T0203576

PRACTICAL GUIDE
TO INDUSTRIAL
SAFETY

PRACTICAL GUIDE TO INDUSTRIAL SAFETY

Methods for Process Safety Professionals

Nicholas P. Cheremisinoff

N&P Consultants, Ltd.
Harpers Ferry, West Virginia

CRC Press
Taylor & Francis Group
Boca Raton London New York

CRC Press is an imprint of the
Taylor & Francis Group, an **informa** business

CRC Press
Taylor & Francis Group
6000 Broken Sound Parkway NW, Suite 300
Boca Raton, FL 33487-2742

First issued in paperback 2019

© 2001 by Taylor & Francis Group, LLC
CRC Press is an imprint of Taylor & Francis Group, an Informa business

No claim to original U.S. Government works

ISBN-13: 978-0-8247-0476-6 (hbk)
ISBN-13: 978-0-367-39803-3 (pbk)

Visit the Taylor & Francis Web site at
http://www.taylorandfrancis.com

and the CRC Press Web site at
http://www.crcpress.com

PREFACE

This volume has been written for process safety professionals. The intent of the volume is to provide supplemental information to assist safety specialists in managing operations in industrial settings. The subject areas covered highlight some of the more important health and safety problems involved in the chemical and allied industries. It is not intended as a handbook, and indeed many safety subjects such as hearing protection, back injuries, and stress fatigue are not covered. Instead, the book concentrates on areas of high-risk personal exposure, with emphasis on inhalation hazards and direct chemical exposure. The volume provides a large number of useful references and points the reader to a number of Web sites where a wide variety of health and safety and emergency preparedness information can be accessed. There are eight chapters, each covering commonplace occupational risks that OSHA standards regulate. Subjects covered include an overview of occupational safety problems (both physical and chemical), specific hazards encountered in selected industry segments within the chemicals and petroleum industries, indoor air quality management issues, the use of personal protective equipment, process safety considerations and emergency preparedness, laboratory safety, bloodborne pathogens, and heat stress.

It is assumed that the reader has a professional background in the area of safety management and engineering, and hence, many commonplace terms are not defined. Instead, the reader may also refer to some of the following books by the same author to facilitate reading and subject understanding:

- *Handbook of Industrial Toxicology and Hazardous Materials*, Marcel Dekker, Inc., 1999;
- *Safety Management Practices for Hazardous Materials*, Marcel Dekker, Inc., 1996;
- *Toxic Properties of Pesticides*, Marcel Dekker, Inc., 1994;

- *Dangerous Properties of Industrial and Consumer Chemicals*, Marcel Dekker, Inc., 1994.

Although the author has relied on well-known and authoritative sources to compile the information in this book, neither he nor the publisher guarantees the designs or procedures that are discussed. Application of safety engineering principles and emergency response practices must closely follow local, state and federal regulations, which are often specific to industry sectors and parts of the country.

The author wishes to thank Marcel Dekker, Inc., for the fine production of this volume.

Nicholas P. Cheremisinoff, Ph.D.

CONTENTS

PRACTICAL GUIDE TO INDUSTRIAL SAFETY

Chapter 1
INTRODUCTION TO INDUSTRIAL SAFETY

INTRODUCTION

Industrial safety deals with the areas of safety engineering and public health that are concerned with the protection of workers' health, through control of the work environment to reduce or eliminate hazards. Industrial accidents and unsafe working conditions can result in temporary or permanent injury, illness, or even death. They also impact on reduced efficiency and loss of productivity. In the United States before 1900 the safety of workers was of little concern to employers. Only with the passage of the Workmen's Compensation Laws and related labor statutes between 1908 and 1948 did U.S. employers start to pay attention to industrial safety; making the work environment safer was less costly than paying compensation. A new national policy was established in 1970 when for the first time all industrial workers in businesses affected by interstate commerce were covered by the Occupational Safety and Health Act. Under this act the National Institute for Occupational Safety and Health (NIOSH) was given responsibility for conducting research on occupational health and safety standards, and the Occupational Safety and Health Administration (OSHA) was charged with setting and enforcing appropriate standards in industry. Various external factors, such as chemical, biological, or physical hazards, can cause work-related injury. Poor working posture or improper design of the workplace often results in muscle strains, sprains, fractures, bruises, and back pain (e.g., Repetitive Stress Injury). In recent years engineers have attempted to develop a systems approach (termed safety engineering) to industrial accident prevention. The systems approach examines all work locations to eliminate or control hazards. It also examines operating methods and practices and the training of employees and supervisors. This first chapter serves as an introduction to the general topics and areas reviewed in various chapters.

ENVIRONMENTAL AND OCCUPATIONAL DISEASES

Environmental and occupational illnesses are caused by exposure to disease-causing agents in the environment, as opposed to illnesses related primarily to an individual's genetic makeup or to immunological malfunctions. In everyday use, the term *environmental disease* is confined to noninfectious diseases and to diseases caused largely by exposures beyond the immediate control of the individual; the latter restriction eliminates diseases related to personal habits such as smoking or to the use or abuse of medications or drugs such as alcohol. Occupational disease, a major category of environmental disease, refers to illness resulting from job-related exposures.

Historically, awareness of environmental diseases began with the recognition of occupational illnesses, because exposures are usually more intense in work settings than in the general environment and therefore can more readily produce overt illnesses. Examples include silicosis, a lung disease of miners, industrial workers, and potters exposed to silica dust; scrotal skin cancer in chimney sweeps exposed to soot; neurological disease in potters exposed to lead glazes; and bone disease in workers exposed to phosphorus in the manufacture of matches. Many such diseases first gained public attention during the Industrial Revolution in the 19th century.

Environmental diseases are caused by chemical agents, radiation, and physical hazards. The effects of exposure, in both natural and work settings, are greatly influenced by the exposure routes: primarily air pollution and water pollution, contaminated food, and direct contact with toxins. Synergistic effects—two or more toxic exposures acting together—are also important, as illustrated by the greatly increased risk of lung cancer in asbestos workers who smoke cigarettes. The potential interaction of multiple hazardous chemicals at toxic waste dumps poses a current public health problem that is of unknown dimensions.

Modern society has introduced or increased human exposure to thousands of chemicals in the environment. Examples are inorganic materials such as lead, mercury, arsenic, cadmium, and asbestos, and organic substances such as polychlorinated biphenyls (PCBs), vinyl chloride, and the pesticide DDT. Of particular concern is the delayed potential for these chemicals to produce cancer, as in the cases of lung cancer and mesothelioma caused by asbestos, liver cancer caused by vinyl chloride, and leukemia caused by benzene. Minamata disease, caused by food contaminated with mercury, and Yusho disease, from food contaminated with chlorinated furans, are examples of acute toxic illnesses occurring in nonoccupational settings. The full toxic potential of most environmental chemicals has not been completely tested. The extent and frequency of an illness are related to the dose of toxin, in degrees depending on the toxin. For chronic or delayed effects such as cancer or adverse reproductive effects, no "safe" dose threshold may exist below which disease is not produced. Thus, the cancer-producing potential of ubiquitous environmental contaminants such as DDT or the PCBs remains undefined.

What has become an even greater concern in recent years is the phenomenon known as *multiple chemical sensitivity*-disorder triggered by exposures to many chemicals in the environment. Synthetic chemicals are all around us. They are in the products we use, in the clothes we wear, in the food we eat, in the air we breathe at work. Because chemicals are everywhere in the environment, it is not possible to escape exposure. For this reason many people have become sensitized to the chemicals around them. In fact, it is estimated that 15% of the population has become sensitized to common household and commercial products. For some people the sensitization is not too serious a problem. They may have what appears to be a minor allergy to one or more chemicals. Other people are much more seriously affected. They may feel tired all the time, and suffer from mental confusion, breathing problems, sore muscles, and a weakened immune system. Such people suffer from a condition known as Multiple Chemical Sensitivity (MCS). MCS is a disorder triggered by exposure to chemicals in the environment. Individuals with MCS can have symptoms from chemical exposure at concentrations far below the levels tolerated by most people. Symptoms occur in more than one organ system in the body, such as the nervous system and the lungs. Exposure may be from the air, from food or water, or through skin contact. The symptoms may look like an allergy because they tend to come and go with exposure, though some people's reactions may be delayed. As MCS gets worse, reactions become more severe and increasingly chronic, often affecting more bodily functions. No single widely available medical test can explain symptoms. In the early stages of MCS, repeat exposure to the substance or substances that caused the initial health effects provokes a reaction. After a time, it takes less and less exposure to this or related chemicals to cause symptoms. As the body breaks down, an ever increasing number of chemicals, including some unrelated to the initial exposure, are found to trigger a reaction.

MCS affects the overall health and feeling of well-being of those with the disorder. It typically impairs many bodily functions including the nervous system and digestion. Each individual affected by MCS has a unique set of health problems. A chemically sensitive person may also have other preexisting health conditions. Many affected people experience a number of symptoms, in relation to their chemical exposures. Typical symptoms include headaches, flu-like symptoms, asthma or other breathing problems, dizziness, increased sensitivity to odors, mental confusion, bloating or other intestinal problems, fatigue and depression, short- and long-term memory loss, chronic exhaustion. People with MCS report many other health conditions such as: persistent skin rashes and sores, inflammation, muscle weakness and joint pains, food allergies, numbness and tingling, visual disturbance, ear, nose and throat problems, autoimmune disorders, cardiovascular irregularities, seizure disorders, genitourinary problems, irritability, persistent infections, behavioral problems, and learning disabilities in children.

MCS may result from a single massive exposure to one or more toxic substances or repeated exposure to low doses. On one hand, some people may become chemically sensitive following a toxic chemical spill at work or in their community or after being sprayed directly with pesticides. On the other, individuals may develop this condition from spending forty hours each week in a poorly ventilated building where they breathe a profusion of chemicals common to our modern way of life. In many cases, MCS has been brought on by a wide array of chemicals found at home and work. Studies show that many people diagnosed with MCS are: industrial workers, teachers, students, office and health care workers in tight buildings, chemical accident victims, people living near toxic waste sites, people whose air or water is highly polluted, people exposed to various chemicals in consumer products, food, and pharmaceuticals, and Gulf War veterans. Not all people with MCS fit these categories. For example, some may have experienced a toxic exposure from flea or roach sprays or from (urea formaldehyde) foam insulation in their home. Other people with MCS cannot identify any situations where they had unusual exposures to chemicals. People with MCS may become partially or totally disabled for several years or for life. This physical condition affects every aspect of their life.

In nonindustrial settings, MCS substances are the cause of indoor air pollution and are the contaminants in air and water. Many of the chemicals which trigger MCS symptoms are known to be irritants or toxic to the nervous system. As an example, volatile organic compounds readily evaporate into the air at room temperature. Permitted airborne levels of such contaminants can still make ordinary people sick. When the human body is assaulted with levels of toxic chemicals that it cannot safely process, it is likely that at some point an individual will become ill. For some, the outcome could be cancer or reproductive damage. Others may become hypersensitive to these chemicals or develop other chronic disorders, while some people may not experience any noticeable health effects. Even where high levels of exposure occur, generally only a small percentage of people become chemically sensitive. The threshold for toxic injury is not the same for everyone because sensitivity varies greatly among individuals. Most chemicals in consumer products remain untested for health effects, such as cancer, reproductive problems, and the impacts of long-term, low level exposure. How these substances affect women, children, and people with existing conditions is also little studied. Once a person's defenses have been broken down and he or she has become hypersensitive, a wide variety of common chemical exposures can trigger a reaction. Just what products and other chemicals which cause problems varies greatly among affected individuals.

Treatment of MCS is difficult for physicians to define and diagnose. There is no single set of symptoms which fit together as a syndrome, nor a single diagnostic test for MCS. Instead, physicians should take a complete patient history which includes environmental and occupational exposures, and act as detectives in diagnosing this problematic condition. After the onset of MCS, a person's health generally continues to deteriorate. It may only begin to improve once the chemical sensitivity condition is uncovered. While a number of treatments may help improve the baseline health status for some patients, at the present time, there is no single "cure" except avoidance. Avoiding the chemicals which may trigger reactions is an essential part of treating MCS. Those with MCS who are able to strictly avoid exposures often experience dramatic improvement in their health over the period of a year or more. Yet the profusion of new and untested synthetic chemicals makes this extremely difficult. Individuals affected by MCS often create a "sanctuary" relatively free from chemical emissions in their home, where they spend as much time as possible. Because of the serious impact of even an accidental unavoidable exposure, MCS sufferers often spend as much time at home as possible and often must choose not to participate in society.

Many traditional allergists and other physicians discount the existence of an MCS diagnosis. They claim that there is not yet sufficient evidence that MCS exists. Research efforts regarding the mechanisms that cause MCS have been inadequate and unfortunately are often financed and supported by the industries which benefit from chemical proliferation. Generally medical doctors have not been trained to understand or seriously investigate conditions such as MCS. In fact, the vast majority of physicians receive very little training in occupational and environmental medicine or in toxicology and nutrition. Therefore, it is not surprising that many affected individuals consult with a large number of specialists. People with MCS are often even diagnosed with serious degenerative diseases. Often baffled doctors tell their patients with MCS that their illness is entirely psychosomatic. And many whose health is impaired by MCS have never heard of the condition. The lack of support and understanding from physicians and the stress created by having no explanation for symptoms tends to produce a high level of anxiety and distress in people with MCS. At this time, conventional medicine offers very few medical treatments for MCS besides avoiding offending products. Unfortunately, medications and other conventional medical treatments offer little or no relief and may even prompt new sets of symptoms. Treatment with anti-depressants masks the underlying condition and can also cause other serious problems. Physicians who clearly recognize the MCS phenomenon include some occupational and environmental health specialists and those MDs who specialize in the new field of clinical ecology. A wide range of new or "alternative" treatments have been utilized by MCS sufferers with varying success. Though some of these treatments are still experimental in nature, they seem to help some individuals with MCS. These treatments may include a combination of the following: nutritional programs, immunotherapy vaccines, food-allergy testing, detoxification regiments through exercise and sweating, chelation for heavy metals, as well as any number of non-Western healing methods. Diagnosis may involve unconventional laboratory tests not customary in conventional medicine, including tests for the presence of chemical contaminants, such as total body burden of accumulated pesticides. Many workers have shown improvement with these treatments, though others have not. Unfortunately, these treatments are not usually reimbursed by insurance plans, since few participating practitioners support alternative approaches. Yet some disabled workers have won reimbursement for such treatments through successful Workers' Compensation claims. MCS is now recognized as a disability. Both the U.S. Department of Housing and Urban Development (HUD) and the Social Security Administration (SSA) have recognized MCS as a disabling condition. People with MCS have won Workers' Compensation cases. A recent human rights lawsuit in Pennsylvania established the right of an affected person to safe living space in subsidized housing. Both the Maryland State Legislature and New Jersey State Department of Health have officially commissioned studies of MCS. The New Jersey study provides an excellent overview of medical and legal issues related to MCS. Just as physical barriers prevent wheelchair access, chemical use and emissions can prevent entry to those with MCS. A new federal law called the Americans with Disabilities Act (ADA) will protect the disabled from many types of discrimination. This law provides for reasonable access to people with disabilities. Reasonable accommodations enable people with MCS to enjoy access to work, public facilities and other necessary settings. Whether an individual developed MCS at work or was already sensitized prior to employment, the right to a safe workplace should be insured. For injured workers who have a right to Workers' Compensation or Disability, it is necessary to find a physician who can diagnose MCS and who will also support the patient's legitimate claims. Finding such a physician is very important in winning such a claim and for gaining reasonable accommodation at work or in rental housing.

Environmental and occupational diseases are by no means limited to chemical exposure. Ionizing and nonionizing radiation can produce both acute and chronic health effects, depending on dose levels. The effects of nonionizing radiation at lower dose levels are uncertain at present. Ionizing radiation at high doses causes both acute disease and delayed effects such as cancer. Victims include workers exposed to various occupational use of X rays or radioactive materials. Although the disease-producing potential of ionizing radiation at low doses is also uncertain, an increase in chromosome damage has been observed in workers in nuclear shipyards.

Major physical hazards include traumatic injuries and noise. Trauma arising from unsafe environments accounts for a large proportion of preventable human illness, and noise in the workplace is responsible for the most prevalent occupational impairment: hearing loss or permanent deafness.

Environmental diseases can affect any organ system of the body. How the diseases are expressed depends on how the particular environmental agent enters the body, how it is metabolized, and by what route it is excreted. The skin, lungs, liver, kidneys, and nervous system are commonly affected by different agents in different settings. Of particular concern is the capacity of many environmental agents to cause various cancers, birth defects or spontaneous abortions (through fetal exposure), and mutations in germ cells, the last-named raising possibilities of environmentally caused genetic diseases in later generations. Environmental illnesses can be mild or severe and can range from transient to chronic, depending on the doses of toxin received. Some diseases occur abruptly after a toxic exposure, whereas the time of onset of other diseases varies after exposure. Environmentally induced cancers, for example, commonly involve latency periods of 15 to 30 years or more. Those illnesses that occur directly after a distinct toxic exposure are usually easily identified as being environmentally or occupationally caused. If the exposure is not clear-cut or illness is delayed, however, the cause is difficult to identify, as clinical features alone are usually nonspecific. In addition, many different causes, environmental or otherwise, may produce identical illnesses. In such instances, epidemiological studies of exposed populations can help relate exposures to the illnesses they cause.

Total frequencies of environmental illness are difficult to measure because of the reasons just described. When causes can be identified, however, scientists observe that frequencies of occurrence of a particular illness vary directly with the severity and extent of exposure. Particularly frequent in the workplace are skin lesions from many different causes and pulmonary diseases related to the inhalation of various dusts, such as coal dust (black lung), cotton dust (brown lung), asbestos fibers (asbestosis), and silica dust (silicosis). Environmental agents can also cause biological effects without overt clinical illness (for example, chromosome damage from irradiation).

REPETITIVE STRESS INJURY

Repetitive Stress Injury (RSI) refers to work-related physical symptoms caused by excessive and repetitive use of the upper extremities. Repetitive stress injuries occur especially when tasks are performed under difficult conditions, using awkward postures and poorly designed equipment. Common tasks that cause RSI include typing for hours on a computer keyboard, cutting meat, or working on a production line. Other terms used to describe RSI include *cumulative trauma disorder, repetitive motion disorder,* and *occupational overuse syndrome.* Recently, the term *upper extremity musculoskeletal disorder* has been adopted. According to the United States Bureau of Labor Statistics, 150,500 cases of RSI were reported in 1997. Eighty percent of these were in businesses such as manufacturing, assembly, and service jobs. The remainder of cases were reported in clerical, sales, professional managerial, secretarial, and data-entry jobs. Because of the nature of their work, musicians and dancers also have a high incidence of RSI, which has led to the creation of a new medical specialty, performing arts medicine. Symptoms of RSI are varied, but they are roughly divided into three categories. Early signs include muscle aches and fatigue in the arms, wrists, or neck occurring during work, which may begin slowly over weeks or months, but which usually disappear with rest. After several months, aching and fatigue persist for longer periods and beyond the work day, diminishing the ability to perform everyday tasks. Advanced symptoms include aching and fatigue while at rest, problems sleeping due to pain, and partial or total disability. As symptoms become more severe, other complications may develop, such as depression resulting from chronic pain. Injury from RSI primarily affects soft tissues, which include muscles, tendons, ligaments, nerves, and connective tissue. These injuries tend to increase in severity over time. Once the chain of injury has begun, RSI tends to progress unless the worker changes the factors that created the stress in the first place. Muscles of the forearm do most of the work of moving the fingers and wrists. These small muscle groups are not designed for extended periods of contraction, so they are vulnerable to injury during repetitive work. Injury is even more likely to occur if the stronger muscles of the back, shoulder, and upper arm are impaired because of poor posture or poorly fitted office equipment, forcing the forearm and hand muscles to do more work. With overuse, forearm muscles contract too often, decreasing the blood and oxygen supply to the muscle.

Lactic acid and other metabolic products build up in the muscle, causing fatigue and pain. The muscle contracts further in response to the pain. The decreased blood supply to the muscles in the arms causes the tendons to tighten, which in turn limits wrist and finger range of motion. When tendons are continually tightened because of muscle injury and contraction, they can be injured by friction as they rub against ligaments or bone, causing the inflammation and pain known as *tendinitis*. As tendons change their angle (as when bending or straightening the wrist), they are kept in place by a variety of sheaths and pulleys similar to the guide rings on a fishing rod. Friction in these areas can cause inflammation and swelling of the sheath and tendon known as *tenosynovitis*. When this occurs at the base of the thumb, it is called deQervain's disease.

Nerves can be squeezed or compressed when surrounding tissues swell. A diminished blood supply can also damage nerve tissue. Friction, or rubbing against inflamed muscle or tendons, injures nerves. Nerves in the spine can be compressed by ruptured discs causing a condition called *spondylosis*. This tends to occur more frequently in the neck and lower back. Nerve root pain caused by bone compression as the nerves exit the spinal column is called *radiculopathy*. As the nerves enter the arms, forearms, and wrists, they encounter a number of tight spots or tunnels where compression or traction (pulling) can occur, including the *cubital tunnel* (ulnar nerve), *radial tunnel* (radial nerve), and *carpal tunnel* (median nerve). Carpal tunnel syndrome, a commonly diagnosed form of RSI, develops when the nerves of the wrist become compressed by inflamed tendons. Symptoms include tingling and numbness in the wrist and hand.

Preventing RSI relies primarily on changes in work style, pacing, conditioning, and training. Sustained keyboard and production line work can lead to a round-shouldered posture with the neck thrust forward. This posture can damage nerves and weakens the shoulder and upper back muscles. Eventually the burden of work shifts to weaker muscles of the forearms and hands. For computer users this is often compounded by wrist rests that encourage excessive forearm and hand muscle use and awkward wrist positions that can result in muscle, tendon, and nerve damage. Poorly designed tools can also cause similar problems.

To minimize these potential strains and injuries, it is necessary to improve the *ergonomics* of the work environment. Ergonomics is the study and practice of arranging furniture and equipment to make work comfortable. For example, computer users benefit from basic ergonomic rules for proper positioning at the keyboard. An adjustable chair is essential, feet should be planted firmly on the ground, and the arms should be free to move without obstruction from arm rests. Production line equipment should accommodate the height of the workers so that they can maintain a healthy posture. While the proper position of equipment is important, how the hands and upper body are used is also critical. Basic computer typing technique that teaches proper hand and arm placement is a critical preventive measure. Production line workers should be taught proper methods of lifting and bending. Pacing work by taking regular breaks, and physical conditioning are also important. Most workers who use their hands are essentially upper body athletes. Like any athlete, they must be in good physical condition to be able to perform their jobs safely. Most forms of RSI involve soft tissue injuries that, with a few exceptions, can be treated conservatively. A medical exam followed by physical or occupational therapy, psychological counseling, and changes in the work situation are the critical components of treatment. A regular program of stretching, strengthening, postural training, and isometric and aerobic exercises done at home and at work is essential. Chronic pain may need the intervention of pain-management specialists who use a variety of techniques, including drug therapy, to lessen pain. Surgery is generally considered as a last resort for most RSI injuries.

Ergonomics

The term *ergonomics* (Greek *ergon*, "work"; *nomos*, "laws"), first appeared in a Polish article published in 1857, but the modern discipline did not take shape until half a century later. The study of human factors did not gain much public attention until World War II (1939-1945). Accidents with military equipment were often blamed on human error, but investigations revealed that some were caused by poorly designed controls. The modern discipline of ergonomics was born in the United Kingdom on July 12, 1949, at an interdisciplinary meeting of those interested in human work problems in the British navy. At another meeting, held on February 16, 1950, the term ergonomics was formally adopted for this growing discipline. Today in the United States, ergonomics professionals belong to the Human Factors and Ergonomics Society (HFES), an organization with over 5,000 members interested in topics ranging from aging and aerospace to computers. The HFES is active in developing national and international technical standards to help improve the design of products and workplaces. Ergonomists also work with the United States Occupational Safety and Health Administration (OSHA) to develop ergonomic guidelines, standards, and regulations to ensure the safety and comfort of American workers. About 40 percent of HFES members have degrees in psychology or an associated behavioral science, about 30 percent have degrees in engineering or design, and others have diverse backgrounds in subjects ranging from computer science to medicine. Over 60 universities in the United States now offer graduate or undergraduate degrees in human factors and ergonomics.

Ergonomics, also known as human engineering or human factors engineering, is the science of designing machines, products, and systems to maximize the safety, comfort, and efficiency of the people who use them. Ergonomists draw on the principles of industrial engineering, psychology, anthropometry (the science of human measurement), and biomechanics (the study of muscular activity) to adapt the design of products and workplaces to people's sizes and shapes and their physical strengths and limitations. Ergonomists also consider the speed with which humans react and how they process information, and their capacities for dealing with psychological factors, such as stress or isolation. Armed with this information of how humans interact with their environment, ergonomists develop the optimum design for products and systems, ranging from the handle of a toothbrush to the flight deck of the space shuttle. Ergonomists view people and the objects they use as one unit, and ergonomic design attempts to marry the best abilities of people and machines. Humans are not as strong as machines, nor can they calculate as quickly and accurately as computers. Unlike machines, humans need to sleep, and they are subject to illness, accidents, or making mistakes when working without adequate rest. But machines are also limited—cars cannot repair themselves, computers do not speak or hear as well as people do, and machines cannot adapt to unexpected situations as well as humans. An ergonomically designed system provides optimum performance because it takes advantage of the strengths and weaknesses of both its human and machine components. Designing with people in mind often requires advanced technology, such as computer-aided design/computer-aided manufacturing (CAD/CAM) programs and robots to simulate human responses. Other ergonomic tools may be relatively simple. Ergonomists frequently use two- or three-dimensional mannequins that represent particular dimensions of the human body, such as seated height, or arm length or reach. Using such tools, ergonomists create products and workstations that fit 90 percent of the possible users. To help evaluate the tools and systems people use in the course of their day, ergonomists use simulations—replicas of workstations, aircraft, and other scenarios together with observations of people operating equipment and products in the replicated environment.

One of the primary goals of ergonomics is prevention of workplace illness and accidents. According to the United States Bureau of Labor Statistics, more than 60 percent of workplace illnesses reported each year are associated with repetitive stress injuries (RSI). These injuries result from continuous repetition of the same motions, for instance screwing or twisting items on an assembly line. The injury may be exacerbated by awkward postures, such as bending or reaching. Carpal tunnel syndrome, for example, is a painful and often debilitating swelling of the tendons in the wrist, which results from overuse of the hands and wrists. It is particularly common in people who must bend or overextend their arms while performing a repetitive task, for instance, typing on a computer keyboard, cutting meat, or tripping knobs and levers. Frequent, unassisted heavy lifting, for example moving hospital patients in and out of beds, is one of the leading causes of work-related back injuries. Noise-induced hearing loss resulting from continuous exposure to excessive noise is another type of RSI, as are headaches and eyestrain due to improper workplace lighting.

Ergonomists work to eliminate these problems by designing workplaces, such as offices or assembly lines, with injury prevention in mind. They position tools and machinery to be accessible without twisting, reaching, or bending. They design adjustable workbenches, desks, and chairs to comfortably accommodate workers of many different sizes, preventing the need to continuously lean or overextend the arms. Ergonomists also determine and design safe workplace environmental conditions, such as correct temperature, lighting, noise, and ventilation to ensure that workers perform under optimal conditions.

Ergonomists also seek to increase worker efficiency and productivity when designing workspaces. They place those pieces of equipment used most frequently in closest proximity to the worker and arrange systems in ways that are convenient and easy to use. Well-designed workspaces ensure that workers perform their jobs in optimal comfort, without experiencing the unnecessary physical and mental fatigue that can slow work performance, reduce accuracy, or cause accidents. Other ergonomists design the individual tools or equipment a worker uses. Specially curved computer keyboards encourage typists to hold their wrists in a position that is less likely to cause carpal tunnel syndrome. To protect the eyes from incessant glare, ergonomically designed computer monitors are equipped with glare reduction screens. Ergonomically designed chairs distribute a person's body weight evenly to avoid back and neck strain. These chairs adjust to a user's height to ensure that the feet rest flat on the ground. In factories and assembly lines, ergonomically designed knobs and levers are positioned appropriately so as not to require reaching, and these knobs and levers also require minimal force to trip. By employing the best possible design for safety systems, ergonomists minimize workplace accidents. Ergonomists consider the way humans interpret information, their reaction speed, and how both of these factors are influenced by the stress of an emergency. Warning signals, such as lights, buzzers, and sirens, must be easy to interpret. Control devices must be easy to identify and use, particularly in workplaces such as aircraft, vehicles, and nuclear power plants, where quick, accurate reactions are imperative to public safety. A poorly designed control panel was a factor during the near meltdown of the nuclear generating station at Three Mile Island, Pennsylvania, in 1979. Some ergonomists practice in the area of job design. These professionals help employers assess both the individual tasks necessary to perform a particular job and the skills needed to accomplish each task. By grouping like tasks and skills, jobs can be redesigned to maximize efficiency. An office telephone receptionist, for example, may perform a number of other tasks as varied as filing, sorting mail, and bookkeeping. Grouping these responsibilities, which can all be performed in the vicinity of the office telephone system, makes use of the receptionist's time when there are no telephone calls. Ergonomists help employers evaluate different ways of organizing workdays to increase worker productivity, ensuring that workers have adequate breaks and rest periods, as well as a well-defined set of tasks. An ergonomist may use similar skill-analysis principles to help an employer identify the best candidate for a particular job. By working with the employer to define the physical, mental, and social skills needed to perform a job, ergonomists can determine the necessary qualifications and help employers with personnel selection. Job task and skill analysis is also used to determine the most effective ways to train employees. Training for astronauts and pilots, for example, may include simulations developed by ergonomists. Training simulations, such as computer virtual reality training, teach trainees how to deal with dangerous scenarios, such as accidents, without exposing them to the dangers of a real accident. Ergonomists also design virtual reality simulations for medical doctors, enabling them to practice diagnostic and surgical skills on computer-simulated patients, thereby not endangering the health of a live patient.

Cognitive ergonomists specialize in information design—the best way to present complex information. These professionals study the way the human brain processes information. Using this knowledge and the principles of graphic design, cognitive ergonomists develop signs, maps, instruction manuals, and even computer programs and Internet sites that are easy to use, or *intuitive*. The work of cognitive ergonomists is particularly evident in public transportation buildings, such as airports or train stations. These buildings are often large, complex, and difficult to navigate. Cognitive ergonomists develop clear, easy-to-understand navigation aids, such as signs and maps, to help people find their way to their gate as simply and efficiently as possible. Color-coded subway maps, for example, help subway riders navigate with relative ease through a complicated maze of interconnected underground tunnels. Cognitive ergonomists also work with manufacturers to design the instruction manuals packaged with consumer products. They evaluate the tasks required to assemble or operate the goods, and present the tasks as a set of sequential, easy to follow instructions. When designing instruction manuals, cognitive ergonomists must consider not only the way the brain processes information, but also the way people expect to receive instructional information. As they develop and learn, humans grow accustomed to receiving different types of information in particular formats. When information does not conform to its customary format, people may find it difficult to follow or understand. Ergonomic design makes consumer products safer, easier to use, and more reliable. In many manufacturing industries, ergonomists work with designers to develop products that fit the bodies and meet the expectations of the people who will use them. An ergonomically designed toothbrush, for example, has a broad handle for easy grip, a bent neck for easier access to back teeth, and a bristle head shaped for better tooth surface contact. The shaving razor has undergone a similar design revolution. The bent-handled, easy-grip models popular today are more comfortable to use and have a better shaving performance than the straight-edged razors of days gone by. Ergonomic design has dramatically changed the interior appearance of automobiles. The steering wheel—once a solid, awkward disk—is now larger and padded for an easier, more comfortable grip. Its center is removed to improve the driver's view of the instruments on the dashboard. Larger, contoured seats, adjustable to suit a variety of body sizes and posture preferences, have replaced the small, upright seats of early automobiles. Equipped with seatbelts and adjustable headrests that prevent the neck from snapping backward in the event of a collision, modern automobile seats are not only more comfortable, they are also safer. The principles of ergonomic design affect other features of the automobile as well. The center-mounted rear windshield brake light, now a required component of all new automobiles, is an ergonomic innovation that saves lives. Perhaps the most compelling ergonomic innovations of our time, improvements to computer user interfaces have changed the way the world uses computers.

Graphical user interface (GUI) is a computer display format that enables the user to choose commands, start programs, and see lists of files and other options by pointing to pictorial representations on the screen. By taking into account the way humans interact with machines, computer scientists developed a GUI that is intuitive and easy to use. This innovation made computers, once the cryptic and complex tool of an elite group of scientists and mathematicians, accessible to almost anyone.

Ergonomic improvements to computer hardware and software are ongoing. The mouse, a hand-shaped input device, enables users to give the computer commands with the click of a button. In the future, keyboards and mice, already ergonomically shaped to reduce the occurrence of carpal tunnel syndrome, may be entirely replaced by voice-activated input systems. Many computer users already use verbal commands, touch the screen, or use pencil-like instruments to enter their commands, rather than typing them on a keyboard or clicking with a mouse.

REGULATIONS, LAWS, AND AGENCIES

The regulation of workplace practices and of potential environmental pollution has evolved as the use of chemicals and human exposure to potential toxins have grown more widespread and complex in modern society. In the United States, numerous laws are directed at protecting occupational and environmental health. Most were passed since 1960, including the Occupational Safety and Health Act (OSH Act) of

1970 and the Resource Conservation and Recovery Act (RCRA) of 1979. Means for the rapid cleanup of toxic waste dumps were provided in the Comprehensive Environmental Response, Compensation, and Liability Act of 1980.

Federal agencies responsible for enforcing such environmental and occupational health laws consist principally of the Environmental Protection Agency and the Occupational Safety and Health Administration (OSHA) within the Department of Labor. The Food and Drug Administration, within the Department of Health and Human Services (HHS), and the Department of Agriculture have regulatory responsibility for preventing the contamination of food supplies. Federal field investigations of potential environmental and occupational hazards are handled through the Center for Environmental Health and the National Institute for Occupational Safety and Health, which are components of the Centers for Disease Control, within HHS. General environmental health research and toxicological testing are directed through the National Institutes of Health and the National Toxicology Program, also within HHS. Comparable regulations and agencies at state and local levels, working with their federal counterparts, play a crucial role as well. International coordination of environmental and occupational control activities in many countries is guided through the World Health Organization. In the developing parts of the world, such activities are of critical importance as modern industrialization proceeds in the face of poverty and growing populations.

Current trends in research in this field focus on the relation of low-dose exposures to human health, the influence of environmental toxins on both male and female reproductive functions, and the potential health implications of subclinical indications of biological damage (for example, genetic or chromosomal damage). In such research, increased emphasis is being placed on delayed or long-term health effects and on a wide range of potential synergistic interactions between environments and hosts.

The *Occupational Safety and Health Administration* (OSHA), is the agency of the U.S. Department of Labor established by an act of Congress in 1970. Its main responsibilities are to provide for occupational safety by reducing hazards in the workplace and enforcing mandatory job safety standards and to implement and improve health programs for workers. OSHA regulations and standards apply to most private businesses in the U.S. From its beginnings, OSHA has been a controversial agency that has drawn much criticism from both business and labor groups. Businesses have charged that the agency's regulations are difficult to understand and often unreasonably rigid; that penalties are unfair, paperwork is excessive, and the cost of compliance is burdensome to small companies. Labor, on occasion, has called OSHA's enforcement procedures weak and complained that the agency has failed to reduce occupational hazards. Since 1977 the agency has made an effort to concentrate on dangerous industries and to eliminate out-of-date regulations. Meanwhile, OSHA is being challenged by some businesses in the courts. The agency, directed by the assistant secretary for occupational safety and health, is headquartered in Washington, D.C. It has ten regional offices located throughout the U.S.

OVERVIEW OF POLLUTION ISSUES

One of many forms of pollution, air pollution occurs inside homes, schools, and offices; in cities; across continents; and even globally. Air pollution makes people sick—it causes breathing problems and promotes cancer—and it harms plants, animals, and the ecosystems in which they live. Some air pollutants return to earth in the form of acid rain and snow, which corrode statues and buildings, damage crops and forests, and make lakes and streams unsuitable for fish and other plant and animal life. It is well recognized that pollution is changing the earth's atmosphere so that it lets in more harmful radiation from the sun. At the same time, our polluted atmosphere is becoming a better insulator, preventing heat from escaping back into space and leading to a rise in global average temperatures. Predictions concerning the temperature increase, referred to as global warming, show that the world food supply may be decreased, sea level will rise, extreme weather conditions may prevail, and increases in the spread of tropical diseases are likely. Most air pollution comes from one human activity: burning fossil fuels—natural gas, coal, and oil—to power industrial processes and motor vehicles. Among the harmful chemical compounds this burning puts into the atmosphere are carbon dioxide, carbon monoxide, nitrogen oxides, sulfur dioxide, and particulate matter. Between 1900 and 1970, motor vehicle use rapidly expanded, and emissions of nitrogen oxides, some of the most damaging pollutants in vehicle exhaust, increased 690 percent. When fuels are incompletely burned, various chemicals called volatile organic chemicals (VOCs) also enter the air. Pollutants also come from other sources. For instance, decomposing garbage in landfills and solid waste disposal sites emits methane gas, and many household products give off VOCs. Refer to Figures 1 and 2 for an explanation of the mechanisms of acid rain and the impact on the environment. Some of these pollutants also come from natural sources. For example, forest fires emit particulate matter and VOCs into the atmosphere. Ultrafine dust particles, dislodged by soil erosion when water and weather loosen layers of soil, increase airborne particulate levels. Volcanoes spew out sulfur dioxide and large amounts of pulverized lava rock known as volcanic ash.

A large volcanic eruption can darken the sky over a wide region and affect the earth's entire atmosphere. The 1991 eruption of Mount Pinatubo in the Philippines, for example, emitted enough volcanic ash into the upper atmosphere to lower global temperatures for the next two years. Unlike pollutants from human activity, however, naturally occurring pollutants tend to remain in the atmosphere for a short time and do not lead to permanent atmospheric change. Once in the atmosphere, pollutants often undergo chemical reactions that produce additional harmful compounds. Air pollution is subject to weather patterns that can trap it in valleys or blow it across the globe to damage pristine environments far from the original sources.

Local and Regional Pollution

Local and regional pollution take place in the lowest layer of the atmosphere, the troposphere, which extends from the earth's surface to about 16 km (about 10 mi).

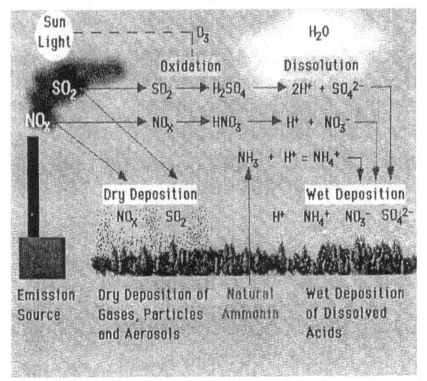

Figure 1. *Mechanisms of acid rain and impact on the environment.*

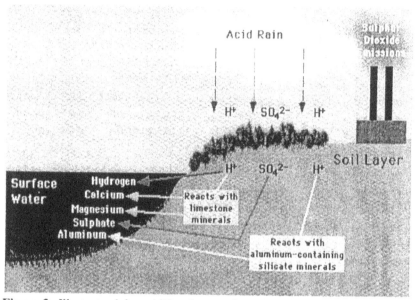

Figure 2. *Illustrates lake acidification resulting from acid rain.*

The troposphere is the region in which most weather occurs. If the load of pollutants added to the troposphere were equally distributed, the pollutants would be spread over vast areas and the air pollution might almost escape our notice. Pollution sources tend to be concentrated, however, especially in cities. In the weather phenomenon known as thermal inversion, a layer of cooler air is trapped near the ground by a layer of warmer air above. When this occurs, normal air mixing almost ceases and pollutants are trapped in the lower layer. Local topography, or the shape of the land, can worsen this effect—an area ringed by mountains, for example, can become a pollution trap.

Smog and Acid Precipitation

Smog is intense local pollution usually trapped by a thermal inversion. Before the age of the automobile, most smog came from burning coal and was so severe that in 19th-century London, street lights were turned on by noon because soot and smog darkened the midday sky. Burning gasoline in motor vehicles is the main source of smog in most regions today. Powered by sunlight, oxides of nitrogen and volatile organic compounds react in the atmosphere to produce photochemical smog. Smog contains ozone, a form of oxygen made up of molecules with three oxygen atoms rather than the normal two. Ozone in the lower atmosphere is a poison—it damages vegetation, kills trees, irritates lung tissues, and attacks rubber. Environmental officials measure ozone to determine the severity of smog. When the ozone level is high, other pollutants, including carbon monoxide, are usually present at high levels as well. In the presence of atmospheric moisture, sulfur dioxide and oxides of nitrogen turn into droplets of pure acid floating in smog. These airborne acids are bad for the lungs and attack anything made of limestone, marble, or metal.

In cities around the world, smog acids are eroding precious artifacts, including the Parthenon temple in Athens, Greece, and the Taj Mahal in Agra, India. Oxides of nitrogen and sulfur dioxide pollute places far from the points where they are released into the air. Carried by winds in the troposphere, they can reach distant regions where they descend in acid form, usually as rain or snow. Such acid precipitation can burn the leaves of plants and make lakes too acidic to support fish and other living things.

Because of acidification, sensitive species such as the popular brook trout can no longer survive in many lakes and streams in the eastern United States. Smog spoils views and makes outdoor activity unpleasant. For the very young, the very old, and people who suffer from asthma or heart disease, the effects of smog are even worse: It may cause headaches or dizziness and can cause breathing difficulties. In extreme cases, smog can lead to mass illness and death, mainly from carbon monoxide poisoning. In 1948 in the steel-mill town of Donora, Pennsylvania, intense local smog killed nineteen people. In 1952 in London over 3000 people died in one of the most notorious smog events known as London Fogs; in 1962 another 700 Londoners died.

With stronger pollution controls and less reliance on coal for heat, today's chronic smog is rarely so obviously deadly. However, under adverse weather conditions, accidental releases of toxic substances can be equally disastrous. The worst such accident occurred in 1984 in Bhopal, India, when methyl isocyanate released from an American-owned factory during a thermal inversion caused at least 3300 deaths.

Global Scale Pollution

Air pollution can expand beyond a regional area to cause global effects. The stratosphere is the layer of the atmosphere between 16 km (10 mi) and 50 km (30 mi) above sea level. It is rich in ozone, the same molecule that acts as a pollutant when found at lower levels of the atmosphere in urban smog. Up at the stratospheric level, however, ozone forms a protective layer that serves a vital function: it absorbs the wavelength of solar radiation known as ultraviolet-B (UV-B). UV-B damages deoxyribonucleic acid (DNA), the genetic molecule found in every living cell, increasing the risk of such problems as cancer in humans. Because of its protective function, the ozone layer is essential to life on earth.

Ozone Depletion

Several pollutants attack the ozone layer. Chief among them is the class of chemicals known as chlorofluorocarbons (CFCs), used as refrigerants (notably in air conditioners), as agents in several manufacturing processes, and formerly as propellants in spray cans. CFC molecules are virtually indestructible until they reach the stratosphere. Here, intense ultraviolet radiation breaks the CFC molecules apart, releasing the chlorine atoms they contain. These chlorine atoms begin reacting with ozone, breaking it down into ordinary oxygen molecules that do not absorb UV-B. The chlorine acts as a catalyst—that is, it takes part in several chemical reactions—yet at the end emerges unchanged and able to react again. A single chlorine atom can destroy up to 100,000 ozone molecules in the stratosphere. Other pollutants, including nitrous oxide from fertilizers and the pesticide methyl bromide, also attack atmospheric ozone.

Scientists are finding that under this assault the protective ozone layer in the stratosphere is thinning. In the Antarctic region, it vanishes almost entirely for a few weeks every year. Although CFC use has been greatly reduced in recent years, CFC molecules already released into the lower atmosphere will be making their way to the stratosphere for decades, and further ozone loss is expected. As a result, experts anticipate an increase in skin cancer, more cataracts (clouding of the lens of the eye), and reduced yields of some food crops.

Global Warming

Humans are bringing about another global-scale change in the atmosphere: the increase in what are called greenhouse gases. Like glass in a greenhouse, these gases admit the sun's light but tend to reflect back downward the heat that is radiated from the ground below, trapping heat in the earth's atmosphere. This process is known as the greenhouse effect. Carbon dioxide is the most significant of these gases—there is 25 percent more carbon dioxide in the atmosphere today than there was a century ago, the result of our burning coal and fuels derived from oil. Methane, nitrous oxide, and CFCs are greenhouse gases as well. Scientists predict that increases in these gases in the atmosphere will make the earth warmer. They expect a global rise in average temperature somewhere between 1.0° and 3.5° C (1.8° and 6.3° F) in the next century. Average temperatures have in fact been rising, and the years from 1987 to 1997 were the warmest ten years on record. Most scientists are reluctant to say that global warming has actually begun because climate naturally varies from year to year and decade to decade, and it takes many years of records to be sure of a fundamental change. There is little disagreement, though, that global warming is on its way.

Global warming will have different effects in different regions. A warmed world is expected to have more extreme weather, with more rain during wet periods, longer droughts, and more powerful storms. Although the effects of future climate change are unknown, some predict that exaggerated weather conditions may translate into better agricultural yields in areas such as the western United States, where temperature and rainfall are expected to increase, while dramatic decreases in rainfall may lead to severe drought and plunging agricultural yields in parts of Africa, for example.

Warmer temperatures are expected to partially melt the polar ice caps, leading to a projected sea level rise of 50 cm (20 in) by the year 2050. A sea level rise of this magnitude would flood coastal cities, force people to abandon low-lying islands, and completely inundate coastal wetlands. If sea levels rise at projected rates, the Florida Everglades will be completely under water in less than 50 years. Diseases like malaria, which at present are primarily found in the tropics, may become more common in the regions of the globe between the tropics and the polar regions, called the temperate zones. For many of the world's plant species, and for animal species that are not easily able to shift their territories as their habitat grows warmer, climate change may bring extinction.

Indoor Air Pollution

Pollution is perhaps most harmful at an often unrecognized site—inside the homes and buildings where we spend most of our time. Indoor pollutants include tobacco smoke; radon, an invisible radioactive gas that enters homes from the ground in some regions; and chemicals released from synthetic carpets and furniture, pesticides, and household cleaners. When disturbed, asbestos, a nonflammable material once commonly used in insulation, sheds airborne fibers that can produce a lung disease called asbestosis.

Pollutants may accumulate to reach much higher levels than they do outside, where natural air currents disperse them. Indoor air levels of many pollutants may be 2 to 5 times, and occasionally more than 100 times, higher than outdoor levels. These levels of indoor air pollutants are especially harmful because people spend as much as 90 percent of their time living, working, and playing indoors. Inefficient or improperly vented heaters are particularly dangerous.

End-of-Pipe Treatment and Pollution Prevention

In the United States, serious effort against local and regional air pollution began with the Clean Air Act of 1970, which was amended in 1977 and 1990. This law requires that the air contain no more than specified levels of particulate matter, lead, carbon monoxide, sulfur dioxide, nitrogen oxides, volatile organic compounds, ozone, and various toxic substances. To avoid the mere shifting of pollution from dirty areas to clean ones, stricter standards apply where the air is comparatively clean.

In national parks, for instance, the air is supposed to remain as clean as it was when the law was passed. The act sets deadlines by which standards must be met. The Environmental Protection Agency (EPA) is in charge of refining and enforcing these standards, but the day-to-day work of fighting pollution falls to the state governments and to local air pollution control districts.

Some states, notably California, have imposed tougher air pollution standards of their own. In an effort to enforce pollution standards, pollution control authorities measure both the amounts of pollutants present in the atmosphere and the amounts entering it from certain sources. The usual approach is to sample the open, or ambient, air and test it for the presence of specified pollutants. The amount of each pollutant is counted in parts per million or, in some cases, milligrams or micrograms per cubic meter. To learn how much pollution is coming from specific sources, measurements are also taken at industrial smokestacks and automobile tailpipes.

Pollution is controlled in two ways: with end-of-the-pipe devices that capture pollutants already created, and by limiting the quantity of pollutants produced in the first place. End-of-the-pipe devices include catalytic converters in automobiles and various kinds of filters and scrubbers in industrial plants. In a catalytic converter, exhaust gases pass over small beads coated with metals that promote reactions changing harmful substances into less harmful ones. When end-of-the-pipe devices first began to be used, they dramatically reduced pollution at a relatively low cost. As air pollution standards become stricter, it becomes more and more expensive to further clean the air. In order to lower pollution overall, industrial polluters are sometimes allowed to make cooperative deals. For instance, a power company may fulfill its pollution control requirements by investing in pollution control at another plant or factory, where more effective pollution control can be accomplished at a lower cost. End-of-the-pipe controls, however sophisticated, can only do so much. As pollution efforts evolve, keeping the air clean will depend much more on preventing pollution than on curing it. Gasoline, for instance, has been reformulated several times to achieve cleaner burning. Various manufacturing processes have been redesigned so that less waste is produced. Car manufacturers are experimenting with automobiles that run on electricity or on cleaner-burning fuels. Buildings are being designed to take advantage of sun in winter and shade and breezes in summer to reduce the need for artificial heating and cooling, which are usually powered by the burning of fossil fuels. The choices people make in their daily lives can have a significant impact on the state of the air. Using public transportation instead of driving, for instance, reduces pollution by limiting the number of pollution-emitting automobiles on the road. During periods of particularly intense smog, pollution control authorities often urge people to avoid trips by car. To encourage transit use during bad-air periods, authorities in Paris, France, make bus and subway travel temporarily free.

Indoor pollution control must be accomplished building by building or even room by room. Proper ventilation mimics natural outdoor air currents, reducing levels of indoor air pollutants by continually circulating fresh air. After improving ventilation, the most effective single step is probably banning smoking in public rooms. Where asbestos has been used in insulation, it can be removed or sealed behind sheaths so that it won't be shredded and get into the air. Sealing foundations and installing special pipes and pumps can prevent radon from seeping into buildings. On the global scale, pollution control standards are the result of complex negotiations among nations. Typically, developed countries, having already gone through a period of rapid industrialization, are ready to demand cleaner technologies. Less developed nations, hoping for rapid economic growth, are less enthusiastic about pollution controls. They seek lenient deadlines and financial help from developed countries to make the expensive changes necessary to reduce pollutant emissions in their industrial processes. Nonetheless, several important international accords have been reached. In 1988, the United States and 24 other nations agreed in the Long-Range Transboundary Air Pollution Agreement to hold their production of nitrogen oxides, a key contributor to acid rain, to current levels. In the Montreal Protocol, adopted in 1987 and strengthened in 1990 and 1992, most nations agreed to stop or reduce the manufacture of CFCs. In 1992 the United Nations Framework Convention on Climate Change negotiated a treaty outlining cooperative efforts to curb global warming. The treaty, which took effect in March 1994, has been legally accepted by 160 of the 165 participating countries. In December 1997 at the Third Conference of the United Nations Framework

Convention on Climate Change in Japan, more than 160 nations formally adopted the Kyoto Protocol. This agreement calls for industrialized nations to reduce their emissions of greenhouse gases to levels 5 percent below 1990 emission levels between 2008 and 2012. The United States, which releases more greenhouse gases than any other nation, has traditionally been slow to support such strong measures. The U.S. Senate may be reluctant to ratify the Kyoto Protocol because it does not require developing countries, such as China and India, to meet similar emissions goals.

All these antipollution measures have helped stem the increase of global pollution emission levels. Between 1970, when the Clean Air Act was passed, and 1995, total emissions of the major air pollutants in the United States decreased by nearly 30 percent. During the same 25-year period, the U.S. population increased 28 percent and vehicle miles traveled increased 116 percent. Air pollution control is a race between the reduction of pollution from each source, such as a factory or a car, and the rapid multiplication of sources. Smog in American cities is expected to increase again as the number of cars and miles driven continues to rise. Meanwhile, developing countries are building up their own industries, and their citizens are buying cars as soon as they can afford them. Ominous changes continue in the global atmosphere. New efforts to control air pollution will be necessary as long as these trends continue.

The reader may wish to refer to some of the following journals and newsletters covering Clean Technology and Environmental Management, Pollution Prevention, Total Quality Environmental Management; which give some detailed information on pollution prevention principles:

Business and the Environment
Cutter Publishing
37 Broadway, Suite 1
Arlington, MA 02174-5552
Tel: +617 641 5125 (client services) +617 641 5123 (editorial)
Fax: +617 648 8707
Email: lovering@cutter.com (customer service), bate@igc.apc.org (editorial)
Web: http://www.cutter.com
Subscription rate: $397 per year; $497 outside North America (subscription includes 12 monthly issues, quarterly Meeting Planner that lists conferences, trade shows, etc.)
Corporate Environmental Strategy
published quarterly by:
PRI Publishing
333 Main Street
Metuchen, NJ 08840

Tel: +908 548 5827 Fax: +908 548 2268
(Editorial offices are in Troy, NY at 518 276 2669 (Fax 2051)
Environment Today
1165 Northchase Parkway NE, Suite 350
Marietta GA 30067
Tel: 404-988-9558
$56/year
Environmental Business Journal
Monthly, $395/yr
Environmental Business International
4452 Park Blvd
San Diego, CA 92116
Tel: 619-295-7685
EBJ's unique strength is its regular economic analysis of the state of the environmental industry, including detailed monthly stock market analysis of 14 sectors of the industry - ranging from solid waste management and asbestos abatement to environmental consulting and environmental energy sources - and discussions of revenue and profitability gains of selected companies. Feature articles present original research on business trends facing these sectors, as well as more general pieces on overall strategic issues.
Environmental Quality Management
Formerly "Total Quality Environmental Management"
Quarterly
John Wiley and Sons Inc.
605 Third Avenue
New York, NY 10158
$176/year, $200 outside North America
Editor-in-Chief: Chris FitzGerald
Editor: John T. Willig
GMI Report: Japan
Monthly
Green Marketing Institute
Aria Ikebukuro Bldg 4F
Minami-Ikebukuro 2-29-12
3 Fukuromachi, Toshima-ku
Tokyo 171
Tel: 81-(0)3-5950-6490
Fax: 81-(0)3-5950-6483
Email: gmi@ppp.bekkoame.or.jp
Web: http://www.mictokyo.co.jp/GMI/

This newsletter offers a unique insight into the substantial pace of environmental innovation in Japan, a country that has set its sights firmly on the world environmental market. While some of the information is specific to Japan, such as environment agency budget requests and consumer surveys, much is readily applicable anywhere, ranging from product news to municipal policy. This makes GMI Report: Japan a stimulating source of ideas even for companies that do not trade in the Japanese market.

Green Business Letter
Tilden Press Inc.
1519 Connecticut Ave. NW
Washington, DC 20036
Tel: +1-202-332-1700
Fax: +1-202-332-3028
Email: gbl@enn.com
Web: http://www.enn.com
1-year introductory subscription (12 monthly issues): US$127. Canada and Mexico: US$132; other countries: US$137. Price includes two free books: "The E-Factor: The Bottom-Line Approach to Environmentally Responsible Business," by Joel Makower; and "50 Simple Things Your Business Can Do to Save the Earth," by The Earthworks Group.

Hazardous & Solid Waste Minimization & Recycling Report
Published by: Government Institute, Inc. (Published monthly)
4 Research Place, Suite 200
Rockville, MD 20850
Tel: (301) 921-2355
Customer Service: 301-921-2323
Publisher: Thomas F.P. Sullivan
Executive Director: Martin Heavner

In Business Magazine for Environmentally-Friendly Products and Entrepenuers
$23/year
JG Press
419 State Avenue
Emmaus, PA 18049
Tel: 610 967-4135

Industrial Environment Newsletter
Worldwide Videotex
PO Box 138
Boston, MA 02157
Tel 508-447-8979, fax 508-477-4236

Industry and Environment

Quarterly + newsletter
United Nations Environment Programme Industry and Environment - UNEP IE
Tour Mirabeau, 39-43 quai Andre-Citroen, 75739
Paris Cedex 15, France
Tel: 33-1-4437-1450
Fax: 33-1-4437-1474
E-mail: unepie@unep.fr.
Director: Jacqueline Aloisi de Larderel
Circulation Address: UN Bookshop/Sales Unit-Palais des Nations- CH 1211
Geneva 10, Switzerland

International Journal of Environmentally Conscious Design and Manufacturing
ECM Press
P.O. Box 20959, Albuquerque, NM
87154-0959, USA
Editors in Chief: Mo Shahinpour; e-mail: Shah@unmb.unm.edu

Journal Of Clean Technology and Environmental Sciences
Princeton Scientific Publishing Co., Inc.
P.O. Box 2155, Princeton
New Jersey 08543
Tel: 609 683-4750
Fax: 609 683-0838
Editors-in-Chief: M.A. Mehlman and Sonia P. Maltezou
Copies of articles are also available through ISI Document Delivery Services c/o
The Genuine Article, 3501 Market Street, Philadelphia, PA 19104

Journal of Cleaner Production
Butterworth-Heinemann Ltd. (Published quarterly)
Linacre House, Jordan Hill, Oxford OX2 8DP,UK
Tel: 44 (0)865-310366
Fax: 44 (0)865-310898
Telex: 83111 BHPOXF G
Publisher: Diane Cogan
Group Editor: Lynne Clayton
UK and overseas orders:
Turpin Distribution Services Ltd.
Blackhorse Road, Letchworth, Herts
SG6 111N, UK
Tel: 44 (0)462 672555
Fax: 44 (0)462 480967
Telex: 825372 TURPIN G
North American orders:

Journals Fulfilment Department
Butterworth-Heinemann, 80 Montvale Avenue
Stoneham, MA 02180
Tel. No: 1 (617) 438-8464
Fax No. 1 (617) 438-1479
Telex: 880052
Journal of Industrial Ecology
(Published quarterly)
MIT Press
55 Hayward Street
Cambridge, MA 02142
Tel: (617) 253-2889
Fax: (617) 527-1545
E-mail: journals-orders@mit.edu
Editor-in-Chief: Reid Lifset
Pollution Prevention Northwest
(Published bimonthly)
Pacific Northwest Pollution Prevention Reseach Center (PPRC)
1326 Fifth Avenue
Suite 650, Seattle, WA 98101
Tel: 206-223-1151
Fax: 206-223-1165
Internet: bsrc_pprc@ccmail.pnl.gov
Managing Editor: Kristi Thorndike
Pollution Prevention Letter
monthly
Charles Knebl
Clarity Publishing
P.O. Box 13315
Silver Spring, MD 20911-3315
Tel: +301 495 7747 Fax: 301 495 7747
Pollution Prevention Review
John Wiley & Sons, Inc. (Published quarterly)
605 Third Avenue, New York, NY 10158
Tel: 212 850 6479
Email: SUBINFO@jwiley.com
Publisher: James F. Slabe
Editor: Ann B. Graham
Pollution Prevention Advisor
U.S. Department of Energy, Office of Defense Programs

Produced by: Systematic Management Services, Inc. (SMS)
McPherson Environmental Resources
109 South Riverside Drive
Elizabethton, TN 37643
Tel: 615-543-5422
Fax: 615-543-4382
E-mail: mer@tricon.net.
Proceedings of the IEEE International Symposia on Electronics and the Environment
Ordering information: Jayne Fitzgerald Cerone, IEEE, 445 Hoes Lane, Piscataway, NJ 08855-1331, Tel: +1 908 562 3908; Fax: +1 908 981 1769; Email: j.cerone@ieee.org
Also phone 1-800-678-4333, outside the U.S. 1-908-981-0060, fax 1-908-981-9667

Proceedings, Corporate Quality Environmental Management Conferences
Global Environmental Management Initiative (GEMI)
2000 L. Street, N.W., Suite 710
Washington, DC 20036
Tel: (202) 296-7449
Fax: (202) 296-7442
Tomorrow Magazine
$48/year
Kim Loughran (managing editor)
Claes Sjoberg (publisher)
Kunsgatan 27, S-111 56 Stockholm
Tel: 46-8-24-34-80
Fax:46-8-24-08-09
email: 100126.3133@compuserve.com
Waste Reduction Tips
(Environmental Newsletters Inc., 11906 Paradise Lane, Herndon, VA 22071-1519, Tel: 703-758-8436)

INJURY AND ILLNESS STATISTICS

In the United States, a total of 5.9 million injuries and illnesses were reported in private industry workplaces during 1998 (most recent OSHA census), resulting in a rate of 6.7 cases per 100 equivalent full-time workers, according to a survey by the Bureau of Labor Statistics, U.S. Department of Labor. Employers reported a 4 percent drop in the number of cases and a 3 percent increase in the hours worked compared with 1997, reducing the case rate from 7.1 in 1997 to 6.7 in 1998. The rate for 1998 was the lowest since the Bureau began reporting this information in the early 1970s. The following tabulation on incidence rates for injuries and illnesses shows the decline in rates per 100 full-time workers since 1994: 1994, 1995, 1996, 1997, 1998- Private industry 8.4, 8.1, 7.4, 7.1, 6.7; Goods-producing 11.9, 11.2, 10.2, 9.9, 9.3; Service-producing 6.9, 6.7, 6.2, 5.9, 5.6. Among goods-producing industries, manufacturing had the highest incidence rate in 1998 (9.7 cases per 100 full-time workers). Within the service-producing sector, the highest incidence rate was reported for transportation and public utilities (7.3 cases per 100 full-time workers), followed by wholesale and retail trade (6.5 cases per 100 workers). This release is the second in a series of three releases covering 1998 from the BLS safety and health statistical series. The first release, in August 1999, covered work-related fatalities from the 1998 National Census of Fatal Occupational Injuries. In April 2000, a third release will provide details on the more seriously injured and ill workers (occupation, age, gender, race, and length of service) and on the circumstances of their injuries and illnesses (nature of the disabling condition, part of body affected, event or exposure, and primary source producing the disability). "More seriously" is defined in this survey as involving days away from work.

About 2.8 million injuries and illnesses in 1998 were lost workday cases, that is, they required recuperation away from work or restricted duties at work, or both. The incidence rate for lost workday cases has declined steadily from 4.1 cases per 100 full-time workers in 1990 to 3.1 cases per 100 workers in 1998. The rate for cases with days away from work has declined for eight years in a row and, at 2.0 cases per 100 full-time workers in 1998, was the lowest on record. By contrast, the rate for cases involving only restricted work activity rose from 0.7 cases per 100 workers in 1990 to 1.2 cases in 1997 and remained at that level in 1998. The latter types of cases may involve shortened hours, a temporary job change, or temporary restrictions on certain duties (for example, no heavy lifting) of a worker's regular job. In 1998, the rate in manufacturing for days-away-from- work cases was lower than the rate for restricted-activity-only cases, 2.3 for days-away-from-work cases and 2.5 for restricted-activity-only cases. In all other divisions, the rate for days-away-from-work cases was higher than the rate for restricted-activity-only cases.

Of the 5.9 million nonfatal occupational injuries and illnesses in 1998, 5.5 million were injuries. Injury rates generally are higher for mid- size establishments (those employing 50 to 249 workers) than for smaller or larger establishments, although this pattern does not hold within certain industry divisions. Eight industries, each having at least 100,000 injuries, accounted for about 1.5 million injuries, or 28 percent of the 5.5 million total. All but one of these industries were in the service-producing sector.

There were about 392,000 newly reported cases of occupational illnesses in private industry in 1998. Manufacturing accounted for three-fifths of these cases. Disorders associated with repeated trauma, such as carpal tunnel syndrome and noise-induced hearing loss, accounted for 4 percent of the 5.9 million workplace injuries and illnesses. They were, however, the dominant type of illness reported, making up 65 percent of the 392,000 total illness cases. Seventy-one percent of the repeated trauma cases were in manufacturing industries.

The Survey of Occupational Injuries and Illnesses is a Federal/State program in which employer reports are collected from about 169,000 private industry establishments and processed by state agencies cooperating with the Bureau of Labor Statistics. Occupational injury and illness data for coal, metal, and nonmetal mining and for railroad activities are provided by the Department of Labor's Mine Safety and Health Administration and the Department of Transportation's Federal Railroad Administration. The survey measures nonfatal injuries and illnesses only. The survey excludes the self-employed; farms with fewer than 11 employees; private households; federal government agencies; and, for national estimates, employees in state and local government agencies. The annual survey provides estimates of the number and frequency (incidence rates) of workplace injuries and illnesses based on logs kept by private industry employers during the year. These records reflect not only the year's injury and illness experience, but also the employer's understanding of which cases are work related under current recordkeeping guidelines of the U.S. Department of Labor. The number of injuries and illnesses reported in any given year also can be influenced by the level of economic activity, working conditions and work practices, worker experience and training, and the number of hours worked. The survey measures the number of new work-related illness cases which are recognized, diagnosed, and reported during the year. Some conditions (for example, long-term latent illnesses caused by exposure to carcinogens) often are difficult to relate to the workplace and are not adequately recognized and reported. These long-term latent illnesses are believed to be understated in the survey's illness measures. In contrast, the overwhelming majority of the reported new illnesses are those that are easier to directly relate to workplace activity (for example, contact dermatitis or carpal tunnel syndrome). Establishments are classified in industry categories based on the 1987 Standard Industrial Classification (SIC) Manual, as defined by the Office of Management and Budget. In the trucking and warehousing and transportation by air industries, SIC coding changes that were introduced with the 1996 BLS Covered Employment and Wages program were incorporated into the estimates for the 1996 survey.

The survey estimates of occupational injuries and illnesses are based on a scientifically selected probability sample, rather than a census of the entire population. Because the data are based on a sample survey, the injury and illness estimates probably differ from the figures that would be obtained from all units covered by the survey. To determine the precision of each estimate, a standard error was calculated. The standard error defines a range (confidence interval) around the estimate. The approximate 95-percent confidence interval is the estimate plus or minus twice the standard error. The standard error also can be expressed as a percent of the estimate, or the relative standard error. For example, the 95-percent confidence interval for an incidence rate of 6.5 per 100 full-time workers with a relative standard error of 1.0 percent would be 6.5 plus or minus 2 percent (2 times 1.0 percent) or 6.37 to 6.63. One can be 95 percent confident that the "true" incidence rate falls within the confidence interval. The 1998 incidence rate for all occupational injuries and illnesses of 6.7 per 100 full-time workers in private industry has an estimated relative standard error of about 0.9 percent.

The statistics clearly indicate that work related injuries and illnesses are still very much a significant issue in the United States. What these figures do not reflect is the cost in productivity as well as out of pocket costs to industry. Proper training and worker incentive programs that ensure safe work environments can have a dramatic impact on operations.

Chapter 2
HAZARDS IN THE CHEMICAL PROCESS INDUSTRIES

INTRODUCTION

The multitude of products and diversity of processes in the petrochemical industries is vast, with numerous situations that may lead to high risk situations both for workers and communities neighboring operations. Perhaps one of the worst chemical disasters of the century was Bhopal, India. The city of Bhopal (1991 pop. 1,063,662), central India, capital of Madhya Pradesh state, was founded in 1728. Bhopal is a railway junction and industrial center, producing electrical equipment, textiles, and jewelry. Landmarks include the old fort (built 1728) and the Taj-ul-Masajid mosque, the largest in India. On Dec. 3, 1984, the worst industrial accident in history occurred there when a toxic gas leak from a Union Carbide insecticide plant killed over 6,400 people and seriously injured 30,000 to 40,000. The Indian government sued on behalf of over 500,000 victims and in 1989 settled for $470 million in damages and exempted company employees from criminal prosecution. The Indian judiciary rejected that exemption in 1991, and the company's Indian assets were seized (1992) after its officials failed to appear to face charges.

The chemical processing industry is so diverse, with products ranging from large volume commodity chemicals to specialty chemicals and products. As such, only an overview of some of the major industry subcategories can be given, with highlights given to air emissions problems. Practices among this industry vary widely throughout the world, however in the United States, OSH standards have demanded strict safety precautions and procedures to protect workers against inhalation hazards.

The chapter provides an overview of the sources of air pollutants, and as such, is the basis for a list of areas to check within a plant operation to assess potential worker overexposure.

The Occupational Safety and Health Act of 1970 emphasizes the need for standards to protect the health and safety of workers. To fulfill this need, the National Institute for Occupational Safety and Health (NIOSH) has developed a strategy for disseminating information that assists employers to protect their workers from workplace hazards. This strategy includes the development of Special NIOSH Hazard Reviews, which support and complement the major standards development and hazard documentation activities of the Institute. These documents deal with hazards that merit research and concern from the scientific community, even though they are not currently suitable for comprehensive review in a criteria document or a Current Intelligence Bulletin. Special NIOSH Hazard Reviews are distributed to the occupational health community at large—industries, trade associations, unions, and members of the academic and scientific communities. Some of the information provided in this chapter is derived from NIOSH Hazard Reviews, and the emphasis is on inhalation hazards.

GENERAL TERMINOLOGY

The following are important terms used throughout this chapter and the volume. The definitions provided are universally recognized, and in many cases the importance of the term in relation to a MSDS (Material Safety Data Sheet) is explained.

Acid: There are several definitions for acid. The Arrhenius definition is a substance that ionizes in water to product H^+ ions. The Bronsted definition is a substance that is a proton (H^+) donor. This does not require the substances to be in aqueous (water) solution. The Lewis definition is a substance that can accept a pair of electrons. This does not require a proton or aqueous solution. There are several other definitions as well. An acidic solution is defined as one that has a pH less than 7.0. The following are examples of strong acids, meaning that they completely dissociate into ions and form H^+ in aqueous (water) solution. For example $HCl \rightarrow H^+ + Cl^-$. All of these will cause severe burns upon skin contact: Perchloric acid ($HClO_4$), Hydroiodic acid (HI), Hydrobromic acid (Hbr), Hydrochloric acid (HCl), Sulfuric acid (H_2SO_4), and Nitric acid (HNO_3). Weak acids do not dissociate completely into ions. Examples of these include acetic acid (a 5% solution of acetic acid in water is called vinegar), formic acid, ammonium cation, and water itself. The strength of acids can be measured using the pH scale. The lower the pH, the greater the acidity of a solution. Just because an acid is weak does not mean that it can't be harmful. For example, HF, hydrofluoric acid, is a weak acid. When you spill it on your hand it doesn't burn...but over the course of hours it migrates to the bones in your fingers and then begins to dissolve them from the inside out (a painful process; amputation can be required). Some common properties of acids are: (1) They have a sour taste. For example, citric acid in lemons and vinegar are both sour; (2) They can react with metals such as magnesium, zinc or iron to corrode them and produce explosive hydrogen gas. Do not store acids in metal containers; (3) Solutions of acids can conduct electricity. It is important to know the pH of substances because they may be corrosive or react with incompatible materials. For example, acids and bases should not be stored or used near each other as their accidental combination could generate a huge amount of heat and energy, possibly resulting in an explosion. pH is also important to know in case you spill the material on your skin or eyes. Whenever a substance enters the eye, flush with water for 15 minutes and get prompt medical attention.

ACGIH-American Conference of Governmental Industrial Hygienists, Inc.: The American Conference of Governmental Industrial Hygienists, Inc., ACGIH, is an organization open to all practitioners in industrial hygiene, occupational health, environmental health, or safety. Their web site is http://www.acgih.org/. ACGIH publishes over 400 titles in occupational and environmental health and safety. They are most famous for their Threshold Limit Values publication which lists the TLV's for over 700 chemical substances and physical agents, as well as 50 Biological Exposure Indices for selected chemicals.

Acute toxicity: Acute toxicity describes the adverse effects resulting from a single exposure to a substance. Acute toxicity helps workers understand the health consequences from a single exposure to a chemical. Acute toxicity differs from chronic toxicity, which describes the adverse health effects from repeated (lower level) exposures to a substance over a longer period (months to years). Human tests for acute toxicity are not performed because of ethical and legal prohibitions. The U.S. EPA describes the following methods for determination of acute toxicity: *Animal testing*: Animal tests are still used where other laboratory protocols are not available. These tests are combined with other assays (lethality, necroscopy, etc.) to minimize the number of animals sacrificed. Evaluation of acute toxicity data should include the relationship, if any, between the exposure of animals to the test substance and the incidence and severity of all abnormalities, including behavioral and clinical abnormalities, the reversibility of observed abnormalities, gross lesions, body weight changes, effects on mortality, and any other toxic effects. *Use of data from structurally related substances or mixtures*: In order to minimize the need for animal testing for acute effects, the

EPA encourages the review of existing acute toxicity information on chemical substances that are structurally related to the agent under investigation. In certain cases it may be possible to obtain enough information to make preliminary hazard evaluations that may reduce the need for further animal testing for acute effects. *Chemical properties*: For example, if a substance is a strong acid then there is really no need to do skin and eye tests as a corrosive material such as this will obviously cause great harm. *In vitro testing* (test tube experiments): Animal rights activists advocate such methods whenever possible. *Limit testing*: A single group of animals is given a large dose of the agent. If no lethality is demonstrated, no further testing is pursued and the substance is classified in a hazard category according to the dose used.

Alopecia: Alopecia is the loss of hair. Acute or chronic exposure to some chemicals may result in the temporary or permanent loss of hair.

Ames Test: The Ames Test is a way of determining whether a compound causes genetic mutations (changes). Animal liver cell extracts are combined with a special form of salmonella bacteria. The mixture is then exposed to the test substance and examined for signs that the bacteria have mutated (a process called mutagenesis). The Ames test does not directly indicate the carcinogenic (cancer-causing) potential of the substance, however there is a good correlation between mutagen strength and carcinogen strength in rodent studies. Avoid the use of mutagens, if at all possible. If you must work with them, be sure to utilize the proper personal protective equipment (PPE) recommended on the MSDS sheet.

Anesthesia: Anesthesia is a loss of sensation or feeling. Anesthesia (or "anesthetics") is often used deliberately by doctors and dentists to block pain and other sensations during surgical procedures. Treatment for pre- or postoperative pain is called analgesia.

Anhydride: An anhydride is a compound that gives an acid or a base when combined with water. Many substances are not themselves acids or bases, but will become such when exposed to water. This does not necessarily require the addition of water (such as from a fire hose). Many anhydrides will react readily with ambient humidity and even the water present in your skin or lungs. The reaction of anhydrides with water is often very violent and exothermic (giving off a great deal of heat energy). The reaction to form sulfuric acid is one step in the reaction of sulfur oxide emissions to form "acid rain". The reaction to form acetic acid is used in certain silicone caulks, leading to the familiar smell of vinegar (vinegar is a 5% solution of acetic acid in water), while the caulk cures.

Anhydrous: An anhydrous material does not contain any water molecules. Many substances occur naturally as hydrates, compounds that have a specific number of water molecules attached to them. This water can often be removed by heating and/or vacuum to give the anhydrous material. Anhydrous materials can absorb water from their surroundings and find use as dessicants. Examples include those packets of silica gel you find in some consumer goods, as well as dehumidifying sachets used in clothes closets. When an anhydrous material reacts with water, this could release a large amount of heat, possibly leading to a heat or pressure buildup that could result in an explosion.

Anorexia: Anorexia is loss of appetite. You may be familiar with the eating disorder, anorexia nervosa, in which the victim restricts dietary intake to starvation levels. Anorexia may be a symptom of acute or chronic exposure to certain chemicals. If you have suffered an unexplained loss of appetite in conjunction with other unusual symptoms, you may want to explore the MSDS's for chemicals that you use in your workplace.

Anosmia: Anosmia is the loss of the sense of smell. Anosmia can be fatal. Certain toxic chemicals have strong detectable odors at low levels. But at higher levels, these saturate your smell receptors ("olfactory fatigue") and you can no longer smell the material. For example, hydrogen sulfide, H_2S, has the distinctive odor of rotten eggs at or below 10 parts per million (ppm) in air. If you were exposed to an H_2S leak and the concentration went above 10 ppm, you might be lulled into thinking that the leak had stopped, because you could not smell it anymore. H_2S can be fatal at concentrations of several hundred ppm.

Anoxia: Anoxia is the absence of oxygen in inspired gases or in arterial blood and/or in the tissues. This is closely related to hypoxia, which is a severe oxygen deficiency in the tissues. One can think of anoxia as the most extreme case of hypoxia.

ANSI-American National Standards Institute: American National Standards Institute, ANSI, is a private, nonprofit membership organization representing over 1,000 public and private organizations, businesses and government agencies. They seek to develop technical, political and policy consensus among various groups. Their web site is http://www.ansi.org/. ANSI does not develop American National Standards (ANSs), but they accredit qualified groups to do so in their area(s) of technical expertise. There are over 14,000 ANSI-approved standards in use today. ANSI-approved standards are voluntary, however, it is possible that some of the content of these standards could be made into law by a governmental body. ANSI is the official U.S. representative to the International Standards Organization (ISO). ANSI standard Z400.1-1998 "Hazardous Industrial Chemicals-Material Safety Data Sheets-Preparation" is the voluntary standard commonly used to construct MSDSs. You can purchase the standard on-line for $100.00 by following links on the ANSI web site. This standard was developed by the Chemical Manufacturers Association (CMA). The ANSI standard MSDS contains 16 sections: (1) Substance identity and company contact information, (2) chemical composition and data on components, (3) hazards identification, (4) First aid measures, (5) Fire-fighting measures, (6) Accidental release measures, (7) handling and storage, (8) exposure controls and personal protection, (9) physical and chemical properties, (10) stability and reactivity, (11) toxicological information, (12) ecological information, (13) disposal considerations, (14) transport information, (15) regulations, (16) other information. MSDSs in ANSI format have a few distinct advantages over those prepared using the standard OSHA Form 174 format. ANSI-format sheets have all of the information required on the OSHA 174 format sheets, as well as additional useful information. The information on an ANSI format sheet is arranged in a consistent format whereas OSHA has no format requirements, only content requirements. ANSI format is likely to be consistent from country to country, possibly permitting one to use the same MSDS in different markets without modification. Most businesses that issue new MSDSs today use the ANSI standard format.

Aqueous: Aqueous refers to a solution in water. A more exact definition is a solution, in which the solute (the substance dissolved) initially is a liquid or a solid and the solvent is water. Aqueous solutions are not usually flammable, but may be able to carry toxic materials into your body through skin contact or ingestion. Be careful with terminology. A solution of ammonia gas (NH_3) in water is often called ammonium hydroxide, NH_4OH, ammonia water, or simply ammonia. Do not confuse this aqueous solution sometimes called ammonia with ammonia gas (anhydrous ammonia). Aqueous solutions fall into three general categories, based on how well they conduct electricity. Strong electrolytes, when dissolved in water, dissociate completely into ions and conduct electricity. For example, sodium chloride, NaCl, dissociates into Na^+ and Cl^- ions in water. Other examples of strong electrolytes are nitric acid (HNO_3) and sodium hydroxide (NaOH). Weak electrolytes, when dissolved in water, do not dissociate to any large extent and, therefore, do not conduct electricity very well. Examples include ammonia (NH_3) and acetic acid (CH_3COOH). Nonelectrolytes do not dissociate to ions in water and do not conduct electricity. Examples include sugar (sucrose; $C_{12}H_{22}O_{11}$), ethanol (CH_3CH_2OH) and methanol (CH_3OH).

Asphysia: Asphysia is a lack of oxygen, which interferes with the oxygenation of the blood. This condition is the result of asphyxiation, which can result from a number of factors, such as (1) suffocation/strangulation or (2) inhalation of an asphyxiant, such as 100% nitrogen gas.

Asphyxiant: An asphyxiant is a substance that can cause unconsciousness or death by suffocation (asphyxiation). Asphyxiation is an extreme hazard when working in enclosed spaces. Be sure you are trained in confined space entry before working in sewers, storage tanks etc. where gases, such as methane, may displace oxygen from the atmosphere. Asphyxiants themselves are not toxic materials. They work by displacing so much oxygen from the ambient atmosphere, that the hemoglobin in the blood can not pick up enough oxygen from the lungs to fully oxygenate the tissues. As a result, the victim slowly suffocates. The normal composition of air is:

Name	Formula	% by volume
Nitrogen	N_2	78.03
Oxygen	O_2	20.99
Argon	Ar	0.94
Carbon dioxide	CO_2	0.033
Neon	Ne	0.0015

According to the Canadian Center for Occupational Health and Safety (CCOHS) the health effects of asphyxiation are:

% O_2 by volume	Symptoms or effects
16 to 12	Breathing and pulse rate increased, muscular coordination slightly disturbed
14 to 10	Emotional upset, abnormal fatigue, disturbed respiration
10 to 6	Nausea and vomiting, collapse or loss of consciousness
Below 6	Convulsive movements, possible respiratory collapse and death

Examples of asphyxiating gases are nitrogen (NH_3), helium (He), neon (Ne), argon (Ar), methane (CH_4), propane ($CH_3CH_2CH_3$), and carbon dioxide (CO_2). All of the above except carbon dioxide are odorless and tasteless. You can be overcome by these gases without realizing they are present. Again, follow OSHA-approved protocols for confined space entry into sewers, storage tanks etc. Related terms are asphysia, asphyxiation.

Asphyxiation: Asphyxiation is the process by which asphysia (lack of oxygen which interferes with the oxygenation of the blood) occurs. Asphyxiation can result from a number of factors such as (1) suffocation/strangulation or (2) inhalation of an asphyxiant such as 100% nitrogen gas.

Asthma: Occupational asthma, one form of asthma, is a lung disease in which the airways overreact to dusts, vapors, gases, or fumes that exist in the workplace. Symptoms include wheezing, a tight feeling in the chest, coughing and shortness of breath. While occupationally-related asthma is usually reversible, chronic exposure to an irritant can result in permanent lung damage. The worker may become generally asthmatic, reacting to molds, allergens, cigarette smoke, dust mites, pet dander etc.

Asymptomatic: Asymptomatic means neither causing nor exhibiting symptoms of disease. Just because one does not display symptoms of a disease or chemical exposure does not necessarily mean that one does not have the disease or was not harmed. Certain symptoms might occur only 50% of the time...or not at all depending on the individual involved. In general, being asymptomatic is a good thing, but it is not a guarantee of health.

Atrophy: Atrophy is a wasting or decrease in size of a bodily organ, tissue, or part owing to disease, injury, or lack of use. You may have heard this term used in reference to accident or paralysis victims: "his muscles atrophied because of non-use". Exposure to certain chemicals can cause internal organs to degrade, weaken and decrease in size, particularly with chronic (long-term) exposure.

Autoignition: The autoignition temperature of a substance is the temperature at or above which a material will spontaneously ignite (catch fire) without anexternal spark or flame. Storing a substance anywhere near its autoignition temperature is a severe safety hazard. Be careful storing substances in hot areas, such as 1) sheds or cabinets exposed to direct sunlight, 2) adjacent to furnaces, hot water heaters or boilers or 3) places where flames or heat are often used. Knowing a substance's autoignition temperatures is also very useful in the event of a fire. The equipment for determining an autoignition temperature is very similar to that used for flash point determinations.

Bradycardia: Bradycardia is a slow heart rate (60 beats per minute or slower) that does not meet the body's metabolic demands. Symptoms of bradycardia include dizziness, extreme fatigue, shortness of breath, or fainting spells. This can be compared to tachycardia, which is an extremely rapid heart rate, usually signified by a pulse of over 100 beats per minute. Adults usually have a resting heart rate of 70-80 beats per minute, although well-trained athletes can have resting rates in the 50's or 60's. Newborn babies have a normal heart rate of 120-160 beats per minute. A slowed heart rate can lead to a variety of other problems. First aid treatment may include administration of oxygen.

Bronchitis: Bronchitis is inflammation of the bronchi (air passages of the lungs consisting of muscle tissue lined with mucous membranes). Chronic bronchitis is defined by the presence of a mucus-producing cough most days of the month, three months of a year for two successive years without other underlying disease to explain the cough. Chronic bronchitis is usually associated with smoking. Certain occupations that involve irritating dust or fumes are at higher risk. The American Lung Association lists coal miners, grain handlers, metal molders, and other workers exposed to dust as being at a higher risk for chronic bronchitis.

Carcinogen: A carcinogen is a substance that causes cancer (or is believed to cause cancer). A material that is carcinogenic is one that is believed to cause cancer. The process of forming cancer cells from normal cells or carcinomas is called carcinogenesis. OSHA's Hazardous Communications standard 1910.1200 accepts the following sources for establishing that a chemical is a known or potential carcinogen: National Toxicology Program (NTP), "Annual Report on Carcinogens" (latest edition); International Agency for Research on Cancer (IARC) "Monographs" (latest editions), part of the World Health Organization (WHO); 29 CFR part 1910, subpart Z, Toxic and Hazardous Substances, Occupational Safety and Health Administration. The "Registry of Toxic Effects of Chemical Substances" published by NIOSH indicates whether a chemical has been found by NTP or IARC to be a potential carcinogen.

Carcinoma: A carcinoma is a malignant (cancerous) growth that arises from the epithelium (the covering of internal and external surfaces of the body, including the lining of vessels and other small cavities). This includes the skin and lining of the organs such as breast, prostate, lung, stomach or bowel. Carcinomas tend to spread (a process called metastasis) through the blood vessels, lymph channels or spinal fluid to other organs, such as the bone, liver, lung or the brain. According to the American Cancer Society, at least 80% of all cancers are carcinomas.

CAS Number - Chemical Abstracts Service Registry Number: A CAS (Chemical Abstracts Service) Registry Number is a unique identifier that tells you, for example, that acetone and dimethyl ketone are actually the same substance. The Chemical Abstracts Service is a division of the American Chemical Society. OSHA only requires certain items on an MSDS and a CAS number is not one of them. However, authors of MSDS's are allowed to add additional information, such as the CAS number, if they desire. The numbers you see on trucks on the highway are not CAS Numbers, but U.S. Department of Transportation (DOT) codes, which are not necessarily specific to each chemical. Their aim is to assist emergency responders.

Chronic Health Effect: A chronic health effect is an adverse health effect resulting from long-term exposure to a substance. The effects could be a skin rash, bronchitis, cancer or any other medical condition. An example would be liver cancer from inhaling low levels of benzene at your workplace over several years. The term is also applied to a persistent (months, years or permanent) adverse health effect resulting from a short-term (acute) exposure. Chronic effects from long-term exposure to chemicals are fairly common. Recognize the PEL (permissible exposure level) for each substance in your workplace and minimize your exposure whenever possible.

Combustible: A combustible material can be a solid or liquid. The U.S. Occupational Health and Safety Administration (OSHA) defines a combustible liquid as "any liquid having a flash point at or above 100 °F (37.8 °C), but below 200 °F (93.3 °C), except any mixture having components with flashpoints of 200 °F (93.3 °C), or higher, the total volume of which make up 99 percent or more of the total volume of the mixture." Compare this definition to flammable, which indicates a liquid that is even easier to ignite (flash point below 100 °F). OSHA divides combustible (and flammable) liquids into several classes. If you'd like to see these, take a look at 29 CFR 1910.106. Combustible solids are those capable of igniting and burning. Wood and paper are examples of such materials. Proper storage and use of combustible materials is absolutely critical in maintaining a safe work place. Avoid placing or using combustible materials near sources of heat or flame.

Conjunctivitis: Conjunctivitis (also known as pink eye because the white part of the eye becomes pink) is inflammation (swelling) of the mucous membrane lining the eye (the conjunctiva). This is often accompanied by itching and watery eyes and sometimes blurred vision, eye pain and sensitivity to light. This inflammation can be caused by physical injury, allergies, or chemical exposure, but is most commonly caused by bacterial or viral infection. These infections can be highly contagious and spread rapidly among schoolchildren or families. Bacterial conjunctivitis can be treated with antibiotic eye drops, prescribed by a physician, whereas viral conjunctivitis usually clears up on its own, if strict hygiene is followed. Exposure to chemicals that irritate the eyes (such as lachrymators) can cause conjunctivitis or make existing conjunctivitis worse. Seek medical treatment, if the condition does not clear up on its own. If you are not certain whether a chemical was splashed in your eye or you are suffering from conjunctivitis, you should consult a physician immediately.

Cutaneous: Cutaneous relates to or affects the skin. The term subcutaneous refers to being below the skin (as in a penetrating injury or injection). Use your MSDS to determine the required personal protective equipment (PPE) that you must use. Protecting the skin (with gloves, aprons, coveralls, face masks etc.) is important. After all, the skin is the largest organ in the human body.

Cyanosis: Cyanosis is an abnormal bluish color of the skin or mucous membranes. The bluish (cyan) or blue-gray color arises from deoxygenated hemoglobin, the oxygen carrier in your bloodstream that carries oxygen from your lungs to your tissues. Hemoglobin and your blood are red when well-oxygenated, but blood appears to be dark red-blue if there is more than 50 g/L of hemoglobin without oxygen. Cyanosis can be seen in cases of anoxia and hypoxia (lack of oxygen) and is a symptom of asphyxiation. Cyanosis is also observed when a chemical agent blocks the ability of hemoglobin to bind oxygen. For example, carbon monoxide, CO, a product of imperfect combustion, binds to hemoglobin approximately 200 times better than oxygen. Victims of carbon monoxide poisoning often have blue lips and fingernails. Treatment for cyanosis includes administration of pure oxygen. In the case of carbon monoxide poisoning, hyperbaric oxygen treatment, placing the victim in a chamber pressurized to 2 or 3 atmospheres of pure oxygen, may be used. Cyanosis is harder to observe in dark-skinned people. The best places to look in this case are in the buccal mucosa (inside of the cheek) and hard palate (roof of the mouth). Cyanosis is an early sign of hypoxia. If you are working in a confined space, where the oxygen content may be less than normal or are working with a chemical substance that interferes with oxygen transport in the body, be sure to recognize the bluish cast of cyanosis.

Dermal Toxicity: Dermal toxicity is the ability of a substance to poison people or animals by contact with the skin. Toxic materials absorb through the skin to various degrees depending on their chemical composition and whether they are dissolved in a solvent. Always wear proper personal protection equipment (PPE), such as gloves and aprons, when working with a toxic (or nontoxic) substance that can be absorbed through the skin.

Dyspnea: Dyspnea is shortness of breath or difficulty in breathing. The victim is usually quite aware of the unusual breathing pattern. Shortness of breath can be an indicator of many physical ailments including simple exertion, a panic attack, a blow to the chest, asthma, cardiac disease, as well as exposure to toxic chemicals. If a person is suffering from shortness of breath, evaluate them for additional symptoms and possible exposures. Keep the victim in a sitting position. Remove the victim to fresh air, if possible, and seek medical attention.

Edema: Edema is an abnormal accumulation of body fluid in tissues. An edema can be as trivial as a blister on your thumb, as life-threatening as a constriction of your airway. As in real estate, the three factors that determine the dangers associated with an edema are location, location, location. Exposure to toxic chemicals can cause a variety of edemas. Pulmonary edema (fluid in the lungs) is particularly dangerous, if not treated. Be sure to avoid inhalation of chemicals whenever possible and to use proper protective measures, as suggested on the MSDS (fume hoods, respirators, etc.).

Emergency Planning and Community Right-To-Know Act (EPCRA): The U.S. Emergency Planning and Community Right-To-Know Act (EPCRA) also known as the Community Right-To-Know Act or SARA, Title III provides for the collection and public release of information about the presence and release of hazardous or toxic chemicals in the nation's communities. The law requires industries to participate in emergency planning and to notify their communities of the existence of, and routine and accidental releases of, hazardous chemicals. The goal is to help citizens, officials, and community leaders to be better informed about toxic and hazardous materials in their communities. To implement EPCRA, Congress required each state to appoint a State Emergency Response Commission (SERC). The SERCs were required to divide their states into Emergency Planning Districts and to name a Local Emergency Planning Committee (LEPC) for each district. Broad representation by fire fighters, health officials, government and media representatives, community groups, industrial facilities, and emergency managers ensures that all necessary elements of the planning process are represented. If you have a major chemical user or manufacturer in your community, plans to deal with emergency releases have already been developed. Consult your local EPA office for more information. A list of over 600 chemicals subject to EPCRA are listed in the Toxics Release Inventory (TRI), which is maintained by the U.S. Environmental Protection Agency (EPA). EPCRA or TRI information is not equivalent to an MSDS, but does provide useful information for people concerned about the presence (or potential presence) of chemicals in their community or environment. The information found in these materials can supplement MSDS information, but is not a substitute for it.

Gastroenteritis: Gastroenteritis is an acute inflammation of the lining of the stomach and intestines. Symptoms include anorexia, nausea, diarrhea, abdominal pain and weakness. Gastroenteritis has many causes, such as bacteria (food poisoning), viruses, parasites, consumption of irritating food or drink, as well as stress. Treatment for the condition depends on the underlying cause.

Hepatic: Hepatic means "pertaining to the liver." For example, hepatitis is inflammation of the liver. Liver disorders are sometimes marked by jaundice, a yellowish coloration to the whites of the eyes and skin. Certain chemicals are hepatotoxins (toxic to the liver), usually as a result of chronic exposure. One example is carbon tetrachloride (CCl_4).

Highly Toxic: A highly toxic material is defined by the U.S. Occupational Health and Safety Administration (OSHA) in CFR 29 1910.1200 Appendix A as a chemical that falls in any of these three categories: A chemical that has a median lethal dose (LD_{50}) of 50 milligrams or less per kilogram of body weight when administered orally to albino rats weighing between 200 and 300 grams each. A chemical that has a median lethal dose (LD_{50}) of 200 milligrams or less per kilogram of body weight, when administered by continuous contact for 24 hours (or less if death occurs within 24 hours) with the bare skin of albino rabbits, weighing between two and three kilograms each. A chemical that has a median lethal concentration (LC_{50}) in air of 200 parts per million by volume or less of gas or vapor, or 2 milligrams per liter or less of mist, fume, or dust, when administered by continuous inhalation for one hour (or less if death occurs within one hour) to albino rats weighing between 200 and 300 grams each. This is the greatest level of toxicity defined in the OSHA Hazard Communication Standard (OSHA does not have an "Extremely toxic" ranking). Accidental release or exposure to a highly toxic chemical can cause serious injury or death. Use proper protective equipment (gloves, safety goggles, fume hoods etc.) when working with highly toxic materials. Know the physical properties of the material, as well as symptoms of exposure and first aid procedures.

Hygroscopic: A hygroscopic material (literally "water seeking") is one that readily absorbs water (usually from the atmosphere). In most cases, the water can be removed from the material by heating (sometimes under vacuum or under a flow of dry gas such as nitrogen). Hygroscopic materials are fairly common. Some may absorb a finite amount of water (such as magnesium sulfate, $MgSO_4$), while others may attract so much water that they form a puddle and dissolve (deliquesce). For example, solid sodium hydroxide (NaOH) pellets will form a small corrosive puddle in less than an hour in moist air. Therefore, always be sure to clean up any spills of hygroscopic materials right away. Also be aware that hygroscopic materials typically release a large amount of heat, when mixed with water. Always store hygroscopic materials in well-sealed containers (or under vacuum or an inert atmosphere). Know their physical properties, so that, if you open a container, you can tell, if the material has been contaminated with water (i.e., that jar of calcium chloride, $CaCl_2$, should be a solid, not a liquid).

Hypergolic: A hypergolic mixture ignites upon contact of the components without any external source of ignition (heat or flame). The only field, in which this is a desirable event, is in rocket fuel research. Accidental mixing of incompatible materials can lead to a fire or explosion. Here is one example provided by the staff at ILPI of what can happen, when incompatibles are mixed. Always read the labels on your bottles (don't assume a chemical's identity by the shape, size or color of the bottle), and know what materials are incompatible with the chemicals that you are using.

Hypoxia: Hypoxia is a deficiency of oxygen in inspired (inhaled) gases or in arterial blood and/or in the tissues. This is closely related to anoxia, which is a complete lack of oxygen in the tissues. One can think of anoxia as the most severe case of hypoxia. Various forms of hypoxia are recognized: anemic hypoxia, which results from a decreased concentration of hemoglobin; hypoxic hypoxia, which results from defective oxygenation of the blood in the lungs; ischemic hypoxia - results from slow peripheral circulation (also called stagnant hypoxia). Not uncommon following congestive cardiac failure; altitude sickness-nosebleed, nausea or pulmonary edema experienced at high altitudes. The most common symptom of hypoxia is cyanosis, a bluish cast to the skin, lips and/or fingernails. If your body isn't getting oxygen, you die. Make sure you recognize cyanosis when you see it. If working in an enclosed space or with an asphyxiant, move to a well-ventilated area if you become light-headed, weak or disoriented. Related terms are anoxia, asphyxiant, cyanosis.

IARC: International Agency for Research on Cancer: The International Agency for Research on Cancer (IARC) is a part of the World Health Organization. IARC's mission is to coordinate and conduct research on the causes of human cancer, the mechanisms of carcinogenesis, and to develop scientific strategies for cancer control. Their web home page is http://www.iarc.fr/. IARC compiles several databases on carcinogenic risk to humans, epidemiology and cancer control. The IARC Monographs series is one of four resources that OSHA uses to list a material as known or probable human carcinogen. You can view a list of over 800 agents covered in the monographs. IARC classifies agents (chemicals, mixtures, occupational exposures etc.) into four basic categories: **Group 1**: The agent (mixture) is carcinogenic to humans. The exposure circumstance entails exposures that are carcinogenic to humans. **Group 2**: The agent (mixture) is probably carcinogenic to humans and **Group 2A**: The exposure circumstance entails exposures that are probably carcinogenic to humans. **Group 2B**: The exposure circumstance entails exposures that are possibly carcinogenic to humans. **Group 3**: The agent (mixture, or exposure circumstance) is unclassifiable as to carcinogenicity in humans. **Group 4**: The agent (mixture, exposure circumstance) is probably not carcinogenic to humans.

Iridocyclitis: Iridocyclitis is an inflammation of the iris (the colored part of the eye) and of the ciliary body (muscles and tissue involved in focusing the eye). This condition is also called "anterior uveitis." The condition can be marked by red eye, pain, photophobia (light sensitivity, literally "fear of light"), watering of the eyes and a decrease in vision. If only one eye is affected, shining light in the good eye can produce pain in the affected eye. This is closely related to conjunctivitis, swelling of the mucous membranes around the eye. Treatments include steroid, atropine, antibiotic or antiviral eyedrops. Exposure to chemicals that irritate the eyes (such as lachrymators) can cause iridocyclitis or aggravate an existing case. Seek medical treatment, if the condition does not clear up on its own. If you are not certain, whether a chemical was splashed in your eye or you are suffering from iridocyclitis, you should consult a physician immediately.

Ketosis: Ketosis is the presence of excess ketones in the body. Ketones are chemicals with a carbonyl unit (a carbon doubly bonded to an oxygen), that has two alkyl or aromatic (hydrocarbon) substituents, bonded to the carbon atom. Ketones are a byproduct of fat metabolism (the breaking down of fat into energy). Normally, your body is efficient at removing these, but when certain enzymes are absent or damaged, the amount of ketones in the body can build up to dangerous levels. Certain individuals are predisposed towards ketosis. For example, those with diabetes have low insulin levels and can not process glucose (sugar) for energy. Therefore, their bodies break down fat, leading to a rise in ketone levels. Ketones can be excreted through the urine and those that are volatile (such as acetone) can be expelled through the lungs. Diabetics can be mistaken for being drunk by the odor on their breath, and acetone being expelled through the lungs can give a false positive result on early model breathalyzers. However, don't expect to escape a ticket as driving while impaired (due to low blood sugar) is still a traffic offense whether you are drunk or not. Ketosis can lead to coma and death, if untreated.

Lachrymator: A lachrymator is an irritant that causes tearing (watering of the eyes). Examples include onions, tear gas, and pepper spray (capsacin). Some typical lachrymating chemicals are thionyl chloride ($SOCl_2$) and acrolein ($CH_2=CH-CHO$). Certain chemicals may say lachrymator on the label so treat these with respect. Use these only in a fume hood. Goggles or safety glasses are not adequate protection for lachrymators, because the fumes can still reach your eyes directly or through inhalation.

LC_{50}, 50% Lethal Concentration: An LC_{50} value is the concentration of a material in air that will kill 50% of the test subjects (animals, hopefully), when administered as a single exposure (typically 1 or 4 hours). This value gives you an idea of the relative toxicity of the material. This value applies to vapors, dusts, mists and gases. Solids and liquids use the closely related LD_{50} value (50% lethal dose). The formula for determination of an LC_{50} is rather complex and can be found in 49 CFR 173.133(b)(1)(i). Both LC_{50} and LD_{50} values state the animal used in the test. This is important, because animal toxicity studies do not necessarily extrapolate (extend) to humans. For example, dioxin (of Love Canal, Times Beach, Sveso, and Agent Orange fame) is highly toxic to guinea pigs and ducklings at extremely low levels, but has never been conclusively linked to a single human death even at very high levels of acute (short term) exposure. However, it is best to err on the safe side, when evaluating animal toxicity studies and assume that most chemicals that are toxic to animals are toxic to humans. Typical units for LC_{50} values are parts per million (ppm) of material in air, micrograms (10^{-6} = 0.000001 g) per liter of air and milligrams (10^{-3} = 0.001 g) per cubic meter of air. Never be exposed to an LC_{50} dose of a hazardous chemical - by definition, there is a 50% chance this will kill you...and if you survive you're not going to be in good shape. Pay close attention to the permissible exposure level (PEL) instead. This is a more realistic determination of the maximum safe exposure to a material and is usually based on the known effects of the chemical on humans, rather than laboratory animals.

LD_{50}, 50% Lethal Dose: An LD_{50} value is the amount of a solid or liquid material that it takes to kill 50% of test animals in one dose. LC_{50} (50% lethal concentration) is a related term used for gases, dusts, vapors, mists, etc. The dose may be administered orally (by mouth), or injected into various parts of the body. The value is usually reported along with the administration method. Both LC_{50} and LD_{50} values state the animal used in the test. This is important, because animal toxicity studies do not necessarily extrapolate (extend) to humans. For example, dioxin (of Love Canal, Times Beach, Sveso, and Agent Orange fame) is highly toxic to guinea pigs and ducklings at extremely low levels, but has never been conclusively linked to a single human death even at very high levels of acute (short term) exposure. However, it is best to err on the safe side when evaluating animal toxicity studies and assume that most chemicals, that are toxic to animals, are toxic to humans. Typical units for LD_{50} values are milligrams or grams of material per kilogram of body weight (mg/kg or g/kg, recall that 1 kg = 2.2 pounds). Never be exposed to an LD_{50} dose of a hazardous chemical- by definition, there is a 50% chance this will kill you...and if you survive you are not going to be in good shape. Pay close attention to the permissible exposure level (PEL) instead. This is a more realistic determination of the maximum safe exposure to a material and is usually based on the known effects of the chemical on humans, rather than laboratory animals.

Metastasis: Metastasis is the spread of a disease (usually cancer) from an original site of infection to other parts of the body. This usually happens when cancer cells break off from the original tumor and travel through the blood vessels to a new site. This kind of cancer is called malignant, meaning that it is life-threatening and usually fatal.

Mutagen: A mutagen is a substance or agent that causes an increase in the rate of change in genes (subsections of the DNA of the body's cells). These mutations (changes) can be passed along as the cell reproduces, sometimes leading to defective cells or cancer. Examples of mutagens include biological and chemical agents as well exposure to ultraviolet light or ionizing radiation. There are many types of mutations, some of which are harmful and others, which have little or no effect on the body's function. See the Life Science Dictionary at the University of Texas, for examples. Mutagens can be identified using the Ames test and other biochemical testing methods. Do not confuse a mutagen with a carcinogen (a substance that causes cancer). Mutagens may cause cancer, but not always. Do not confuse a mutagen with a teratogen (a substance that causes change or harm to a fetus or embryo). Whenever you work with a mutagen be sure to wear proper protective equipment (PPE) and minimize your exposure.

Narcosis: Narcosis is a state of deep stupor or unconsciousness, produced by a chemical substance, such as a drug or anesthesia. Inhalation of certain chemicals can lead to narcosis. For example, diethyl ether and chloroform, two common organic solvents, were among the first examples of anesthesia known. Many other chemicals that you would not suspect can also cause narcosis. For example, even though nitrogen gas comprises 80% of the air we breathe and is considered chemically inert (unreactive) it can cause narcosis under certain conditions. Always work with adequate inhalation and avoid inhaling chemical fumes, mists, dusts etc. whenever possible. Use fume hoods and respirators as necessary.

Necrosis: Necrosis is the death of cells or tissue due to disease, injury, exposure to chemical agents, radiation etc. Necrotic tissue can be found anywhere in the body, but is generally a localized phenomenon. When extremely large areas of tissue are involved, the condition is generally called gangrene. Corrosive materials may cause painful chemicals burns and necrosis of the skin or eyes. Always wear proper personal protection equipment (PPE) such as gloves and goggles when dealing with such materials.

Neoplasm: A neoplasm is an abnormal growth of tissue that has no useful function. A synonym is "tumor." Neoplasms may be benign (no ability to spread to other parts of the body) or malignant (cancerous). Chronic (long-term) exposure to certain chemicals can result in the formation of neoplasms. While not all of these tumors are cancerous, benign tumors have the potential to interfere with vital body functions or become malignant. Limit your exposure to chemicals that are known to cause neoplasms (tumors). This includes reducing the usage of such chemicals in your workplace as well as using proper personal protective equipment (PPE) such as gloves, respirators and fume hoods. Related terms are carinogen, carcinoma, malignant, mutagen, and teratogen.

NIOSH - National Institute for Occupational Safety and Health: The National Institute for Occupational Safety and Health, NIOSH, is part of the U.S. federal government's Centers for Disease Control and Prevention (CDC). NIOSH's web site is http://www.cdc.gov/niosh/homepage.html. NIOSH is the only federal Institute responsible for conducting research and making recommendations for the prevention of work-related illnesses and injuries. NIOSH was created by the Occupational Health and Safety (OHS) Act of 1970. OSHA is responsible for creating and enforcing workplace safety and health regulations while NIOSH is in the Department of Health and Human Services and is a research agency. Examples of NIOSH activities include: investigating potentially hazardous working conditions as requested by employers or employees; evaluating hazards in the workplace, ranging from chemicals to machinery; creating and disseminating methods for preventing disease, injury, and disability; conducting research and providing scientifically valid recommendations for protecting workers; providing education and training to individuals preparing for or actively working in the field of occupational safety and health; creating new ways to prevent workplace hazards is the job of NIOSH. Many NIOSH-approved standards appear on MSDSs. Other examples of NIOSHs involvement with MSDSs: NIOSH cooperates with other world agencies in providing International Chemical Safety Cards which, while not strictly MSDSs, provide workers with succinct information about the hazards of chemicals;   NIOSH plays a role in establishing PEL, TLV and STELs for a variety of chemicals; the NIOSH Pocket Guide to Chemical Hazards (NPG); NIOSH publishes Occupational Health Guidelines for Chemical Hazards listed by chemical; The Registry for Toxic Effects of Chemical Substances (RTECS), a toxicology database containing over 140,000 chemicals substances.

NTP- National Toxicology Program: The National Toxicology Program (NTP) is a unit of the U.S. Department of Health and Human Services. NTP coordinates toxicology research and testing activities within the Department and provides information about potentially toxic chemicals to regulatory and research agencies and the public. NTP is the world leader in designing, conducting, and interpreting animal assays for toxicity and carcinogenicity. Their web home page is http://ntp-server.niehs.nih.gov/. The NTP consists of relevant toxicology activities of the U.S. National Institutes of Health's National Institute of Environmental Health Sciences (NIH/NIEHS), the U.S. Centers for Disease Control and Prevention's National Institute for Occupational Safety and Health (CDC/ NIOSH), and the U.S. Food and Drug Administration's National Center for Toxicological Research (FDA/NCTR). The NTP's Annual Reports on Carcinogens (see the latest edition) are one of the four sources used by OSHA to declare substances as carcinogens. NTP is one of your most trusted sources when it comes to information about toxic chemicals or potential carcinogens.

Nystagmus: Nystagmus is a rapid, involuntary, motion of the eyeball (side to side, up and down, rotating and/or oscillating). Nystagmus can be caused by tumors or drugs (such as barbiturates). It can also be a congenital condition (present at birth). Exposure to certain chemicals can cause nystagmus. This condition is very easy for other persons to observe. Be certain you read the MSDS's for all chemicals that you are working with and note whether exposure to any of these can cause this condition.

OSHA- Occupational Health and Safety Administration: The U.S. Occupational Health and Safety Administration, OSHA, is a federal government agency in the U.S. Department of Labor. OSHA's web site is http://www.osha.gov/. The primary goals of OSHA are to save lives, prevent injuries and protect the health of America's workers. OSHA employs over 2,000 inspectors to ensure job site safety. OSHA was created by the Occupational Health and Safety (OHS) Act of 1970. While many see OSHA as an intrusive government agency intent on enforcing arcane rules, the fact is that OSHA saves lives. And if that is not enough for you, their Voluntary Protection Plan (VPP) saves money by reducing the cost of injuries, accidents, downtime and litigation. OSHA's Hazard Communication Standard 1910.1200 requires employers to establish hazard communication programs to transmit information on the hazards of chemicals to their employees by means of labels on containers, material safety data sheets, and training programs. Implementation of these hazard communication programs will ensure all employees have the "right-to-know" the hazards and identities of the chemicals they work with, and will reduce the incidence of chemically-related occupational illnesses and injuries. The OSHA HazCom standard specifies the required elements that must be on an MSDS among other important data. It is a very readable document, and it is suggested that anyone involved with MSDS management print out a hard copy for future reference. OSHA has a suggested format for MSDS's, Form 174 (OMB #1218-0072). You can download this form in HTML or PDF format from the U.S. Department of Labor's Occupational Safety and Health Administration world wide web site, if you wish. While this format is nonmandatory, it is a frequently utilized format. An MSDS can contain more information than that required by OSHA, but not less. Form 174 has the following sections: Chemical Identity-The identity of the substance as it appears on the label. Section I. Manufacturer's Name and Contact Information: Manufacturer's name, address, telephone number and emergency telephone number. Date the MSDS was prepared and an optional signature of the preparer. Section II. Hazardous Ingredients/Identity Information - Lists the hazardous components by chemical identity and other common names. Includes OSHA PEL (Permissible Exposure Limit), ACGIH TLV (Threshold Level Value) and other recommended exposure limits. Percentage listings of the hazardous components is optional. Section III. Physical/Chemical Characteristics - Boiling point, vapor pressure, vapor density, specific gravity, melting point, evaporation rate, solubility in water, physical appearance and odor. Section IV. Fire and Explosion Hazard Data - Flash point (and method used to determine it), flammability limits, extinguishing media, special firefighting procedures, unusual fire and explosion hazards. Section V. Reactivity Data - Stability, conditions to avoid, incompatibility (materials to avoid), hazardous decomposition or byproducts, hazardous polymerization (and conditions to avoid). VI. Health Hazard Data - Routes of entry (inhalation, skin, ingestion), health hazards (acute = immediate and chronic = build up over time), carcinogenicity (NTP, IARC monographs, OSHA regulated), signs and symptoms of exposure, medical conditions generally aggravated by exposure, emergency, and first aid procedures.VII. Precautions for Safe Handling and Use - Steps to be taken in case material is released or spilled, waste disposal method, precautions to be taken in handling or storage, and other precautions. VIII. Control Measures - Respiratory protection (specify type), ventilation (local, mechanical exhaust, special or other), protective gloves, eye protection, other protective clothing or equipment, work/hygienic practices. A competing format, the ANSI format is emerging as the standard format for MSDS. This format contains all of the information found on Form 174, but includes additional information/categories and has a consistent organization.

Permissible Exposure Limit (PEL): A Permissible Exposure Limit (PEL) is the maximum amount or concentration of a chemical that a worker may be exposed to under OSHA regulations. PEL can be defined in two different ways as discussed in the OSHA regulation on air contaminants 1910.1000: Ceiling values at no time should this exposure limit be exceeded. *8-hour Time Weighted Averages (TWA):* This is an average value of exposure over the course of an 8 hour work shift. TWA levels are usually lower than ceiling values. Thus, a worker may be exposed to a level higher than the TWA for part of the day (but still lower than the ceiling value) as long as he is exposed to levels below the TWA for the rest of the day. See 1910.1000 for the formulas used in the calculations. PELs are defined by OSHA in 3 Tables: Table Z-1 Limits for Air Contaminants, Table Z-2 Acceptable maximum peak above the acceptable ceiling level for an 8 hour shift, Table Z-3 Mineral dusts. In general, PELs refer to substances that may be inhaled, although some can be absorbed through the skin or eyes. When working with materials that have a PEL or TWA listed use proper precautions to minimize the generation of a vapor or dust in the first place. Always use appropriate personal protective equipment (PPE) such as gloves, dust masks, and respirators to limit your exposure to chemicals. Remember, exposure limits are not some magic threshold that define the border between safe and dangerous. A PEL that was acceptable in 1950 may be recognized as dangerously high today. Therefore, always do everything reasonable to limit the airborne release of chemicals or dusts in the first place. Chemical Sampling Information at OSHA lists the PELs and/or TWAs for many substances, health effects, and equipment/manufacturers that can monitor concentration for PEL/TWA compliance.

Poison: A poison is a substance that adversely affects one's health by causing injury, illness, or death. A gas poisonous by inhalation is defined by 49 CFR 173.115 as "material which is a gas at 20 °C (68 °F) or less and a pressure of 101.3 kPa (14.7 psi) (a material which has a boiling point of 20 °C (68 °F) or less at 101.3 kPa (14.7 psi)) and which: (1) Is known to be so toxic to humans as to pose a hazard to health during transportation, or (2) In the absence of adequate data on human toxicity, is presumed to be toxic to humans because when tested on laboratory animals it has an LC_{50} value of not more than 5000 ml/m^3 ... LC_{50} values for mixtures may be determined using the formula in Sec. 173.133(b)(1)(i) of this subpart." A poisonous substance (other than a gas) is defined by 49 CFR 173.132 as a material "which is known to be so toxic to humans as to afford a hazard to health during transportation, or which, in the absence of adequate data on human toxicity: Is presumed to be toxic to humans because it falls within any one of the following categories when tested on laboratory animals (whenever possible, animal test data that has been reported in the chemical literature should be used): *Oral Toxicity*: A liquid with an LD_{50} for acute oral toxicity of not more than 500 mg/kg or a solid with an LD_{50} for acute oral toxicity of not more than 200 mg/kg. *Dermal Toxicity*: A material with an LD_{50} for acute dermal toxicity of not more than 1000 mg/kg. *Inhalation Toxicity*: A dust or mist with an LC_{50} for acute toxicity on inhalation of not more than 10 mg/L; or A material with a saturated vapor concentration in air at 20 °C (68 °F) of more than one-fifth of the LC_{50} for acute toxicity on inhalation of vapors and with an LC_{50} for acute toxicity on inhalation of vapors of not more than 5000 ml/m^3 or Is an irritating material, with properties similar to tear gas, which causes extreme irritation, especially in confined spaces." Treat poisonous materials with respect! Read the MSDS sheet to find out what ways a poison can kill you - is it by inhalation, ingestion and/or skin contact? What precautions should you take with the material? What kinds of personal protective equipment are recommended?

Registry of Toxic Effects of Chemical Substances (RTECS): The Registry of Toxic Effects of Chemical Substances (RTECS) is a toxicology database of over 140,000 chemicals compiled, maintained, and updated by the U.S. National Institute of Occupational Safety and Health (NIOSH). Its goal is to include "all known toxic substances... and the concentrations at which... toxicity is known to occur." RTECS was mandated by the same act that created the U.S. Occupational Health and Safety Administration (OSHA). It was originally called the Toxic Substances List. RTECS data is obtained from the open scientific literature and maintained by NIOSH. Unlike many other government databases, RTECS is only available from vendors (if you dislike this, which you should, consult your congressional representatives). NIOSH has additional info and a vendor list if you would like access to the database. The price is roughly $300 U.S. for a one-year CD-ROM subscription and $250 for renewals. RTECS data can be used to help you construct an MSDS. Six types of toxicity data are included in the file primary irritation, mutagenic effects, reproductive effects, tumorgenic effects, acute toxicity, and other multiple dose toxicity, including LD_{50} and LC_{50} values.

SARA: Superfund Amendments and Reauthorization Act: The U.S. Superfund Amendments and Reauthorization Act, SARA is an amendment and reauthorization of CERCLA, the Comprehensive Environmental Response, Compensation & Liability Act (CERCLA) of 1980, better known as the SuperFund Act. Both CERCLA and SARA have the goals of identifying, remediating and preventing the release of hazardous substances to the environment. SARA not only extended the life of CERCLA, but made several important changes to provide new tools for enforcement, remedies, funding, and both state and individual input. SARA also resulted in a revision of the U.S. EPA's (Environmental Protection Agency) Hazard Ranking System to assess the degree of hazard to humans and the environment. The Emergency Planning and Community Right-To-Know Act (EPCRA), also known as the Community Right-To-Know Law, is also known as Title III of SARA. This provides specific plans for preparing for, preventing, and responding to the release of over 600 chemicals listed in the Toxics Release Inventory. Any release of one or more of the roughly 800 CERCLA or 360 EPCRA hazardous substances that equals or exceeds a reportable quantity (RQ) must be reported to the National Response Center (NRC). RQs are adjusted to one of five levels: 1, 10, 100, 1,000, or 5,000 pounds. EPA bases adjustments to the RQs on the intrinsic characteristics of each hazardous substance, such as the aquatic toxicity, acute and chronic toxicity, ignitability, reactivity, and potential carcinogenicity. An RQ value is established for each of these characteristics of a hazardous substance, with the most stringent RQ value (i.e., the lowest quantity) becoming the final RQ or reporting trigger for that hazardous substance.

Sensitizer: A sensitizer is defined by OSHA as "a chemical that causes a substantial proportion of exposed people or animals to develop an allergic reaction in normal tissue after repeated exposure to the chemical." The condition of being sensitized to a chemical is also called chemical hypersensitivity. Certain chemicals have no immediate health effect. But if you are exposed to them several times, they can make you allergic or sensitive to other chemicals. A classic example is formaldehyde (HCHO). Typical reactions to sensitizers can include skin disorders such as eczema. When working with sensitizers, always use proper protective equipment such as gloves, respirators, etc. Once you are sensitized to a particular chemical, even minute amounts will cause symptoms. Sensitization is usually a life-long effect.

Short Term Exposure Limit (STEL): A Short Term Exposure Limit (STEL) is defined by ACGIH as the concentration to which workers can be exposed continuously for a short period of time without suffering from: irritation, chronic or irreversible tissue damage, narcosis of sufficient degree to increase the likelihood of accidental injury, impair self-rescue or materially reduce work efficiency. STELs are generally used only when toxic effects have been reported from high acute (short-term) exposures in either humans or animals. A STEL is not a separate independent exposure limit, but supplements time-weighted average limits where there are recognized acute effects from a substance whose toxic effects generally chronic (long-term) in nature. For example, one can not be exposed to a STEL concentration if the TLV-TWA (time weighted average for an 8 hour shift; see Permissible Exposure Limit (PEL)) would be exceeded. Workers can be exposed to a maximum of four STEL periods per 8 hour shift, with at least 60 minutes between exposure periods. In general, PELs and TlV-STELs refer to substances that may be inhaled, although some can be absorbed through the skin or eyes (STELs will often have "-skin" after them, when skin exposure is possible). When working with materials that have listed exposure limits, use proper precautions to minimize the generation of a vapor or dust in the first place. Always use appropriate personal protective equipment (PPE) such as gloves, dust masks, and respirators to limit your exposure to chemicals. Remember, exposure limits are not some magic threshold that define the border between safe and dangerous. A PEL or STEL that was acceptable in 1950 may be recognized as dangerously high today. Therefore, always do everything reasonable to limit the airborne release of chemicals or dusts in the first place.

Target Organ Effects: Target organ effects indicate which bodily organs are most likely to be affected by exposure to a substance. Some terms used when describing target organ effects are defined as follows:

Class and Definition	Signs/Symptoms	Examples
Hepatotoxins - produce hepatic (liver) damage	Jaundice, liver enlargement	Carbon tetrachloride, nitrosamines
Nephrotoxins- produce kidney damage	Edema, proteinuria	Halogenated hydrocarbons, uranium

Class and Definition	Signs/Symptoms	Examples
Neurotoxins- produce their primary toxic effects on the nervous system	Narcosis, behavioral changes, decrease in motor functions	Mercury, carbon disulfide
Hematopoietic agents- act on the blood or hematopoietic system, decrease hemoglobin function, deprive the body tissues of oxygen	Cyanosis, loss of consciousness	Carbon monoxide, cyanides
Agents which damage the lung- these irritate or damage pulmonary (lung) tissue	Cough, tightness in chest, shortness of breath	Silica, asbestos
Reproductive toxins- affect the reproductive capabilities including chromosomal damage (mutations) and effects on fetuses (teratogenesis)	Birth defects, sterility	Lead, DBCP
Cutaneous hazards- affect the dermal layer (skin) of the body	Defatting of the skin, rashes, irritation	Ketones, chlorinated compounds
Eye hazards- affect the eye or visual capacity	Conjunctivitis, corneal damage	Organic solvents, acids

When working with chemicals that have target organ effects it is critical to prevent exposure. This is especially true if you have a pre-existing condition, disease or injury to that particular organ. Read the MSDS to find out what the most effective personal protection equipment (PPE) for dealing with the chemical and be certain to minimize release of the chemical in the first place.

Teratogen: A teratogen is an agent that can cause malformations of an embryo or fetus. This can be a chemical substance, a virus or ionizing radiation. Pregnant women should avoid all contact with teratogens, particularly during the first three months of pregnancy, as this can result in damage to the developing child. For example, alcohol is a teratogen and drinking during pregnancy can lead to a child born with fetal alcohol syndrome. Many drugs can also have an adverse effect on developing fetuses, the most infamous example being thalidomide. Always minimize the use and release of teratogens (or believed teratogens) in the workplace. Women who are of child-bearing age should pay particular attention to teratogenic materials because they could be pregnant without knowing it and expose their fetus. Teratogens typically cause their most severe damage during the first three months of pregnancy when many pregnancies are not yet known. Many teratogens cause effects at very low exposure levels.

Toxics Release Inventory (TRI): The Toxics Release Inventory (TRI) contains information concerning waste management activities and the release of over 600 toxic chemicals by facilities that manufacture, process, or otherwise use such materials. Using this information, citizens, businesses, and governments can work together to protect the quality of their land, air and water. Section 313 of the Emergency Planning and Community Right-To-Know Act (EPCRA) and section 6607 of the Pollution Prevention Act (PPA), mandate that a publicly accessible toxic chemical database be developed and maintained by the U.S. Environmental Protection Agency (EPA). The TRI database includes information on: What chemicals were released into the local environment during the preceding year; How much of each chemical went into the air, water, and land in a particular year; How much of the chemicals were transported away from the reporting facility for disposal, treatment, recycling, or energy recovery; How chemical wastes were treated at the reporting facility; The efficiency of waste treatment; Pollution prevention and chemical recycling activities. TRI applies to companies that utilize 25,000 pounds of the approximately 600 designated chemicals or 28 chemical categories specified in the regulations, or uses more than 10,000 pounds of any designated chemical or category. TRI information is not equivalent to an MSDS, but does provide useful information for people concerned about the presence (or potential presence) of chemicals in their community or environment.

Toxic: Toxic is defined by OSHA 29 CFR 1910.1200 App A as a chemical which falls in any of these three categories: (1) A chemical that has a median lethal dose (LD_{50}) of more than 50 milligrams per kilogram, but not more than 500 milligrams per kilogram of body weight when administered orally to albino rats weighing between 200 and 300 grams each; (2) A chemical that has a median lethal dose (LD_{50}) of more than 200 milligrams per kilogram, but not more than 1,000 milligrams per kilogram of body weight, when administered by continuous contact for 24 hours (or less, if death occurs within 24 hours) with the bare skin of albino rabbits weighing between two and three kilograms each; (3) A chemical that has a median lethal concentration (LC_{50}) in air of more than 200 parts per million, but not more than 2,000 parts per million by volume of gas or vapor, or more than two milligrams per liter, but not more than 20 milligrams per liter of mist, fume, or dust, when administered by continuous inhalation for one hour (or less, if death occurs within one hour) to albino rats weighing between 200 and 300 grams each. Highly toxic is defined by OSHA as: (1) A chemical that has a median lethal dose (LD_{50}) of 50 milligrams or less per kilogram of body weight, when administered orally to albino rats, weighing between 200 and 300 grams each; (2) A chemical that has a median lethal dose (LD_{50}) of 200 milligrams or less per kilogram of body weight, when administered by continuous contact for 24 hours (or less, if death occurs within 24 hours) with the bare skin of albino rabbits, weighing between two and three kilograms each; (3) A chemical that has a median lethal concentration (LC_{50}) in air of 200 parts per million by volume or less of gas or vapor, or 2 milligrams per liter or less of mist, fume, or dust, when administered by continuous inhalation for one hour (or less, if death occurs within one hour) to albino rats, weighing between 200 and 300 grams each. Toxicology is the study of the nature, effects, detection, and mitigation of poisons and the treatment or prevention of poisoning. Treat all toxic materials with great respect. Avoid their use whenever possible, but, if you do you use them, take responsible measures to limit their use and minimize hazards. Always wear appropriate personal protective equipment (PPE), such a gloves, fume hoods, respirators etc.

POLYMER PRODUCTION

A polymer or resin can be defined as a solid or semisolid, water-insoluble, organic substance, with little or no tendency to crystallize. Resins are the basic components of plastics and are important components of surface-coating formulations. There are two types of resins -- natural and synthetic. The natural resins are obtained directly from sources such as fossil remains and tree sap. Synthetic resins can be classified by physical properties as thermoplastic or thermosetting. Thermoplastic resins undergo no permanent change upon heating. They can be softened, melted, and molded into shapes they retain upon cooling, without change in their physical properties. Thermosetting resins, on the other hand, can be softened, melted, and molded upon heating, but upon continued heating, they harden or set to a permanent, rigid state and cannot be remolded. Each basic resin type requires many modifications both in ingredients and techniques of synthesis in order to satisfy end uses and provide desired properties. Not all these variations, however, will be discussed, since not all present individual air pollution problems. Thermosetting resins are obtained from fusible ingredients that undergo condensation and polymerization reactions under the influence of heat, pressure, and a catalyst and form rigid shapes that resist the actions of heat and solvents. These resins, including phenolic, amino, polyester, and polyurethane resins, owe their heat resisting qualities to cross-linked molecular structures.

Phenolic Resins: Phenolic resins can be made from almost any phenolic compound and an aldehyde. Phenol and formaldehyde are by far the most common ingredients used, but others include phenol-furfural, resorcinol-formaldehyde, and many similar combinations. A large proportion of phenolic-resin production goes into the manufacture of molding materials. Phenol and formaldehyde, along with an acid catalyst (usually sulfuric, hydrochloric, or phosphoric acid), are charged to a steam-jacketed or otherwise indirectly heated resin kettle that is provided with a reflux condenser and is capable of being operated under vacuum. Heat is applied to start the reaction, and then the exothermic reaction sustains itself for a while without additional heat. Water formed during the reaction is totally refluxed to the kettle. After the reaction is complete, the upper layer of water in the kettle is removed by drawing a vacuum on the kettle. The warm, dehydrated resin is poured onto a cooling floor or into shallow trays and then ground to powder after it hardens. This powder is mixed with other ingredients to make the final plastic material. Characteristics of the molding powder, as well as the time and rate of reaction, depend upon the concentration of catalyst used, the phenol-formaldehyde ratio used, and the reaction temperature maintained.

Amino Resins: Among the most important amino resins are the urea-formaldehyde and melamine-formaldehyde resins. The urea-formaldehyde reaction is simple: 1 mole of urea is mixed with 2 moles of formaldehyde as 38 percent solution. The mixture is kept alkaline with ammonia pH 7. 6 to 8. The reaction is carried out at 77 °F at atmospheric pressure without any reflux. The melamine resins are made in much the same manner except that the reactants must be heated to about 176 °F initially, in order to dissolve the melamine. The solution is then cooled to 77 °F to complete the reaction. The equipment needed for the synthesis of the amino resins consists of kettles for the condensation reaction (usually nickel or nickel-clad steel), evaporators for concentrating the resin, and some type of dryer. The amino resins are used as molding compounds, adhesives, and protective coatings, and for treating textiles and paper.

Polyester and Alkyd Resins: By chemical definition, the product obtained by the condensation reaction between a polyhydric alcohol and a polybasic acid, whether or not it is modified by other materials, is properly called a polyester. All polyesters can then be divided into three basic classes: Unsaturated polyesters, saturated polyesters, and alkyds. Unsaturated polyesters are formed when either of the reactants (alcohol and acid) contains, or both contain, a double-bonded pair of carbon atoms. The materials usually used are glycols of ethylene, propylene, and butylene and unsaturated dibasic acids such as maleic anhydride and fumaric acid. The resulting polyester is capable of crosslinking and is usually blended with a polymerizable material, such as styrene. Under heat or a peroxide catalyst, or both, this blend copolymerizes into a thermosetting resin. It has recently found extensive use in the reinforced plastics field where it is laminated with fibrous glass. It is also molded into many forms for a variety of uses. Saturated polyesters are made from saturated acids and alcohols. The polyesters formed are long-chain, saturated materials not capable of crosslinking. Several of these are used as plasticizers. A special type made from ethylene glycol and terephthalic acid is made into fiber and film. Still others of this type with lower molecular weights are used with di-isocyanates to form polyurethane resins. Alkyd resins differ from other polyesters as a result of modification by additions of fatty, monobasic acids. This is known as oil modification since the fatty acids are usually in the form of naturally occurring oils such as linseed, tung, soya, cottonseed, and, at times, fish oil. The alkyds, thinned with organic solvents, are used predominantly in the protective coating industry in varnishes, paints, and enamels. The most widely used base ingredients are phthalic anhydride and glycerol. Smaller quantities of other acids such as maleic, fumaric, and others and alcohols, such as pentaerythritol, sorbitol, mannitol, ethylene glycol, and others are used. These are reacted with the oils already mentioned to form the resin. The oils, as they exist naturally, are predominantly in the form of triglycerides and do not react with the polybasic acid. They are changed to the reactive monoglyceride by reaction with a portion of the glycerol or other alcohol to be used. Heat and a catalyst are needed to promote this reaction, which is known as alcoholysis. The resin is then formed by reacting this monoglyceride with the acid by agitation and sparging with inert gas, until the condensation reaction product has reached the proper viscosity. The reaction takes place in an enclosed resin kettle, equipped with a condenser and usually a scrubber, at temperatures slightly below 500 °F. The alcoholysis can be accomplished first and then the acid and more alcohol can be added to the kettle.

Polyurethane: The manufacture of the finished polyurethane resin differs from the others described in that no heated reaction in a kettle is involved. One of the reactants, however, is a saturated polyester resin, as already mentioned, or, a polyether resin. To form a flexible foam product, the resin, typically a polyether, such as polyoxypropylenetriol, is reacted with tolylene diisocyanate and water, along with small quantities of an emulsifying agent, a polymerization catalyst, and a silicone lubricant. The ingredients are metered to a mixing head that deposits the mixture onto a moving conveyor. The resin and tolylene diisocyanate (TDI) polymerize and cross-link to form the urethane resin. The TDI also reacts with the water, yielding urea and carbon dioxide. The evolved gas forms a foam-like structure. The product forms as a continuous loaf. After room temperature curing, the loaf can be cut into desired sizes and shapes, depending upon required use. The flexible foams have found wide use in automobile and furniture upholstery and in many other specialty items for many years.

Thermoplastic Resins: As already stated, thermoplastic resins are capable of being reworked after they have been formed into rigid shapes. The subdivisions in this group that are discussed here are the vinyls styrenes, and petroleum base resins.

Polyvinyl Resins: The polyvinyl resins are those having a vinyl ($CH=CH_2$) group. The most important of these are made from the polymerization of vinyl acetate and vinyl chloride. Vinyl acetate monomer is a clear liquid made from the reaction between acetylene and acetic acid. The monomer can be polymerized in bulk, in solution, or in beads or emulsion. In the bulk reaction, only small batches can be safely handled because of the almost explosive violence of the reaction once it has been catalyzed by a small amount of peroxide. Probably the most common method of preparation is in solution. In this process, a mixture of 60 volumes vinyl acetate and 40 volumes benzene is fed to a jacketed, stirred resin kettle equipped with a reflux condenser. A small amount of peroxide catalyst is added and the mixture is heated, until gentle refluxing is obtained. After several hours, approximately 80 to 90 percent is polymerized, and the run is transferred to another kettle, where the solvent and unreacted monomer are removed by steam distillation. The wet polymer is then dried. Polyvinyl acetate is used extensively in water-based paints, and for adhesives, textile finishes, and production of polyvinyl butyral. Vinyl chloride monomer under normal conditions is a gas that boils at -14 °C. It is usually stored and reacted as a liquid under pressure. It is made by the catalytic combination of acetylene and hydrogen chloride gas or by the chlorination of ethylene, followed by the catalytic removal of hydrogen chloride. It is polymerized in a jacketed, stirred autoclave. Since the reaction is highly exothermic and can result in local overheating and poor quality, it is usually carried out as a water emulsion to facilitate more precise control. To ensure quality and a properly controlled reaction, several additives are used. These include an emulsifying agent such as soap, a protective colloid such as glue a pH control, such as acetic acid or other moderately weak acid (2.5 is common), oxidation and reduction agents such as ammonium persulfate and sodium bisulfite, respectively, to control the oxidation-reduction atmosphere, a catalyst or initiator, like benzoyl peroxide, and a chain length controlling agent, such as carbon tetrachloride. The reaction is carried out in a completely enclosed vessel with the pressure controlled to maintain the unreacted vinyl chloride in the liquid state. As the reaction progresses, a suspension of latex or polymer is formed. This raw latex is removed from the kettle, and the unreacted monomer is removed by evaporation and recovered by compression and condensation. Another form of the emulsion reaction is known as suspension polymerization. In this process, droplets of monomer are kept dispersed by rapid agitation in a water solution of sodium sulfate or in a colloidal suspension such as gelatin in water.

During the reaction, the droplets of monomer are converted to beads of polymer that are easily recovered and cleaned. This process is more troublesome and exacting than the emulsion reaction but eliminates the contaminating effects of the emulsifying agent and other additives. Other vinyl-type resins are polyvinylidene chloride Saran polytetrafluoroethylene (fluoroethene), polyvinyl alcohol, polyvinyl butyral, and others. The first two of these are made by controlled polymerization of the monomers in a manner similar to that previously described for polyvinyl chloride. Polyvinyl alcohol has no existing monomer and is prepared from polyvinyl acetate by hydrolysis. Polyvinyl alcohol is unique among resins in that it is completely soluble in both hot and cold water. Polyvinyl butyral is made by the condensation reaction of butyraldehyde and polyvinyl alcohol. All have specific properties that make these materials superior for certain applications.

Polystyrene: Polystyrene, discovered in 1831, is one of the oldest resins known. Because of its transparent, glasslike properties, its practical application was recognized even then. Two major obstacles prevented its commercial development, the preparation of styrene monomer itself, and some means of preventing premature polymerization. These obstacles were not overcome until nearly 100 years later. Styrene is a colorless liquid that boils at 145 °C. It is prepared commercially from ethylbenzene, which, in turn, is made by reaction of benzene with ethylene in presence of a Fridel-Crafts catalyst such as aluminum chloride. During storage or shipment the styrene must contain a polymerization inhibitor such as hydroquinone and must be kept under a protective atmosphere of nitrogen or natural gas. Styrene can be polymerized in bulk, emulsion, or suspension by using techniques similar to those previously described. The reaction is exothermic and has a runaway tendency unless the temperature is carefully controlled. Oxygen must be excluded from the reaction since it causes a yellowing of the product and affects the rate of polymerization. Polystyrene is used in tremendous quantities for many purposes. Because of its ease of handling, dimensional stability, and unlimited color possibilities, it is used widely for toys, novelties, toilet articles, houseware parts, radio and television parts, wall tile, and other products. Disadvantages include limited heat resistance, brittleness, and vulnerability to attack by organic solvents such as kerosine and carbon tetrachloride.

Manufacturing Equipment and Inhalation Exposures

Most resins are polymerized or otherwise reacted in a stainless steel, jacketed, indirectly heated vessel, which is completely enclosed, equipped with a stirring mechanism, and generally contains an integral reflux condenser. Since most of the reactions previously described are exothermic, cooling coils are usually required. Some resins, such as the phenolics, require that the kettle be under vacuum during part of the cycle. This can be supplied either by a vacuum pump or by a steam or water jet ejector. Moreover, for some reactions, that of polyvinyl chloride for example, the vessel must be capable of being operated under pressure. This is necessary to keep the normally gaseous monomer in a liquid state.

The size of the reactor vessels varies from a few hundred to several thousand gallons capacity. Because of the many types of raw materials, ranging from gases to solids, storage facilities vary accordingly: ethylene, a gas, is handled as such; vinyl chloride, a gas at standard conditions, is liquefied easily under pressure. It is stored, therefore, as a liquid in a pressurized vessel. Most of the other liquid monomers do not present any particular storage problems. Some, such as styrene, must be stored under an inert atmosphere to prevent premature polymerization. Some of the more volatile materials are stored in cooled tanks to prevent excessive vapor loss.

Some of the materials have strong odors, and care must be taken to prevent emission of odors to the atmosphere. Solids, such as phthalic anhydride, are usually packaged and stored in bags or fiber drums. Treatment of the resin after polymerization varies with the proposed use.

Resins for moldings are dried and crushed or ground into molding powder. Resins, such as the alkyd resins, to be used for protective coatings are normally transferred to an agitated thinning tank, where they are thinned with some type of solvent and then stored in large steel tanks equipped with water-cooled condensers to prevent loss of solvent to the atmosphere. Still other resins are stored in latex form as they come from the kettle.

The major sources of possible air contamination are the emissions of raw materials or monomer to the atmosphere, emissions of solvent or other volatile liquids during the reaction, emissions of sublimed solids such as phthalic anhydride in alkyd production, emissions of solvents during thinning of some resins, and emissions of solvents during storage and handling of thinned resins. Table 1 lists probable types and sources of air contaminants from various operations.

In the formulation of polyurethane foam, a slight excess of tolylene diisocyanate is usually added. Some of this is vaporized and emitted along with carbon dioxide during the reaction. The TDI fumes are extremely irritating to the eyes and respiratory system and are a source of local air pollution. Since the vapor pressure of TDI is small, the fumes are minute in quantity and, if exhausted from the immediate work area and discharged to the outside atmosphere, are soon diluted to a nondetectible concentration. No specific controls have been needed to prevent emission of TDI fumes to the atmosphere.

The finished solid resin represents a very small problem - chiefly some dust from crushing and grinding operations for molding powders. Generally the material is pneumatically conveyed from the grinder or pulverizer through a cyclone separator to a storage hopper. The fines escaping the cyclone outlet are collected by a baghouse type dust collector. Many contaminants are readily condensable. In addition to these, however, small quantities of noncondensable, odorous gases similar to those from varnish cooking may be emitted.

These are more prevalent in the manufacture of oil-modified alkyds where a drying oil such as tung, linseed, or soya is reacted with glycerin and phthalic anhydride. When a drying oil is heated, acrolein and other odorous materials are emitted at temperatures exceeding about 350 °F. The intensity of these emissions is directly proportional to maximum reaction temperatures.

Table 1. Principal Air Contaminants and Sources of Emissions.

Resin	Air contaminant	Possible sources of emission
Phenolic	Aldehyde odor	Storage, leaks, condenser outlet, vacuum pump discharge
Amino	Aldehyde odor	Storage, leaks
Polyester and alkyds	Oil-cooking odors Phthalic anhydride fumes, Solvents	Uncontrolled resin kettle discharge Kettle or condenser discharge
Polyvinyl acetate	Vinyl acetate odor, Solvent	Storage, condenser outlet during reaction, condenser outlet during steam distillation to recover solvent and unreacted monomer
Polyvinyl chloride	Vinyl chloride odor	Leaks in pressurized system
Polystyrene	Styrene odor	Leaks in storage and reaction equipment
Polyurethane resins	Tolylene diisocyanate	Emission from finished foam resulting from excess TDI in formulation

Control of monomer and volatile solvent emissions during storage before the reaction and of solvent emissions during thinning and storage after the polymerization of the polymer is relatively simple. It involves care in maintaining gas-tight containers for gases or liquefied gases stored under pressure, and condensers or cooling coils on other vessels handling liquids that might vaporize. Since most resins are thinned at elevated temperatures near the boiling point of the thinner, resin-thinning tanks, especially, require adequate condensers. Aside from the necessity for control of air pollution, these steps are needed to prevent the loss of valuable products. Heated tanks used for storage of liquid phthalic and maleic anhydrides should be equipped with condensation devices to prevent losses of sublimed material. One traditional device is a water-jacketed, vertical condenser with provisions for admitting steam to the jacket and provisions for a pressure relief valve at the condenser outlet set at perhaps 4 ounces of pressure. During storage the tank is kept under a slight pressure of about 2 ounces, an inert gas making the tank completely closed. During filling, the displaced gas, with any sublimed phthalic anhydride, is forced through the cooled condenser, where the phthalic is deposited on the condenser walls. After filling is completed, the condensed phthalic is remelted by passing steam through the condenser jacket. The addition of solids, such as phthalic anhydride to other ingredients that are mentioned above, the sublimation temperature of the phthalic anhydride causes temporary emissions, that violate most air pollution standards regarding opacity of smoke or fumes. These emissions subside somewhat as soon as the solid is completely dissolved, but remain in evidence at a reduced opacity, until the reaction has been completed. The emissions can be controlled fairly easily with simple scrubbing devices. Various types of scrubbers can be used. A common system that has been proved effective consists of a settling chamber, commonly called a resin slop tank, followed by an exhaust stack equipped with water sprays. The spray system should provide for at least 2 gallons per 1,000 scf at a velocity of 5 fps. The settling chamber can consist of an enclosed vessel partially filled with water, capable of being circulated with gas connections from the reaction vessel and to the exhaust stack. Some solids and water of reaction are collected in the settling tank, the remainder being knocked down by the water sprays in the stack.

Many resin polymerization reactions, for example polyvinyl acetate by the solution method, require refluxing of ingredients during the reaction. Thus, all reactors for this or other reactions involving the vaporization of portions of the reactor contents must be equipped with suitable reflux- or horizontal-type condensers or a combination of both. The only problems involved here are proper sizing of the condensers and maintaining the cooling medium at the temperature necessary to effect complete condensation.

When noncondensable, odor-bearing gases are emitted during the reaction, especially with alkyd resin production, and these gases are in sufficient concentration to create a nuisance, more extensive air pollution control equipment is necessary. These include equipment for absorption and chemical oxidation, adsorption, and combustion, both catalytic and direct-flame type.

There are a variety of air pollution control equipment that are used to minimize emissions, however, not all are effective, depending upon the problem. Scrubbing and condensing equipment is not capable of controlling some odors adequately because some of the objectionable material is in the form of noncondensable or insoluble gas or vapor, or is particulate matter of very small size. Scrubbers are, however, valuable adjuncts when used as precleaners. The scrubbing equipment used upstream from the final collection device is generally a spray tower, a plate tower, a chamber or tower with a series of baffles and water curtains, an agitated tank, or a water jet scrubber. The spray tower is probably the most efficient because the high degree of atomization that can be obtained in the scrubbing liquid by the sprays allows for maximum contact between the scrubbing water and particulate matter. A major disadvantage of the spray tower, however, is the excessive maintenance required to keep the spray nozzle free from clogging and in proper operation if the scrubbing media are recirculated. From an economic point of view, the baffled water curtain scrubber is better but is less efficient.

Packed towers are not usually used in polymer operations, because of the gummy condensed fumes that rapidly plug the tower. In practically all the scrubbing devices, the flow of vapor is countercurrent to that of the scrubbing liquid. Although various liquids have been used as scrubbing media, for example, acids, bases, various oils, and solvents, all are too expensive to be used in large-scale operations. Water is generally used and is usually not recirculated. Wetting agents are at times added to the water to increase its efficiency as an absorbent.

Adsorbing equipment, especially activated-carbon filters, are very efficient in removing solvents and odors from gas streams. To maintain this efficiency, however, the gas stream entering the carbon filters must be almost completely free of solids and entrained oil droplets.

Unfortunately, polymerization process effluent is not always free of, these materials. Without some highly efficient precleaning device, an activated-carbon filter serving a polymerization reactor would rapidly become clogged and inoperative. If used downstream from an efficient filter or precipitator, an activated- carbon unit could control solvents and odors effectively. The economy of this combined system is, however, questionable, compared to that of combustion in terms of both the original cost and the operating cost.

Combustion has proven to be an acceptable control method for many years. The other methods listed individually remove varying amounts of the contaminants, but a properly designed afterburner can do the job alone. In some instances, using a scrubber as a precleaner may be desirable from an operational point of view.

Afterburners that have been used for controlling emissions have been predominantly of the direct-flame type, effluent gases and flame entering tangentially, or the axially fired type. The burners are normally designed to be capable of reaching a temperature of 1,400 °F under maximum load conditions. For most operations, however, 1,200 °F completely controls all visible emissions and practically all odors. At temperatures appreciably below 1,200 °F incomplete combustion results in intermediate products of combustion, and highly odorous materials are emitted. The afterburner should be designed to have the maximum possible flame contact with the gases to be controlled and it should be of sufficient size to have a gas retention time of at least 0.3 seconds. The length-to-diameter ratio should be about 2.5 to 4:1. In order to prevent flashback and serious fire hazard, the fumes must enter the afterburner at a velocity faster than the flame propagation rate in the reverse direction. An even more positive fire control is a flame trap or barrier between the afterburner and the kettle. This could be a simple scrubber. Catalytic afterburners have been used to control emissions.

RUBBER PRODUCTS MANUFACTURING INDUSTRY

Excess deaths from bladder, stomach, lung, hematopoietic, and other cancers have occurred among workers involved in the manufacture of rubber products. These workers may also risk adverse respiratory effects, dermatologic effects, reproductive effects, injuries, and repetitive trauma disorders. The adverse health effects cannot be attributed to a single chemical or group of chemicals, because workplace exposures vary greatly and chemical formulations change frequently. Epidemiologic, toxicologic, and industrial hygiene studies are needed to assess the risk of cancer and other adverse health effects for rubber products workers. Many epidemiologic studies have reported excess deaths from bladder, stomach, lung, hematopoietic, and other cancers among tire and non-tire rubber products workers. Most of these excess deaths cannot be attributed to a specific chemical, because (1) workplace exposures involve many individual chemicals and combinations, and (2) changes occur in chemical formulations. Most of the chemicals found in these industries have not been tested for carcinogenicity or toxicity, nor do they have Occupational Safety and Health Administration (OSHA) permissible exposure limits (PELs) or National Institute for Occupational Safety and Health (NIOSH) recommended exposure limits (RELs).

Rubber products, such as automobile tires, automotive and appliance moldings, rubber bands, rubber gloves, and prophylactics are an important part of modern life. However, production of these items involves subjecting heterogeneous mixtures of hundreds of chemicals to heat, pressure, and catalytic action during a variety of manufacturing processes. As a result, the work environment may be contaminated with dusts, gases, vapors, fumes, and chemical byproducts (e.g., Nitrosamines). Workers may be exposed to these hazards through inhalation and skin absorption during rubber processing and product manufacturing. Physical hazards, such as noise, repetitive motion, and lifting may also be present.

The rubber products manufacturing industry employs a considerable number of workers. For example, in 1989, approximately 54,600 U.S. workers were employed in tire and inner tube production, and 132,500 workers were employed in the manufacture of non-tire rubber products. Although the products and byproducts of tire and non-tire rubber manufacturing contain hundreds of chemicals, only a small proportion of them are covered by applicable Federal occupational health standards [29 CFR 1910].

Historically, cancer has been the chronic disease, most frequently reported in rubber products workers. In the late 1940s, British rubber workers were reported to be at increased risk of bladder cancer from exposure to an antioxidant that contained 1-naphthylamine(alpha-naphthylamine) and 2-naphthylamine (beta-naphthylamine). In the United States, early investigations revealed excess cancer deaths among rubber products workers employed in 1938 and 1939; these investigators recommended additional studies of U.S. rubber workers. In 1970, the United Rubber, Cork, Linoleum, and Plastic Workers of America (URW) joined with six major American rubber companies to establish a joint occupational health program. A contract was negotiated with the Schools of Public Health at Harvard University and the University of North Carolina to conduct epidemiologic studies of rubber workers that emphasized cancer incidence and mortality. A large number of published and unpublished reports were produced as a result of these studies until the program was discontinued in 1980. The principal adverse health effects reported were cancer and respiratory effects (e.g., reductions in pulmonary function, chest tightness, shortness of breath, and other respiratory symptoms). Currently, the risks for cancer and other chronic diseases in rubber products workers are unknown because of the lack of substantial epidemiologic and industrial hygiene research in the past decade. Toxicity data are also lacking for many chemical formulations found in tire and non-tire manufacturing. Categories of rubber compounding additives may include the following list of ingredients:

Accelerators	Oils (process and extender)
Antioxidants	Organic vulcanizers
Antiozonants	Pigment blends
Antitack agents	Plasticizers
Chemical byproducts	Reinforcing agents
Curing fumes	Resins
Extenders	Solvents
Fillers	

Most studies of cancer among rubber products workers have been conducted as retrospective cohort or case control mortality studies of workers employed in the tire and nontire industries between 1940 and 1975. Many of these studies have reported statistically significant numbers of excess deaths from bladder cancer and stomach cancer. Excess deaths from colon cancer, prostate cancer, liver and biliary cancer, and esophageal cancer have been noted in individual studies. Occupational exposure data do not exist for most of these studies and have been estimated historically. The uncertainty of these exposure estimates is exacerbated by chemical formulations that differ with each plant or process. In 1982, the International Agency for Research on Cancer (IARC) published a rubber industry monograph that evaluated the available epidemiologic, toxicologic, and industrial hygiene data [IARC 1982]. In that monograph, IARC concluded that "sufficient" evidence existed to associate leukemia with occupational solvent exposure in the rubber industry; however, "no clear evidence" existed to indicate an excess of bladder cancer in British rubber workers first employed after 1950 or in U.S. rubber workers. IARC also concluded in the monograph that evidence was "limited" for associating stomach, lung, and skin cancer with occupational exposures in the rubber industry, and "inadequate" for associating lymphoma and colon, prostate, brain, thyroid, pancreatic, and esophageal cancer with these exposures. NIOSH recently recommended measures to reduce worker exposures to otoluidine and aniline (chemicals used as intermediates in the manufacture of rubber antioxidants and accelerators) to the lowest feasible concentrations.

The epidemiologic evidence reported by NIOSH associated occupational exposure to o-toluidine and aniline with an increased risk of bladder cancer among workers at a plant that manufactured rubber antioxidants and accelerators. However, it is unknown whether a similar risk exists for workers involved in the manufacture of rubber products.

The studies, conducted by Harvard University and the University of North

Carolina, report an increase in adverse respiratory effects (i.e., chest tightness, shortness of breath, reductions in pulmonary functions, and other respiratory symptoms) among rubber products workers. The University of North Carolina researchers also investigated chronic disabling pulmonary disease among rubber products workers. Emphysema was the primary condition responsible for early retirement that was due to pulmonary disease among rubber products workers who retired from a large Ohio company during the period 1964–73. Respiratory effects and chronic disabling pulmonary diseases were reported to be more prevalent among rubber workers in the curing, processing (premixing, weighing, mixing, and heating of raw ingredients), and finishing and inspection areas of tire and non-tire plants.

The prevalence is unknown for respiratory effects and diseases resulting from current occupational exposures in the rubber products manufacturing industry. The results from one study identified naphthalene diisocyanate (NDI) as the cause of respiratory irritation among workers in a Swedish tire plant.

In another study, however, no agent could be identified as the cause of acute respiratory illnesses, recurrent bronchitis with loss of lung function, or peripheral eosinophilia among 30 workers involved in a synthetic chloroprene rubber thermoinjection process. A 1987 study by NIOSH at nine U.S. tire plants found no significant increase in pneumoconiosis when investigators examined 987 chest X-rays of workers who were at least 40 years old. Contact dermatitis has been reported frequently among rubber workers and even more frequently among users of rubber products. A cross-sectional survey of 999 workers in an Australian tire plant reported a prevalence rate of 37 cases of occupational contact dermatitis per 1,000 workers. Since 1972, several NIOSH health hazard evaluations (HHEs) have reported contact dermatitis in tire and nontire plants, but most of the evaluations could not identify a specific chemical as the causative agent.

The Bureau of Labor Statistics (BLS) reported that in the rubber and miscellaneous plastics industry, the 1991 incidence rate for skin diseases or disorders (contact dermatitis, eczema, or rash caused by primary irritants and sensitizers) was 19.0 cases per 10,000 full-time workers. The relationship between dermatitis and the chemicals used in this industry has not been well defined, but some chemicals documented to cause dermatitis are no longer used. Many other chemicals used in the industry today have not been evaluated and are not regulated.

Lack of information about sources of worker exposure (including direct contact with bulk chemicals, processed stocks, and machinery contaminated with chemicals) has contributed to the difficulty in determining the association between contact dermatitis and specific chemicals. Table 2 provides a partial list of additives that can cause dermatitis. Refer also to Table 3. Although pregnancy outcomes have been studied among Swedish and Finnish rubber workers, no conclusions can be drawn because of equivocal results and lack of occupational exposure data. NIOSH has investigated reports of spontaneous abortions among American rubber workers, but none of these could be attributed to exposures in the work environment. Investigators of reproductive impairment and occupation have rarely studied the possibility of synergistic effects occurring in industries where workers encounter a multitude of exposures.

The BLS annual survey reports injury incidence rates for industries by total cases, lost workday cases, nonfatal cases without lost workdays, and lost workdays. The annual survey is based on a stratified random sample of employers and does not include self-employed individuals, farms with fewer than 11 workers, employers regulated by Federal safety and health laws other than the Occupational Safety and Health Act, or Federal, State, or local government agencies. A review of the data indicates that the injury incidence rate for the rubber and miscellaneous plastics industry was about 15.0 injuries per 100 full-time workers in 1988, 1989, and 1990.

Table 2. Agents that Cause Contact Dermatitis in Rubber Products Workers

Chemical	Process	Product
2(2'-4'Dinitrophenylthio) benzothiazole, which was contaminated with dinitrochlorbenzene (DNCB)	All areas	Tires
4,4'-Dithiodimorpholine, 1%	Not specified	Tires
n-Isopropyl-n'- phenylparaphenylenediamine (IPPD)	Assembly, maintenance, compounding	Tires
n-Dimethyl-1,3 butyl-n'- phenylparaphenylenediamine (DMPPD)	Assembly, maintenance, compounding	Tires
para-Phenylenediamine compounds	Not specified	Tires, footwear
Ethylene thiourea (ETU)	Sewing	Nontire
Resorcinal (sic)	Not specified	Tires

Table 3. NIOSH Surveys For Contact Dermatitis in Tire And Nontire Plants

Chemical	Product	Recommendations
Solution containing zinc oxide, resorcinol, formaldehyde, sodium hydroxide, ammonia; rayon fibers in cord	Reinforced rubber hose	Use ventilation and barrier cream.
Various, not specified	Bladders for aircraft fuel cells	Use gloves and wash hands after contact.
Various, not specified (present in dust in banbury, milling, and pigment blending areas)	Tires, tubes, flaps, bladders	Use clean rubber stock liners; train workers in personal hygiene; relocate sensitized workers; maintain surveillance; implement a dust control program.
2,2'Dithioaniline (bis-2)	Urethane rubber bumper pads	Use disposable gloves and disposable smock; shower before leaving the worksite.

NS = Not specified.

Disorders associated with repeated trauma is a broad BLS category that includes noise-induced hearing loss, synovitis, tenosynovitis, carpal tunnel syndrome, and other conditions resulting from repeated motion, vibration, or pressure. No data show the number of rubber and plastics workers affected by each of these disorders, but the overall incidence in 1991 was 80.5 cases per 10,000 fulltime workers. Data also indicate that this incidence rate has increased yearly since 1988, when the incidence rate was 56.5 per 10,000. In the early 1980s, NIOSH conducted industrial hygiene surveys of the tire and rubber industry and recommended that engineering controls be implemented and that substitute chemical formulations be introduced to reduce worker exposure to toxic agents. However, few studies have been conducted since then to determine whether these recommendations have been instituted. Researchers in Italy sampled volatile emissions in three locations: the vulcanization area of a shoe sole factory, the vulcanization and extrusion areas of a tire retreading factory, and the extrusion area of an electrical cable insulation plant. Approximately 100 different chemicals were identified, but the health effects associated with exposure to these chemicals have not been studied. In addition, it is not known whether the chemicals identified are representative of the emissions in other plants.

The National Occupational Exposure Survey (NOES) lists many chemical and other possible hazards (e.g., elevated temperature, whole body vibration, infrared and microwave radiation, impact noise, handwrist manipulations, etc.) found in 37 tire and non-tire plants surveyed from 1981 to 1983. The survey involved approximately 19,500 workers. Most of the NOES surveys were conducted in the work areas of fabricated rubber products plants, where more than 1,000 potential chemical hazards were identified. Although the NOES results estimate the number and distribution of workers potentially exposed to these hazards, they provide no quantitative exposure data.

Citations can be issued by OSHA for chemical exposures that exceed the PEL, failure to abate a hazard, lack of adequate engineering controls and personal protective equipment, lack of adequate training, lack of medical surveillance, or failure to meet other occupational safety and health standards. Of the hazards typically cited, only continuous or intermittent noise was noted. However, because not all facilities were inspected and only a small proportion of the workers were observed, these data do not provide adequate information about the health and safety risks for this industry. Also, the substances monitored during OSHA inspections are usually those that had OSHA PELs and sampling methods [29 CFR 1910]. Thus, the substances monitored may not have been the substances that represented the greatest hazards. In the absence of current epidemiologic and occupational exposure data, information about chemical formulations, and specific injury analyses, it is impossible to assess the risk posed to rubber products workers by cancer, other occupational diseases, and certain injuries. The hazards that exist today may be different from those in the past because of changes in chemical formulations and the introduction of automated processes. To quantify the health risks for rubber products workers, the following types of research are needed:

- Evaluation of hazardous exposures
- Assessment of control measures
- Epidemiologic research
- Collection of injury data
- Evaluation of health and safety programs
- Identification of compounds or groups of substances

Industrial hygiene characterization studies should be conducted in tire and non-tire plants to further evaluate hazardous agents. These data should be used to reconstruct historical exposures (e.g., airborne and dermal). The effectiveness of control measures (e.g., engineering controls and personal protective equipment and clothing) should be assessed in tire and non-tire plants. Such an assessment should include the transfer of new and innovative technologies across industries.

For example, control measures for bag dumping and powder weighout operations have been studied by NIOSH in other industries where powdered materials are handled. Investigations are needed to determine the type and amount of personal hygiene information being communicated to rubber products workers (e.g., the information described by OSHA [1980]).

Practices recommended by OSHA include (1) removal of chemical contaminants before eating, drinking, smoking, or using cosmetics, (2) refraining from eating, drinking, or smoking in work areas, (3) showering before going home, and (4) leaving protective clothing at work. The application of these practices by rubber products workers should also be surveyed.

From 1980 to the present, a gap exists in the epidemiologic research among rubber products workers. Only three retrospective cohort mortality studies were published after those done by Harvard University and the University of North Carolina. Updated retrospective analyses and other epidemiologic designs are needed to determine whether current workers in this industry risk cancer and other adverse health effects such as respiratory effects, skin disorders, and reproductive impairments. Cross-sectional surveys should be performed throughout the rubber products industry to determine the prevalence of skin diseases and repetitive trauma disorders. Detailed injury data (e.g., data on the types and severity of injuries within a particular job process) should be collected to identify areas where preventive measures, such as worker education and training could be implemented. The BLS Annual Survey reports injury incidence rates, but not types of injuries. Current health and safety programs in the rubber products manufacturing industry should be evaluated to determine whether they are adequately addressing the need for injury reduction. Although toxicologic studies have been conducted on chemicals, used or formed during rubber products manufacturing, this research has included only a small fraction of all chemicals used. Investigators have suggested that instead of focusing on individual chemicals, research in the rubber industry should address compounds or groups of substances such as accelerators or curing fumes.

This research should include systematic identification and quantitation of the organic vapors and gases produced from the heating and curing of rubber compounds. Epidemiologic, toxicologic, and industrial hygiene studies are needed to assess the risk of cancer and other diseases from current occupational exposures. Detailed analyses of worker injuries are also needed to identify areas for worker training and other preventive measures. The results of future studies should be closely monitored to determine whether recommendations to Federal regulatory agencies should be developed for the rubber products manufacturing industry.

SULFURIC ACID MANUFACTURING

Sulfuric acid is used as a basic raw material in an extremely wide range of industrial processes and manufacturing operations. Table 4 lists the properties and chemical hazards of sulfuric acid. Basically, the production of sulfuric acid involves the generation of sulfur dioxide (SO_2), its oxidation to sulfur trioxide (SO_3), and the hydration of the SO_3 to form sulfuric acid. The two main processes are the chamber process and the contact process. The chamber process uses the reduction of nitrogen dioxide to nitric oxide as the oxidizing mechanism to convert the SO_2 to SO_3. The contact process, using a catalyst to oxidize the SO_2 to SO_3, is the more modern and the more commonly encountered. For these reasons further discussion will be restricted to the contact process of sulfuric acid manufacture. Combustion air is drawn through a silencer, or a filter when the air is dust laden, by either a single-stage, centrifugal blower or a positive pressure type blower. Since the blower is located on the upstream side, the entire plant is under a slight pressure. The combustion air is passed through a drying tower before it enters the sulfur burner. In the drying tower, moisture is removed from the air by countercurrent scrubbing with 98 to 99 percent sulfuric acid at temperatures from 90 ° to 120 °F. The drying tower has a topside internal-spray eliminator, located just below the air outlet to minimize acid mist carryover to the sulfur burner. Molten sulfur is pumped to the burner, where it is burned with the dried combustion air to form SO_2. The combustion gases together with excess air leave the burner at about 1,600 °F and are cooled to approximately 800 °F in a water tube-type waste-heat boiler. The combustion gases then pass through a hot-gas filter into the first stage or "pass" of the catalytic converter to begin the oxidation of the SO_2 to SO_3. If the molten sulfur feed has been filtered at the start of the process, the hot-gas filter may be omitted. Because the conversion reaction is exothermic, the gas mixture from the first stage of the converter is cooled in a smaller waste-heat boiler. Gas cooling after the second and third converter stages is achieved by steam superheaters. Gas, leaving the fourth stage of the converter, is partially cooled in an economizer. Further cooling takes place in the gas duct before the gas enters the absorber. The extent of cooling required depends largely upon whether or not oleum is to be produced.

The total equivalent conversion from SO_2 to SO_3 in the four conversion stages is about 98 percent. The cooled SO_3 combustion gas mixture enters the lower section of the absorbing tower, which is similar to the drying tower. The SO_3 is absorbed in a circulating stream of 99 percent sulfuric acid. The nonabsorbed tail gases pass overhead through mist removal equipment to the exit gas stack.

Table 4. Physical and Hazardous Properties of Sulfuric Acid

Chemical Designations - *Synonyms*: Battery acid; Chamber acid; Fertilizer acid; Oil of vitriol; *Chemical Formula*: H_2SO_4.

Observable Characteristics - *Physical State (as shipped)*: Liquid; *Color*: Colorless (pure) to dark brown; *Odor*: Odorless unless hot, then choking.

Physical and Chemical Properties - *Physical State at 15 °C and 1 atm.*: Liquid; *Molecular Weight*: 98.08; *Boiling Point at 1 atm.*: 644 °F, 340 °C, 613 °K; *Freezing Point*: Not pertinent; *Critical Temperature*: Not pertinent; *Critical Pressure*: Not pertinent; *Specific Gravity*: 1.84 at 20 °C (liquid); *Vapor (Gas) Specific Gravity*: Not pertinent.

Health Hazards Information - *Recommended Personal Protective Equipment*: Safety shower; eyewash fountain; safety goggles; face shield; approved respirator (self-contained or air-line); rubber safety shoes; rubber apron; *Symptoms Following Exposure*: Inhalation of vapor from hot, concentrated acid may injure lungs. Swallowing may cause severe injury or death. Contact with skin or eyes causes severe burns; *General Treatment for Exposure*: Call a doctor. INHALATION: observe victim for delayed pulmonary reaction. INGESTION: have victim drink water if possible; do NOT induce vomiting. EYES AND SKIN: wash with large amounts of water for at least 15 min.; do not use oils or ointments in eyes; treat skin burns; *Toxicity by Inhalation (Threshold Limit Value)*: 1 mg/m³; *Short-Term Inhalation Limits*: 10 mg/m³ for 5 min.; 5 mg/m³ for 10 min.; 2 mg/m³ for 30 min.; 1 mg/m³ for 60 min.; *Toxicity by Ingestion*: No effects except those secondary to tissue damage; *Late Toxicity*: None; *Vapor (Gas) Irritant Characteristics*: Vapors from hot acid (77-98%) cause moderate irritation of eyes and respiratory system. The effect is temporary; *Liquid or Solid Irritant Characteristics*: 77-98% acid causes severe second- and third-degree burns on short contact and is very injurious to the eyes; *Odor Threshold*: Greater than 1 mg/m³.

A contact process plant, intended mainly for use with various concentrations of hydrogen sulfide (H_2S) as a feed material, is known as a wet-gas plant. The wet-gas plant's combustion furnace is also used for burning sulfur or dissociating spent sulfuric acid.

A common procedure for wet-gas plants located near petroleum refineries is to burn simultaneously H_2S, molten sulfur, and spent sulfuric acid from the alkylation processes at the refineries. In some instances a plant of this type may produce sulfuric acid by using only H_2S or spent acid. In a wet-gas plant, the H_2S gas, saturated with water vapor, is charged to the combustion furnace along with atmospheric air. The SO_2 formed, together with the other combustion products, is then cooled and treated for mist removal. Gas may be cooled by a waste-heat boiler or by a quench tower followed by updraft coolers. Mist formed is removed by an electrostatic precipitator. Moisture is removed from the SO_2 and airstream with concentrated sulfuric acid in a drying tower. A centrifugal blower takes suction on the drying tower and discharges the dried SO_2 and air to the converters. The balance of the wet-gas process is essentially the same as that of the previously discussed sulfur-burning process.

The most significant source of air contaminant discharge from a contact sulfuric acid plant is the tail gas discharge from the SO_3 absorber. While these tail gases consist primarily of innocuous nitrogen, oxygen, and some carbon dioxide, they also contain small concentrations of SO_2 and smaller amounts of SO_3 and sulfuric acid mist. Tail gases that contain SO_3, owing to incomplete absorption in the absorber stack, hydrate and form a finely divided mist upon contact with atmospheric moisture. The optimum acid concentration in the absorbing tower is 98. 5 percent. This concentration has the lowest SO_3 vapor pressure. The partial pressure of SO_3 increases, if the absorbing acid is too strong, and SO_3 passes out with the tail gases. If a concentration of absorbing acid less than 98.5 percent is used, the beta phase of SO_3, which is less easily absorbed, is produced. A mist may also form, when the process gases are cooled before final absorption, as in the manufacture of oleum.

Physical and hazardous properties of oleum are summarized in Table 5. Water-based mists can form as a result of the presence of water vapor in the process gases fed to the converter. This condition is often caused by poor performance of the drying tower. Efficient performance should result in a moisture loading of 5 milligrams or less per cubic foot. In sulfur-burning plants, mists may be formed from water resulting from the combustion of hydrocarbon impurities in the sulfur. Mists formed in the wet-purification systems of an acid sludge regeneration plant are not completely removed by electrostatic precipitation. The mists pass through the drying tower and are volatilized in the converter. The mist reforms, however, when the gases are cooled in the absorption tower. Water-based mists can also form from any steam or water leaks into the system. The SO_3 mist presents the most difficult problem of air pollution control, since it is generally of the smallest particle size. The particle size of these acid mists ranges from submicron to 10 microns and larger. Acid mist, composed of particles of less than 10 microns in size, is visible in the absorber tail gases, if present in amounts greater than 1 milligram of sulfuric acid per cubic foot of gas.

Table 5. Physical and Hazardous Properties of Oleum

Chemical Designations - *Synonyms*: Fuming sulfuric acid; *Chemical Formula*: SO_3-H_2SO_4
Observable Characteristics - *Physical State (as shipped)*: Liquid; *Color*: Colorless to cloudy; *Odor*: Sharp penetrating; choking
Physical and Chemical Properties - *Physical State at 15 °C and 1 atm.*: Liquid; *Molecular Weight*: Not pertinent; *Boiling Point at 1 atm.*: Decomposes; *Freezing Point*: Not pertinent; *Critical Temperature*: Not pertinent; *Critical Pressure*: Not pertinent; *Specific Gravity*: 1.91-1.97 at 15 °C (liquid); *Vapor (Gas) Specific Gravity*: Not pertinent; *Ratio of Specific Heats of Vapor (Gas)*: Not pertinent; *Latent Heat of Vaporization*: Not pertinent; *Heat of Combustion*: Not pertinent; *Heat of Decomposition*: Not pertinent
Health Hazards Information - *Recommended Personal Protective Equipment*: Respirator approved by U.S. Bureau of Mines for acid mists; rubber gloves; splashproof goggles; eyewash fountain and safety shower; rubber footwear; face shield; *Symptoms Following Exposure*: Acid mist is irritating to eyes, nose and throat. Liquid causes severe burns of skin and eyes; *General Treatment for Exposure*: INGESTION: have victim drink water or milk; do NOT induce vomiting. EYES: flush with plenty of water for at least 15 min.; call a doctor. SKIN: flush with plenty of water; *Toxicity by Inhalation (Threshold Limit Value)*: 1 mg/m³; *Short-Term Inhalation Limits*: 5 mg/m³ for 5 min.; 3 mg/m³ for 10 min.; 2 mg/m³ for 30 min.; 1 mg/m³ for 60 min.; *Toxicity by Ingestion*: Severe burns of mouth and stomach; *Late Toxicity*: None; *Vapor (Gas) Irritant Characteristics*: Vapors causes a severe irritation of the eye and throat and can cause eye and lung injury. They cannot be tolerated even at low concentrations; *Liquid or Solid Irritant Characteristics*: Severe skin irritant. Causes second- and third-degree burns on short contact; very injurious to the eyes; *Odor Threshold*: 1 mg/m³.

As the particle size decreases, the plume becomes more dense, because of the greater light-scattering effect of the smaller particles. Maximum light scattering occurs when the particle size approximates the wave length of light. Thus, the predominant factor in the visibility of an acid plant's plume is particle size of the acid mists rather than the weight of mist discharged. Acid particles larger than 10 microns are probably present as a result of mechanical entrainment. These larger particles deposit readily on duct and stack walls and contribute little to the opacity of the plume.

Water scrubbing of the SO_3 absorber tail gases can remove 50 to 75 percent of the SO_2 content. Scrubbing towers using 3-inch or larger stacked rings are often employed. On startups, when SO_2 concentrations are large, soda ash solution is usually used in place of straight water. Water scrubbing is feasible, where disposal of the acidic waste water does not present a problem. Tail gases may be scrubbed with soda ash solution to produce marketable sodium bisulfite. The traditional process for removal of SO_2 from a gas stream is scrubbing with ammonia solution. Single- and two-stage absorber systems reduce SO_2 concentrations in tail gases. Two-stage systems are designed to handle SO_2 gas concentrations as great as 0.9 percent. Large SO_2 concentrations resulting from acid plant startups and upsets could be handled adequately by a system, such as this.

Electrostatic precipitators are widely used for removal of sulfuric acid mist from the cold SO_2 gas stream of wet-purification systems. The wet-lead-tube type is used extensively in this service. Tube-type and plate-type precipitators have also been used for treating tail gases from SO_3 absorber towers. Dry gas containing SO_2, carbon dioxide, oxygen, nitrogen, and acid mist enters two inlet ducts to the precipitator.

The gas flows upward through distribution tiles to the humidifying section. This section contains several feet of single-spiral tile irrigated by weak sulfuric acid. The conditioned gas then flows to the ionizing section. Ionized gas then flows to the precipitator section where charged acid particles migrate to the collector plate electrodes.

Acid migrating to the plates flows down through the precipitator and is collected in the humidifying section. The gas from the precipitator section flows to a lead-lined stack that discharges to the atmosphere above grade. The high-voltage electrode wires are suspended vertically by sets of insulators. Horizontal motion is eliminated by diagonally placed insulators, which are isolated from the gas stream by oil seals. All structural material in contact with the acid mist is usually lead clad. Electrical wires are stainless steel cores with lead cladding. Voltage is supplied from a generator.

Packed-bed separators employ sand, coke, or glass or metal fibers to intercept acid mist particles. The packing also causes the particles to coalesce by reason of high turbulence in the small spaces between packing. In the past, glass fiber filters have not been very effective in mist removal because of a tendency on the part of the fiber to sag and mat. The threshold concentration for mist visibility after scrubbing is about 3.6×10^{-4} gram SO_3 per cubic foot. Wire mesh mist eliminators are usually constructed in two stages. The lower stage of wire mesh may have a bulk density of about 14 pounds per cubic foot, while the upper stage is less dense. The two stages are separated by several feet in a vertical duct. The highdensity lower stage acts as a coalescer. The re-entrained coalesced particles are removed in the upper stage. Typical gas velocities for these units range from 11 to 18 fps.

The kinetic energy of the mist particle is apparently too low to promote coalescence at velocities less than 11 fps, and re-entrainment becomes a problem at velocities greater than 18 fps. The tail gas pressure drop through a wire mesh mist installation is approximately 3 inches water column. Exit sulfuric acid mist loadings of less than 5 milligrams per cubic foot of gas are normally obtained from wire mesh units serving plants making 98 percent acid. No type of mechanical coalescer, however, has satisfactorily controlled acid mists from oleum-producing plants.

Corrosion possibilities from concentrated sulfuric acid must be considered in selecting wire mesh material. The initial cost of wire mesh equipment is modest. Porous ceramic filter tubes have proved successful in removing acid mist. The filter tubes are usually several feet in length and several inches in diameter with a wall thickness of about 3/8 inch. The tubes are mounted in a horizontal tube sheet, with the tops open and the bottoms closed. The tail gases flow downward into the tubes and pass out through porous walls. Appreciably more filtering area is required for the ceramic filter than for the wire mesh type.

The porous ceramic filter is composed of small particles of alumina or similar refractory material fused with a binder. The maintenance costs for ceramic tubes is considerably higher, than those for wire mesh filters because of tube breakage. Initial installation costs are also considerably higher, than those for wire mesh. A pressure drop of 8 to 10 inches water column is required to effect mist removal equivalent to that of a wire mesh filter. Thus, operating costs would also be appreciable.

Another approach is sonic agglomeration. The principle of sonic agglomeration is also used to remove acid particles from waste-gas streams. Sound waves cause smaller particles in an aerosol to vibrate and thereby coalesce into larger particles. Conventional cyclone separators can then be used for removal of these larger particles. A nuisance factor must be taken into consideration, however, since some of the sound frequencies are in the audible range. Simple baffles and cyclone separators are not effective in collecting particles smaller than 5 microns in size.

A considerable amount of the larger size acid mist particles may be removed; however, the visibility of the stack plume is not greatly affected, since the smallest particle size contributes most to visibility. Vane-type separators operate at relatively high gas velocities and thus make better use of the particles' kinetic energy. They have been found to be moderately effective for contact plants having wet-purification systems in reducing stack plume opacities.

PHOSPHORIC ACID MANUFACTURING

Phosphorus-containing chemical fertilizers, phosphoric acid, and phosphate salts and derivatives are widely used throughout the world. In addition to their very large use in fertilizers, phosphorus derivatives are widely used in food and medicine, and for treating water, plasticizing in the plastic and lacquer industries, flameproofing cloth and paper, refining petroleum, rustproofing metal, and for a large number of miscellaneous purposes. Most of the phosphate salts are produced for detergents in washing compounds. With the exception of the fertilizer products, most phosphorus compounds are derived from orthophosphoric acid, produced by the oxidation of elemental phosphorus. The physical and hazardous properties of phosphoric acid are summarized in Table 6. Generally, phosphoric acid is made by burning phosphorus to form the pentoxide and reacting the pentoxide with water to form the acid. Specifically, liquid phosphorus (melting point 112 °F) is pumped into a refractory-lined tower where it is burned to form phosphoric oxide, P_4O_{10}, which is equivalent algebraicly to two molecules of the theoretical pentoxide, P_2O_5, and is, therefore, commonly termed phosphorus pentoxide. An excess of air is provided to ensure complete oxidation so that no phosphorus trioxide (P_2O_3) or yellow phosphorus is coproduced. The reaction is exothermic, and considerable heat must be removed to reduce corrosion. Generally, water is sprayed into the hot gases to reduce their temperature before they enter the hydrating section. Additional water is sprayed countercurrently to the gas stream, hydrating the phosphorus pentoxide to orthophosphoric acid and diluting the acid to about 75 to 85 percent. The hot phosphoric acid discharges continuously into a tank, from which it is periodically removed for storage or purification. The tail gas from the hydrator is discharged to a final collector where most of the residual acid mist is removed before the tail gas is vented to the air. The raw acid contains arsenic and other heavy metals. These impurities are precipitated as sulfides. A slight excess of hydrogen sulfide, sodium hydrosulfide, or sodium sulfide is added and the treated acid is filtered. The excess hydrogen sulfide is removed from the acid by air blowing. The entire process is very corrosive, and special materials of construction are required. Stainless steel, carbon, and graphite are commonly used for this severe service. Special facilities are required for handling elemental yellow phosphorus since it ignites spontaneously on contact with air and is highly toxic.

Table 6. Physical and Hazardous Properties of Phosphoric Acid

---

Chemical Designations - *Synonyms*: Orthophosphoric acid; *Chemical Formula*: H_3PO_4

Observable Characteristics - *Physical State (as shipped)*: Liquid; *Color*: Colorless; *Odor*: Odorless

---

Physical and Chemical Properties - *Physical State at 15 °C and 1 atm.*: Liquid; *Molecular Weight*: 98.00; *Boiling Point at 1 atm.*: $>266, >130, >403$; *Freezing Point*: Not pertinent; *Critical Temperature*: Not pertinent; *Critical Pressure*: Not pertinent; *Specific Gravity*: 1.892 at 25 °C (liquid); *Vapor (Gas) Specific Gravity*: Not pertinent; *Ratio of Specific Heats of Vapor (Gas)*: Not pertinent; *Latent Heat of Vaporization*: Not pertinent; *Heat of Combustion*: Not pertinent; *Heat of Decomposition*: Not pertinent

---

Health Hazards Information - *Recommended Personal Protective Equipment*: Goggles or face shield; rubber gloves and protective clothing; *Symptoms Following Exposure*: Burns on mouth and lips, sour acrid taste, severe gastrointestinal irritation, nausea, vomiting, bloody diarrhea, difficult swallowing, severe abdominal pains, thirst, acidemia, difficult breathing, convulsion, collapse, shock, death; *General Treatment for Exposure*: INGESTION: do NOT induce vomiting; give water, milk, or vegetable oil. SKIN OR CONTACT: flush with water for at least 15 min.; *Toxicity by Inhalation (Threshold Limit Value)*: 1.0 mg/m³; *Short-Term Inhalation Limits*: Not pertinent; *Toxicity by Ingestion*: Grade 3, LD_{50} 50 to 500 mg/kg; *Late Toxicity*: None; *Vapor (Gas) Irritant Characteristics*: Not volatile; *Liquid or Solid Irritant Characteristics*: Fairly severe skin irritant; may cause pain and second-degree burns after a few minutes contact; *Odor Threshold*: Not pertinent.

Phosphorus is always shipped and stored under water to prevent combustion. The tank car of phosphorus is heated by steam coils to melt the water-covered phosphorus. Heated water at about 135 °F is then pumped into the tank car and displaces the phosphorus, which flows into a storage tank. A similar system using hot displacement water is frequently used to feed phosphorus to the burning tower.

A number of air contaminants, such as phosphine, phosphorus pentoxide, hydrogen sulfide, and phosphoric acid mist, may be released by the phosphoric acid process. Phosphine (PH_3) is a very toxic gas, and may be formed by the hydrolysis of metallic phosphides that exist as impurities in the phosphorus. When the tank car is opened, the phosphine usually ignites spontaneously but only momentarily. Phosphorus pentoxide (P_4O_{10}), created when phosphorus is burned with excess air, forms an extremely dense fume. The fumes are submicron in size and are 100 percent opaque.

Except for military use, phosphorus pentoxide is never released to the atmosphere unless phosphorus is accidentally spilled and exposed to air. Since handling elemental phosphorus is extremely hazardous, stringent safety precautions are mandatory, and phosphorus spills are very infrequent. Hydrogen sulfide (H_2S) is released from the acid during treatment with NaHS to precipitate sulfides of antimony and arsenic and other heavy metals.

Removal of these heavy metals is necessary for manufacture of food grade acid. H_2S is highly toxic and flammable. The odor threshold is less 0.2 ppm and has the characteristic odor of rotten eggs. Refer to Table 7. In practice, however, H_2S is blown from the treating tank and piped to the phosphorous-burning tower where it is burned to SO_2. Evolution of H_2S is also minimized by restricting the amount of NaHS in excess of that needed to precipitate arsenic and antimony and other heavy metals.

The manufacture of phosphoric acid cannot be accomplished in a practical way by burning phosphorus and bubbling the resultant products through either water or dilute phosphoric acid. When water vapor comes into contact with a gas stream that contains a volatile anhydride, such as phosphorus pentoxide, an acid mist consisting of liquid particles of various sizes is formed almost instantly.

The particle size of the phosphoric acid aerosol is small, about 2 microns or less, and it has a median diameter of 1.6 microns, with a range of 0.4 to 2.6 microns. The tail gas discharged from the phosphoric acid plant is saturated with water vapor and produces a 100 percent opaque plume. The concentration of phosphoric acid in this plume may be kept small with a well-designed plant. This loss amounts to 0.2 percent or less of the phosphorus charged to the combustion chamber as phosphorus pentoxide. All the reactions involved take place in closed vessels. The phosphorus-burning chamber and the hydrator vessel are kept under a slight negative pressure by the fan that handles the effluent gases. This is necessary to prevent loss of product as well as to prevent air pollution. The hydrogen sulfide generated during the acid purification treatment must be captured and collected, and sufficient ventilation must be provided to prevent an explosive concentration, for hydrogen sulfide has a lower explosive limit of 4.3 percent. The sulfiding agent must be carefully metered into the acid to prevent excessively rapid evolution of hydrogen sulfide. The hydrogen sulfide can be removed by chemical absorption or by combustion. Weak solutions of caustic soda or soda ash sprayed countercurrently to the gas stream react with the hydrogen sulfide and neutralize it. The hydrogen sulfide may also be oxidized in a suitable afterburner. The phosphoric acid mist in the tail gas is commonly removed by an electrostatic precipitator, a venturi scrubber, or a fiber mist eliminator. All are very effective in this service.

Table 7. Physical and Hazardous Properties of Hydrogen Sulfide

Chemical Designations - *Synonyms*: Sulfuretted hydrogen; *Chemical Formula*: H_2S
Observable Characteristics — *Physical State (as shipped)*: Liquid under pressure; *Color*: Colorless; *Odor*: Offensive odor, like rotten eggs
Physical and Chemical Properties - *Physical State at 15 °C and 1 atm.*: Gas; *Molecular Weight*: 34.08; *Boiling Point at 1 atm.*: -76.7 °F, -60.4 °C, 212.8 °K; *Freezing Point*: -117, -82.8, 190.4; *Critical Temperature*: 212.7, 100.4, 373.6; *Critical Pressure*: 1300 psia; *Specific Gravity*: 0.916 at -60 °C (liquid); *Vapor (Gas) Specific Gravity*: 1.2; *Ratio of Specific Heats of Vapor (Gas)*: 1.322; *Latent Heat of Vaporization*: 234 But/lb; *Heat of Combustion*: -6552, -3640, -152.4; *Heat of Decomposition*: Not pertinent
Health Hazards Information - *Recommended Personal Protective Equipment*: Rubber-framed goggles; approved respiratory protection; *Symptoms Following Exposure*: Irritation of eyes, nose and throat. If high concentrations are inhaled, hyperpnea and respiratory paralysis may occur. Very high concentrations may produce pulmonary edema; *General Treatment for Exposure*: INHALATION: remove victim from exposure; if breathing has stopped, give artificial respiration; administer oxygen if needed. EYES: wash with plenty of water; *Toxicity by Inhalation (Threshold Limit Value)*: 10 ppm; *Short-Term Inhalation Limits*: 200 ppm for 10 min.; 100 ppm for 30 min. and 50 ppm for 60 min.; *Toxicity by Ingestion*: Hydrogen sulfide is present as a gas at room temperature, so ingestion not likely; *Late Toxicity*: Data not available; *Vapor (Gas) Irritant Characteristics*: Vapor is moderately irritating such that personnel will not usually tolerate moderate or high vapor concentrations; *Liquid or Solid Irritant Characteristics*: Minimum hazard. If spilled on clothing and allowed to remain, may cause smarting and reddening of skin; *Odor Threshold*: 0.0047 ppm.

INSECTICIDE MANUFACTURE

The substances used commercially as insecticides can be classified according to method of action, namely: (1) Stomach poisons, which act in the digestive system; (2) contact poisons, which act by direct external contact with the insect at some stage of its life cycle; and (3) fumigants, which attack the respiratory system. A few of the commonly used insecticides, classified according to method of action, are listed in Table 8.

The classification is somewhat arbitrary in that many poisons, such as nicotine, possess the characteristics of two or three classes. Human threshold limit values of various insecticides are shown in Table 9. They represent conditions under which it is believed that nearly all workers may be repeatedly exposed day after day, without adverse effect. A number of the insecticides listed in tables 8 and 9 are now banned from use in some countries. In the United States, the insecticide industry has had a constant battle over the last three decades to try and find safe substitutes for many of these materials.

The amount by which these figures may be exceeded for short periods without injury to health depends upon factors such as:

1.   the nature of the contaminant,

2.   whether large concentrations over short periods produce acute poisoning,

3.   whether the effects are cumulative,

4.   the frequency with which large concentrations occur, and

5.   the duration of these periods.

Table 8. Common Insecticides Classified According to Method of Action

Stomach poisons	Contact poisons	Fumigants
Paris green	DDT	Sulfur dioxide
Lead arsenate	Pyrethrum	Nicotine
Calcium arsenate	Sulfur	Hydrocyanic acid
Sodium fluoride	Lime-sulfur	Naphthalene
Cryolite	Nicotine sulfate	P-dichloro-benzene
Rotenone	Methoxychlor	Ethylene oxide

Table 9. Threshold Limit Values (mg/m^3) of Various Insecticides

Substance	TLV
Aldrin (1, 2, 3, 4, 10, 10-hexachloro- 1,4,4a,5,8,8a-hexahydro-1,4,5,8-dimethanonaphthalene)	0.25
Arsenic	0.5
Calcium arsenate	1
Chlordane (1, 2, 4, 5, 6, 7, 8, 8-octachloro-3a, 4, 7, 7a-tetrahydro-4, 7-methanoindane)	0.5
Chlorinated camphene, 60%	0.5
2, 4-D (2, 4-dichlorophenoxyacetic acid	10
DDT (2, 2-bis(p-chlorophenyl) - 1, 1, 1 -trichloroethane)	1
Dieldrin (1, 2, 3, 4, 10, 10-hexachloro-6,7, epoxy - 1, 4, 4a, 5, 6, 7, 8, 8a - octahydro - 1, 4, 5, 8 -dimethano -naphthalene)	0.25
Dinitro-o-cresol	0.2
Ferbam (ferric dimethyl dithiocarbamate)	15
Lead arsenate	0.15
Lindane (hexachlorocyclohexane gamma isomer)	0.5
Malathion (0, O-dimethyl dithiophosphate of diethyl mercaptosuccinate)	15
Methoxychlor (2, 2-di-p-methoxyphenyl-1, 1, 1- trichloroethane)	15
Nicotine	0.5
Parathion (0, 0-diethyl-0-p-nitrophenyl thiophosphate)	0.1
Pentachlorophenol	0.5
Phosphorus pentasulfide	1
Picric acid	0.1
Pyrethrurn	5
Rotenone	5
TEDP (tetraethyl dithionopyrophosphate)	0.2
TEPP (tetraethyl pyrophosphate)	0.05
Thiram (tetramethyl thiuram disulfide)	5
Warfarin (3-(a-acetonylbenzyl) 4- hydroxycoumarin)	0.1

Tables 10 through 15 provide a summary of the physical and hazardous properties of some of the common insecticides.

Table 10. Physical and Hazardous Properties of Parathion, Liquid

Chemical Designations - *Synonyms*: O,O-Diethyl O-(p-nitrophenyl) phoshorothioate; O,O-Diethyl O-(p-nitrophenyl) thiophosphate; Ethyl Parathion; Phosphorothioic acid; O,O-diethyl O-p-nitrophenyl ester; *Chemical Formula*: $(C_2H_5O)_2PSOC_6H_4NO_2$;

Observable Characteristics - *Physical State (as shipped)*: Liquid; *Color*: Deep brown to yellow; *Odor*: Characteristic

Physical and Chemical Properties - *Physical State at 15 °C and 1 atm.*: Liquid; *Molecular Weight*: 291.3; *Boiling Point at 1 atm.*: Very high; decomposes; *Freezing Point*: 43 °F, 6 °C, 279 °K; *Specific Gravity*: 1.269 at 25 °C (liquid); *Heat of Combustion*: -9,240 Btu/lb.

Health Hazards Information - *Recommended Personal Protective Equipment*: Neoprene-coated gloves; rubber work shoes or overshoes; latex rubber apron; goggles; respirator or mask approved for toxic dusts and organic vapors; *Symptoms Following Exposure*: Inhalation of mist, dust, or vapor (or ingestion, or absorption through the skin) cause dizziness, usually accompanied by constriction of the pupils, headache, and tightness of the chest. Nausea, vomiting, abdominal cramps, diarrhea, muscular twitchings, convulsions and possibly death may follow. An increase in salivary and bronchial secretions may result which simulate severe pulmonary edema. Contact with eyes causes irritation; *General Treatment for Exposure*: Call a doctor for all exposures. INHALATION: remove victim from exposure immediately; have physician treat with atropine injections until full atropinization; 2-PAM may also be administered by physician. EYES: flush with water immediately after contact for at least 15 min. SKIN: remove all clothing and shoes; quickly wipe off the affected area with a clean cloth; follow immediately with a shower, using plenty of soap. If a complete shower is impossible, wash the affected skin repeatedly with soap and water. INGESTION: if victim is conscious, induce vomiting and repeat until vomit fluid is clear; make victim drink plenty of milk or water; have him lie down and keep warm; *Toxicity by Inhalation (Threshold Limit Value)*: 0.01 mg/m³; *Short-Term Inhalation Limits*: 0.5 mg/m³ for 30 min.; *Toxicity by Ingestion*: Grade 4, oral LD_{50} = 2 mg/kg (rat); *Late Toxicity*: Birth defects in chick embryos; *Odor Threshold*: 4.04 ppm.

Table 11. Physical and Hazardous Properties of Pentachlorophenol

Chemical Designations - *Synonyms*: Dowicide 7; Penta; Santophen 20; *Chemical Formula*: C_6Cl_5OH
Observable Characteristics - *Physical State (as shipped)*: Solid; *Color*: Colorless to light brown; *Odor*: Very weak
Physical and Chemical Properties - *Physical State at 15 °C and 1 atm.*: Solid; *Molecular Weight*: 266.35; *Boiling Point at 1 atm.*: 590 °F, 310 °C, 583 °K; *Freezing Point*: 370, 188, 461; *Specific Gravity*: 1.98 at 15 °C (solid).
Health Hazards Information - *Recommended Personal Protective Equipment*: Respirator for dust; goggles; protective clothing; *Symptoms Following Exposure*: Dust or vapor irritates skin and mucous membranes, causing coughing and sneezing. Ingestion causes loss of appetite, respiratory difficulties, anesthesia, sweating coma. Overexposure can cause death; *General Treatment for Exposure*: Call a doctor! INGESTION: induce vomiting at once. EYES: flush with water for 15-30 min. SKIN: wash well with soap and water; *Toxicity by Inhalation (Threshold Limit Value)*: 0.5 mg/m^3; *Short-Term Inhalation Limits*: Data not available; *Toxicity by Ingestion*: Grade 3, LD_{50} 50 to 500 mg/kg; *Late Toxicity*: Data not available; *Vapor (Gas) Irritant Characteristics*: Vapor is moderately irritating such that personnel will not usually tolerate moderate or high vapor concentrations; *Liquid or Solid Irritant Characteristics*: Causes smarting of the skin and first-degree burns on short exposure; may cause secondary burns on long exposure; *Odor Threshold*: Data not available.

Production of the toxic substances used in insecticides involves the same operations employed for general chemical processing. Similarly, chemical-processing equipment, that is, reaction kettles, filters, heat exchangers, and so forth, are the same as discussed in other sections of this chapter. Emphasis is given, therefore, to the equipment and techniques encountered in the compounding and blending of commercial insecticides to achieve specific chemical and physical properties. Most commercial insecticides are used as either dusts or sprays. Insecticides employed as dusts are in the solid state in the 0.5 to 10 micron size range. Insecticides employed as sprays may be manufactured and sold as either solids or liquids. The solids are designed to go into solution in an appropriate solvent or to form a colloidal suspension; liquids may be either solutions or water base emulsions.

Table 12. Physical and Hazardous Properties of Lead Arsenate

Chemical Designations - *Synonyms*: Lead arsenate, acid; Plumbous arsenate; *Chemical Formula*: PbHASO₄
Observable Characteristics - *Physical State (as shipped)*: Solid; *Color*: White; *Odor*: None
Physical and Chemical Properties - *Physical State at 15 °C and 1 atm.*: Solid; *Molecular Weight*: 347.12; *Boiling Point at 1 atm.*: Decomposes; *Freezing Point*: Not pertinent; *Critical Temperature*: Not pertinent; *Critical Pressure*: Not pertinent; *Specific Gravity*: 5.79 at 15 °C (solid).
Health Hazards Information - *Recommended Personal Protective Equipment*: Dust respirator; protective clothing to prevent accidental inhalation or ingestion of dust; *Symptoms Following Exposure*: Inhalation or ingestion causes dizziness, headache, paralysis, cramps, constipation, collapse, coma. Subacute doses cause irritability, loss of weight, anemia, constipation. Blood and urine concentrations of lead increase; *General Treatment for Exposure*: A specific medical treatment is used for exposure to this chemical; call a physician immediately! Give victim a tablespoon of salt in glass of warm water and repeat until vomit is clear. Then give two tablespoon of epsom salt or milk of magnesia in water, and plenty of milk and water. Have victim lie down and keep quiet; *Toxicity by Inhalation (Threshold Limit Value)*: (dust) 0.15 mg/m³; *Short-Term Inhalation Limits*: Not pertinent; *Toxicity by Ingestion*: Grade 4, LD₅₀ below 50 mg/kg (rabbit, rat).

No matter what physical state or form is involved, insecticides are usually a blend of several ingredients in order to achieve desirable characteristics. A convenient means of classifying equipment and their related processing techniques is to differentiate them by the state of the end product. Equipment used to process insecticides where the end product is a solid is designated solid insecticide-processing equipment. Equipment used to process insecticides where the end product is a liquid is designated as liquid-insecticide-processing equipment. Solid mixtures of insecticides may be compounded by either (1) adding the toxicant in liquid state to a dust mixture or (2) adding a solid toxicant to the dust mixture. If the toxicant is in liquid state, it is sprayed into a dust mixture during the blending process. After leaving the rotary sifter, the solid raw materials are carried by elevator to the upper mixer where the liquid toxicant is introduced by means of spray nozzles.

Table 13. Physical and Hazardous Properties of Aldrin

Chemical Designations - *Synonyms*: endo-, exo-,1,2,3,4,10,10-Hexachloro-1,4,4a,5,8,8a-Hexahydro-1,4,5,8-Dimethanonaphthalene, HHDN; *Chemical Formula*: $C_{12}H_8Cl_6$
Observable Characteristics - *Physical State (as normally shipped)*: Solid; *Color*: Tan to dark brown; *Odor*: Mild chemical
Physical and Chemical Properties - *Physical State at 15 °C and 1 atm.*: Solid; *Molecular Weight*: 364.93; *Boiling Point at 1 atm.*: Not pertinent; *Freezing Point*: 219, 104, 377; *Critical Temperature*: Not pertinent; *Critical Pressure*: Not pertinent; *Specific Gravity*: 1.6 at 20°C (solid); *Vapor (Gas) Density*: Not pertinent; *Ratio of Specific Heats of Vapor (Gas)*: Not pertinent; *Latent Heat of Vaporization*: Not pertinent; *Heat of Combustion*: Not pertinent; *Heat of Decomposition*: Not pertinent
Health Hazards Information - *Recommended Personal Protective Equipment*: During prolonged exposure to mixing and loading operations, wear clean synthetic rubber gloves and mask or respirator of the type passed by the U.S. Bureau of Mines for aldrin protection; *Symptoms Following Exposure*: Ingestion, inhalation, or skin absorption of a toxic dose will induce nausea, vomiting, hyperexcitability, tremors, epileptiform convulsions, and ventricular fibrillation. Aldrin may cause temporary reversible kidney and liver injury. Symptoms may be seen after ingestion of less than 1 gram in an adult; ingestion of 25 mg has caused death in children; *General Treatment for Exposure*: SKIN CONTACT: wash with soap and running water. If material gets into eyes, wash immediately with running water for at least 15 min.; get medical attention. INGESTION: call physician immediately; induce vomiting. Repeat until vomit fluid is clear. Never give anything by mouth to an unconscious person. Keep patient prone and quiet. PHYSICIAN: administer barbiturates as anti-convulsant therapy. Observe patient carefully because repeated treatment may be necessary; *Toxicity by Inhalation (Threshold Limit Value)*: 0.25 mg/m³; *Short-Term Exposure Limits*: 1 mg/m³ for 30 min.; *Toxicity by Ingestion*: Grade 3; LD 50 to 500 mg/kg (rat); *Late Toxicity*: Chronic exposure produces benign tumors in mice; *Vapor (Gas) Irritant Characteristics*: Vapors cause slight smarting of the eyes or respiratory system if present in high concentration. Effects is temporary; *Liquid or Solid Irritant Characteristics*: Minimum hazard. If spilled on clothing and allowed to remain, may cause smarting and reddening if the skin; *Odor Threshold*: Data not available.

There can be discharge gates at each end of the upper mixer, which permit the wetted mixture to be introduced either directly into a second mixer or into the high-speed fine-grinding pulverizer and then into the second mixer.

Table 14. Physical and Hazardous Properties of Chlordane

Chemical Designations - *Synonyms*: Chlordan, 1,2,4,5,6,7,8,8-Octachloro-2,3,3a,4,7,7a-Hexahydro-4,7-Methanoindene, Texichlor; *Chemical Formula*: $C_{10}H_6Cl_8$

Observable Characteristics - *Physical State (as normally shipped)*: Liquid; *Color*: Brown; *Odor*: Penetrating; aromatic; slightly pungent, like chlorine

---

Physical and Chemical Properties - *Physical State at 15 °C and 1 atm.*: Liquid; *Molecular Weight*: 409.8; *Boiling Point at 1 atm.*: Decomposes; *Freezing Point*: Not pertinent; *Critical* Temperature: Not pertinent; *Critical Pressure*: Not pertinent; *Specific Gravity*: 1.6 at 25 °C (liquid); *Vapor (Gas) Density*: Not pertinent; *Ratio of Specific Heats of Vapor (Gas)*: Not pertinent; *Latent Heat of Vaporization*: Not pertinent; *Heat of Combustion*: -4,000, -2,200, -93; *Heat of Decomposition*: Not pertinent

---

Health Hazards Information - *Recommended Personal Protective Equipment*: Use respirators for spray, fogs, mists or dust; goggles; rubber gloves; *Symptoms Following Exposure*: Moderately irritating to eyes and skin. Ingestion, absorption through skin, or inhalation of mist or dust may cause excitability, convulsions, nausea, vomiting, diarrhea, and local irritation of the gastrointestinal tract; *General Treatment for Exposure*: INHALATION: Administer the victim oxygen and give fluid therapy; do not give epinephrine, since it may induce ventricular fibrillation; enforce complete rest. EYES: flush with water for at least 15 minutes. SKIN: wash off skin with large amounts of fresh running water and wash thoroughly with soap and water. Do not scrub infected area of skin. INGESTION: induce vomiting and follow with gastric lavage and administration of saline cathartics; ether and barbiturates may control convulsions; oxygen and fluid therapy are also recommended. Do not give epinephrine. Since no specific antidotes are known, symptomatic therapy must be accompanied by complete rest.; *Toxicity by Inhalation (Threshold Limit Value)*: 0.5 mg/m³; *Short-Term Exposure Limits*: 2 mg/m³ for 30 min; *Toxicity by Ingestion*: oral LD_{50} = 283 mg/kg (rat); *Late Toxicity*: Possible liver damage; loss of appetite or weight.; *Vapor (Gas) Irritant Characteristics*: No data; *Liquid or Solid Irritant Characteristics*: No data; *Odor Threshold*: No data.

Table 15. Physical and Hazardous Properties of Benzene Hexachloride (Lindane)

Chemical Designations - *Synonyms*: BHC, 1,2,3,4,5,6-Hexachlorocyclohexane Lindane; *Chemical Formula*: $C_6H_6Cl_6$
Observable Characteristics - *Physical State (as normally shipped)*: Solid; *Color*: Light tan to dark brown; *Odor*: Characteristic
Physical and Chemical Properties - *Physical State at 15 °C and 1 atm.*: Solid; *Molecular Weight*: 290.83; *Boiling Point at 1 atm.*: Not pertinent; *Freezing Point*: Not pertinent; *Critical Temperature*: Not pertinent; *Critical Pressure*: Not pertinent; *Specific Gravity*: 1.891 at 19 °C (solid); *Vapor (Gas) Density*: Not pertinent; *Ratio of Specific Heats of Vapor (Gas)*: Not pertinent; *Latent Heat of Vaporization*: Not pertinent; *Heat of Combustion*: Not pertinent; *Heat of Decomposition*: Not pertinent
Health Hazards Information - *Recommended Personal Protective Equipment*: Respiratory protection; ensure handling in a well ventilated area.; *Symptoms Following Exposure*: Hyperirritability and central nervous system excitation; notably vomiting, restlessness, muscle spasms, ataxia, clonic and tonic convulsions. Occasional dermatitis and urticaria.; *General Treatment for Exposure*: Gastric lavage and saline cathartics (not oil laxatives because they promote abortion). Sedatives: pentobarbital or phenobarbitol in amounts adequate to control convulsions. Calcium gluconate intravenously may be used in conjunction with sedatives to control convulsions. Keep patient quiet. Do not use epinephrine because ventricular fibrillation may result; *Toxicity by Inhalation (Threshold Limit Value)*: 0.5 mg/m^3; *Short-Term Exposure Limits*: 1 mg/m^3 for 30 minutes; *Toxicity by Ingestion*: LD_{50} 0.5 ~ 5 g/kg (Technical Mixture); LD_{50} 50 ~ 500 mg/kg (rat) (Gamma Isomer - Lindane); *Late Toxicity*: Mutagen to human lymphocytes; *Vapor (Gas) Irritant Characteristics*: Moderately irritating. Workers will not usually tolerate moderate to high concentrations.; *Liquid or Solid Irritant Characteristics*: Minimum hazard. If spilled on clothing and allowed to remain, the chemical may cause smarting or reddening of skin.; *Odor Threshold*: No data.

From the second mixer, a discharge gate with a built-in feeder screw conveys the mixture to a second elevator for transfer to the holding bin where the finished batch is available for packaging. For more extensive information and additional chemicals, the reader should refer to the *Handbook of Industrial Toxicology and Hazardous Materials* (Marcel Dekker Publishers, 1999). Additional references and resources are cited at the end of this chapter where the reader can find a multitude of safety information and sites to link to on the Internet.

CONCEPTS OF INDUSTRIAL HYGIENE

From the point of view of industrial hygiene, what are the hazardous work processes? Within each process, what are the chemical exposures? The first step in a preventive program in occupational health is the identification or recognition of potential health hazards. Mastery of this process is based on thorough knowledge of industrial materials and processes. Without this knowledge it is difficult to identify those industrial processes that have the potential to cause occupational disease.

Burgess has developed a basis for linking industrial processes to hazardous chemicals. Tables 16 and 17 provide examples, developed by Burgess, based on "Recognition of Health Hazards in Industry". They list the following unit processes and links them to unit operations and hazards for the following:

- Metal Preparation (abrasive blasting)
- Product Fabrication (forging, melting & pouring, machining, and welding)
- Finishing (electroplating and metal thermal spraying).

The Hazard Score is used to rate each of the 989 chemicals and biological agents in the Agents table of Haz-Map. First, how great is the risk of industrial exposure? Each chemical gets points for the number of common industrial processes in which exposure could occur from using, mishandling or spilling. Industrial chemicals enter the body usually through either the skin or inhalation routes.

Each chemical gets 3 points if it has the ACGIH skin designation and 3 points if it is classified as "PIH" by the North American Emergency Response Guidebook. PIH stands for *Poison Inhalation Hazard*, and these volatile and toxic chemicals present the highest inhalation danger to spill responders. Second, how toxic is the chemical in terms of potency, persistence in the body, carcinogenic potential and injury to target organs?

Each chemical gets 2 points for potency if it has a low TLV (< or = to 2 ppm or 5 mg/m^3) and 1 point for potency if it has a medium TLV (< or = to 10 ppm or 100 mg/m^3).

Examples are shown for the eight chemicals with the highest hazard scores: lead, carbon monoxide, mercury, chromium, nickel, formaldehyde, ammonia, and cadmium in Table 18.

Table 16. Exposures in Metal Preparation, Fabrication and Finishing: Dusts and Metal Fumes *(after Burgess).* [a]

	Silica: **Si**; Fluorides, Inorganic: **Fl**		Nickel: **Ni**; Cadmium: **Cd**; Lead: **Pb**; Chromium: **Cr**; Manganese: **Mn**					Beryllium: **Be**; Cobalt: **Co**; Aluminum: **Al**; Barium, Soluble: **Ba**; Zinc: **Zn**				
	Si	Fl	Ni	Cd	Pb	Cr	Mn	Be	Co	Al	Ba	Zn
Abrasive Blasting	x		x	x	x	x						
Forging												
Melting & Pouring	x	x	x	x	x	x	x	x		x		x
Metal Machining			x			x		x	x			
Metallizing			x	x	x	x			x			
Electroplating			x	x		x			x			
Welding		x	x	x	x	x	x			x	x	x
Gas Cutting & Welding				x	x							x

(a)- Metal exposures depend upon the specific metals used in the process.

Table 17. Exposures in Metal Preparation, Fabrication and Finishing: Gases, Vapors and Particulate Matter

	CO	NO$_2$	O$_3$	SO$_2$	AsH$_3$	CN	H$_2$S	HF	HCl	H$_2$SO$_4$	PAH
Abrasive Blasting											
Forging	x										x
Melting & Pouring	x			x							x
Metal Machining											
Metallizing											
Electroplating		x			x	x	x	x	x	x	
Welding	x	x	x								
Gas Welding, Cutting	x	x									

In Table 17, the chemical symbols are: Carbon Monoxide: **CO**; Nitrogen Dioxide: **NO₂**; Ozone: **O₃**; Sulfur Dioxide: **SO₂**; Arsine: **AsH₃**; Cyanide: **CN**; Hydrogen Sulfide: **H₂S**; Hydrogen Fluoride: **HF**; Hydrogen Chloride: **HCl**; Sulfuric Acid: **H₂SO₄**; Polycyclic Aromatic Hydrocarbons: **PAH**.

Table 18. Illustrates the Hazard Score Approach.

Criteria	Points	Pb	CO	Hg	Ni	HCHO	NH₃	Cr	Cd
Associated Processes	count	12	16	5	9	10	12	12	9
Skin Absorption	3			3					
Inhalation (PIH)	3		3				3		
Bioaccumulates	3	3		3	3			3	3
Low TLV	2	2		2	2	2		2	2
Medium TLV	1								
Organohosphate	3								
Organochlorine	3								
Carbamate	2								
Chemical Asphyxiant	3		3						
Simple Asphyxiant	2								
IARC Known Carcinogen	3				3				3
IARC Probable	2					2			
IARC Possible	1	1							
Reproductive Toxin	3	3	3	3					
Skin Burns	2					2	2		
Allergic Dermatitis	2			2	2	2		2	
Chloracne	3								
Interstitial Fibrosis	3								
Asthma	3				3	3		3	
Pneumonitis	3			3	3	3	3		3
Chronic Bronchitis	3						3		
MetHgb, Primary	3								
MetHgb, Secondary	1								
Aplastic Anemia	3								

Table 18. Continued

Criteria	Points	Pb	CO	Hg	Ni	HCHO	NH$_3$	Cr	Cd
Hemolytic Anemia	3	3							
Liver, Primary	3								
Liver, Secondary	1								
Kidney Damage	2	2		2					2
Peripheral Neuropathy	3	3		3					
Parkinson's Syndrome	3		3						
CNS Solvent Syndrome	2								
Totals		29	28	26	25	24	23	22	22

We focus on chromium, both by way of an example, but also because of its high toxicity. The score of 22 for chromium is for chromium metal and Cr III compounds. The International Agency for Research on Cancer considers these compounds to be "not classifiable as to carcinogenicity in humans." Chromium VI compounds are classified by IARC as human carcinogens (Class 1). Hexavalent chromium compounds (Cr VI) include:

• Water-soluble compounds: chromium trioxide (chromic acid), and monochromates and dichromates of sodium, potassium, ammonium, lithium, cesium and rubidium

• Water-insoluble compounds: zinc chromate, strontium chromate and sintered chromium trioxide

"NIOSH considers all Cr(VI) compounds (including chromic acid, tert-butyl chromate, zinc chromate, and chromyl chloride) to be potential occupational carcinogens." [NIOSH Pocket Guide Appendix] "Compounds of Cr^{III} do not cause chrome ulcerations and do not generally initiate allergic dermatitis without prior sensitization by Cr^{VI} compounds." [ILO Encyclopedia] Chromates, the most common cause of allergic contact dermatitis, are released as hexavalent chromium from chrome-plated metal tools and machine parts.

Animal experiments have associated chromium with birth defects, but there are no studies implicating it as a cause of birth defects in humans. Chronic exposure to hexavalent chromium may produce evidence of kidney and liver injury. Asthma reported in printer, plater, welder and tanner (chromium and nickel); Allergic contact dermatitis in agricultural workers, construction workers, mechanics and printers.

SOURCES OF INFORMATION

Journals & Monographs

1. Aetiological Agents in Occupational Asthma, Chan-Yeung M, Malo J. (Eur Respir J. 1994;7:346-371).

2. Occupational Sentinel Health Events: An Up-Dated List for Physician Recognition and Public Health Surveillance, Mullan RJ, Murthy LI. (Am J Ind Med. 1991;19:775-799).

3. ATSDR Case Studies in Environmental Medicine, Lum, M, Ortyl D, Project Officers, published by the Agency for Toxic Substances and Disease Registry.

Books

1. Contact and Occupational Dermatology, 2nd ed., Marks JG, DeLeo VA. (Mosby, 1997).

2. Control of Communicable Diseases Manual, 16th ed., Benenson AS, ed. (APHA, 1995).

3. Environmental & Occupational Medicine, 3nd ed., Rom WN, ed.. (Lippincott-Raven, 1998).

4. Hawley's Condensed Chemical Dictionary, 12th ed., Lewis, RJ. (Van Nostrand Reinhold Company, 1993).

5. Hazardous Materials Toxicology. Sullivan J, Krieger G, eds.(Williams & Wilkins, 1992).

6. Occupational & Environmental Medicine, 2nd ed., LaDou J, ed. (Appleton & Lange, 1997).

7. Occupational and Environmental Respiratory Diseases, Harber P, Schenker MB, Balmes JR, eds. (Mosby, 1996).

8. Occupational Health: Recognizing and Preventing Work-Related Disease, 3rd ed., Levy BS, Wegman DH, eds. (Little Brown and Company; 1995).

9. Preventing Occupational Disease and Injury, Weeks JL, Levy BS, Wagner GR, eds. (APHA, 1991).

10. Textbook of Clinical Occupational and Environmental Medicine, Rosenstock L, Cullen MR, eds. (W.B. Saunders, 1994).

11. Recognition and Management of Pesticide Poisonings, 5th ed. Reigart JR, Roberts JR, eds.(EPA, 1999).

12. The Merck Index, An Encyclopedia of Chemicals, Drugs, and Biologicals, 11th ed., Budavari S, O'Neil MJ, Smith A, Heckelman PE, eds., (Merck & Co., Inc., 1989).

13. The Merck Manual, 16th ed. Berkow, R., ed (Merck & Co., Inc., 1992).

14. Recognition of Health Hazards in Industry, 2nd ed., Burgess WA. (John Wiley & Sons, 1995).

Computer Databases

1. ATSDR's Toxicological Profiles on compact disc. (Lewis Publishers, 1997 CRC Press, Inc.)

2. Encyclopaedia of Occupational Health and Safety, 4th ed., CD-ROM. Stellman JM, ed. (ILO, 1998)

3. Hazardous Substances Data Bank (HSDB) on compact disc from the Canadian Center for Occupational Health and Safety (can buy at CCOHS web site). "The HSDB(R) (Hazardous Substances Data Bank(R)) database contains data profiles on 4,500 potentially toxic chemical substances. It is created and updated by specialists at the U.S. National Library of Medicine. Compiled from an extensive range of authoritative sources, HSDB is widely recognized as a reliable and practical source of health and safety information. Much of the data is peer reviewed.

4. NIOSHTIC on compact disc from the Canadian Center for Occupational Health and Safety (can buy at CCOHS web site). "NIOSHTIC(R) is a bibliographic database which provides comprehensive international coverage of documents on occupational health and safety, as well as related fields. It contains detailed summaries of over 200,000 articles, reports and publications, spanning over 100 years. NIOSHTIC(R) sources include over 160 scientific and technical journals, NIOSH reports (published and unpublished), NIOSH research bibliographies, abstracts from CIS Abstracts, and personal files from respected professionals on selected topics.

5. Quick Guide: The Electronic NIOSH Pocket Guide to Chemical Hazards, Version 4.0.

6. TLVs and other Occupational Exposure Values—1995 on compact disc from the American Conference of Governmental Industrial Hygienists

7. BC Cancer Agency: Carcinogens CDC National Center for Infectious Diseases

8. ChemIDplus at the National Library of Medicine is a database of 56,645 chemical structures. Code of Federal Regulations

Chapter 3
INHALATION HAZARDS IN REFINERIES

INTRODUCTION

Operations of the petroleum industry can logically be divided into production, refining, and marketing. Production includes locating and drilling oil wells, pumping and pretreating the crude oil, recovering gas condensate, and shipping these raw products to the refinery or, in the case of gas, to commercial sales outlets. Refining, which extends to the conversion of crude to a finished salable product, includes oil refining and the manufacture of various chemicals derived from petroleum. The chemical manufacturing segment is referred to as the petrochemical industry. Marketing involves the distribution and the actual sale of the finished products. To prevent unsafe operating pressures in process units during shutdowns and startups and to handle miscellaneous hydrocarbon leaks, a refinery must provide a means of venting hydrocarbon vapors safely. Either a properly sized elevated flare using steam injection or a series of venturi burners actuated by pressure increases is generally considered satisfactory, although there is less of a tendency to rely on flaring practices in the United States today. Good instrumentation and properly balanced steam-to-hydrocarbon ratios are prime factors in the design of a safe, smokeless flare. Refinery operations are complex systems, involving many unit operations ranging from high pressure systems, wide temperature variations, and products and intermediates that vary from toxic to highly flammable. There are numerous dangers associated with inhalation hazards, chemical exposure by workers, and fire and explosion. This chapter provides an overview of the basic operations of a refinery, and the hazards associated during operation and emergency situations, along with standard safety precautions and designs. It is important to note that no two refineries are exactly alike. Designs and the sequential use of certain equipment can vary significantly, and hence the operation and specific safety requirements are a matter of careful process safety design that must be established by a local technical staff that understands the operations.

INHALATION AND FIRE HAZARDS

Sources of Inhalation Hazards

Inhalation hazards from operations can be associated with air emissions from very specific equipment within refineries. The following are common equipment where emissions occur.

Pressure Relief Valves: In refinery operations, process vessels are protected from overpressure by relief valves. These pressure-relieving devices are normally springloaded valves. Corrosion or improper reseating of the valve seat results in leakage. Proper maintenance through routine inspections, or use of rupture discs, or manifolding the discharge side to vapor recovery or to a flare minimizes air contamination from this source.

Storage Vessels: Tanks used to store crude oil and volatile petroleum distillates are a large potential source of hydrocarbon emissions. Hydrocarbons can be discharged to the atmosphere from a storage tank as a result of diurnal temperature changes, filling operations, and volatilization. Control efficiencies of 85 to 100 percent can be realized by using properly designed vapor recovery or disposal systems, floating-roof tanks, or pressure tanks.

Bulk-Loading Facilities: The filling of vessels used for transport of petroleum products is potentially a large source of hydrocarbon emissions that workers may be exposed to. As the product is loaded, it displaces gases containing hydrocarbons to the atmosphere. An adequate method of preventing these emissions consists of collecting the vapors by enclosing the filling hatch and piping the captured vapors to recovery or disposal equipment. Submerged filling and bottom loading also reduces the amount of displaced hydrocarbon vapors.

Catalyst Regenerators: Modern refining processes include many operations using solid-type catalysts. These catalysts become contaminated with coke buildup during operation and must be regenerated or discarded. For certain processes to be economically feasible, for example, catalytic cracking, regeneration of the catalyst is a necessity and is achieved by burning off the coke under controlled combustion conditions. The resulting flue gases may contain catalyst dust, hydrocarbons, and other impurities originating in the charging stock, as well as the products of combustion. The dust problem encountered in regeneration of moving-bed-type catalysts requires control by scrubbers and cyclones, cyclones and precipitators, or high-efficiency cyclones, depending upon the type of catalyst, the process, and the regenerator conditions. Hydrocarbons, carbon monoxide, ammonia, and organic acids can be controlled effectively by incineration in carbon monoxide waste-heat boilers. The waste-heat boiler offers a secondary control feature for plumes emitted from fluid catalytic cracking units. This type of visible plume, whose degree of opacity is dependent upon atmospheric humidity, can be eliminated by using a carbon monoxide waste-heat boiler as an example. Other processes in refining operations employ liquid or solid catalysts, Regenerating some of these catalysts at the unit is feasible. Other catalysts are consumed or require special treatment by their manufacturer. Where regeneration is possible, a closed system can be effected to minimize the release of any air contaminants by venting the regenerator effluent to the firebox of a heater.

Effluent-Waste Disposal: Waste water, spent acids, spent caustic and other waste liquid materials are generated by refining operations and present disposal problems. The waste water is processed through clarification units or gravity separators. Unless adequate control measures are taken, hydrocarbons contained in the waste water are emitted to the atmosphere. Acceptable control is achieved by venting the clarifier to vapor recovery and enclosing the separator with a floating roof or a vapor-tight cover. In the latter case, the vapor section should be gas blanketed to prevent explosive mixtures and fires. Spent waste materials can be recovered as acids or phenolic compounds, or hauled to an acceptable disposal site.

Pumps and Compressors: Pumps and compressors required to move liquids and gases in the refinery can leak product at the point of contact between the moving shaft and stationary casing. Properly maintained packing glands or mechanical seals minimize the emissions from pumps. Compressor glands can be vented to a vapor recovery system or smokeless flare. In older refineries, internal combustion engines are used to drive the compressors. These are fueled by natural or refinery process gas. Even with relatively high combustion efficiency and steady load conditions, some fuel can pass through the engine unburned. Nitrogen oxides, aldehydes, and sulfur oxides can also be found in the exhaust gases.

Air-Blowing Operations: Venting the air used for "brightening" and agitation of petroleum products or oxidation of asphalt results in a discharge of entrained hydrocarbon vapors and mists, and malodorous compounds. Mechanical agitators that replace air agitation can reduce the volumes of these emissions. For the effluent fumes from asphalt oxidation, incineration gives effective control of the hydrocarbons and malodors.

Pipeline Valves and Flanges, Blind Changing, Process Drains: Liquid and vapor leaks can develop at valve stems as a result of heat, pressure, friction, corrosion, and vibration. Regular equipment inspections followed by adequate maintenance can keep losses at a minimum. Leaks at flange connections are negligible if the connections are properly installed and maintained. Installation or removal of pipeline blinds can result in spillage of some product. A certain amount of this spilled product evaporates regardless of drainage and flushing facilities. Special pipeline blinds are used to reduce the amount of spillage. In refinery operation, condensate water and flushing water must be drained from process equipment. These drains also remove liquid leakage or spills and water used to cool pump glands. Modern refining designs provide wastewater-effluent systems with running liquid-sealed traps and liquid-sealed and covered junction boxes. These seals keep the amount of liquid hydrocarbons exposed to the air at a minimum and thereby reduce hydrocarbon losses.

Cooling Towers: The large amounts of water used for cooling are conserved by recooling the water in towers Cooling is accomplished by evaporating part of this water. Any hydrocarbons that might be entrained or dissolved in the water as a result of leaking heat exchange equipment are readily discharged to the atmosphere. Proper design and maintenance of heat exchange equipment minimizes this loss. Fin-fan cooling equipment has also replaced the need of the conventional cooling tower in many instances. Process water that has come into contact with a hydrocarbon stream or has otherwise been contaminated with odorous material should not be piped to a cooling tower.

Vacuum Jets and Barometric Condensers: Some process equipment is operated at less than atmospheric pressure. Steam-driven vacuum jets and barometric condensers are used to obtain the desired vacuum. The lighter hydrocarbons that are not condensed are discharged to the atmosphere unless controlled. These hydrocarbons can be completely controlled by incinerating the discharge. The barometric hot well can also be enclosed and vented to a vapor disposal system. The water of the hot well should not be turned to a cooling tower.

Table 1 provides a list of typical sources of inhalation hazards that workers face in petroleum refining operations. Some of the health risks associated with exposure to specific inhalation hazards are discussed later in this chapter. There are three common methods of minimizing the risk to workers from inhalation hazards. These are engineering controls, managerial controls, and personal respiratory protection. Managerial controls and respiratory protection are discussed later in this volume.

Engineering Control Methods

Engineering controls are often specific to the process, and hence are discussed in relation to oil refining practices. The control of air contaminants can be accomplished by process change, installation of control equipment, improved housekeeping, and better equipment maintenance. Some combination of these often proves the most effective solution. Table 2 indicates various methods of controlling air pollution sources encountered in an oil refinery. These techniques are also applicable to petrochemical operations. Most of these controls result in some form of economic saving as well, and can be thought of in some instances as pollution prevention measures

In addition to refining operations, there are an extensive network of pipelines, terminals, truck fleets, marine tankers, and storage and loading equipment must be used to deliver the finished petroleum product to the user.

Hydrocarbon emissions from the distribution of products derive principally from storage vessels and filling operations. Additional hydrocarbon emissions may occur from pump seals, spillage, and effluent-water separators. Table 3 lists practical methods of minimizing these emissions from this part of the industry.

Table 1. Sources of Inhalation Hazards from Oil Refining

Type of emission	Potential source
Hydrocarbons	Air blowing, barometric condensers, blind changing, blowdown systems, boilers, catalyst regenerators, compressors, cooling towers, recoking operations, flares, heaters, incinerators, loading facilities, processing vessels, pumps, sampling operations, tanks, turnaround operations, vacuum jets, waste-effluent handling equipment
Sulfur oxides	Boilers, catalyst regenerators, decoking operations, flares, heaters, incinerators, treaters, acid sludge disposal
Carbon monoxide	Catalyst regenerators, compressor engines, coking operations, incinerators
Nitrogen oxides	Boilers, catalyst regenerators, compressor engines, flares
Particulate matter	Boilers, catalyst regenerators, coking operations, heaters, incinerators
Odors	Air blowing, barometric condensers, drains, process vessels, steam blowing, tanks, treaters, waste-effluent handling equipment
Aldehydes	Catalyst regenerators, compressor engines
Ammonia	Catalyst regenerators

Table 2. Control Measures for Reduction of Air Contaminants

Source	Control method	Source	Control method
Storage vessels	Vapor recovery systems; floating-roof tanks; pressure tanks; vapor balance; paint tanks white	Effluent-waste disposal	Enclosing separators; covering sewer boxes and using liquid seal; liquid seals on drains
Catalyst regenerators	Cyclones, precipitator, CO boiler; cyclones, water scrubber; multiple cyclones	Bulk-loading facilities	Vapor collection with recovery or incineration; submerged or bottom loading
Accumulator vents	Vapor recovery; vapor incineration	Acid treating	Continuous -type agitators with mechanical mixing; replace with catalytic hydrogenation units; incinerate vented cases
Blowdown systems	Smokeless flares; gas recovery	Acid sludge storage	Caustic scrubbing; incineration; vapor return
Pumps and compressors	Mechanical seals; vapor recovery; sealing glands by oil pressure; maintenance	Spent-caustic handling	Incineration; scrubbing
Vacuum jets	Vapor incineration	Doctor treating	Steam strip spent doctor solution to hydrocarbon recovery before air regeneration; replace treating unit with other, less objectionable units
Equipment valves	Inspection and maintenance	Sour-water treating	Use sour-water oxidizers and gas incineration; conversion to ammonium sulfate
Pressure relief valves	Vapor recovery; vapor incineration; rupture discs; inspection and maintenance	Mercaptan disposal	Conversion to disulfides; adding to catalytic cracking charge stock; incineration; using organics in synthesis
Asphalt blowing	Incineration; water scrubbing	Shutdowns, turnarounds	Depressure and purge to vapor recovery

Table 3. Sources and Control of Hydrocarbon Losses from Marketing Equipment

Source	Control method
Storage vessels	Floating-roof tanks; vapor recovery; vapor disposal; vapor balance; pressure tanks; painting tanks white
Bulk-loading facilities	Vapor collection with recovery or incineration; submerged loading, bottom loading
Service station delivery	Vapor return; vapor incineration
Automotive fueling	Vapor return
Pumps	Mechanical seals; maintenance
Separators	Covers; use of fixed-roof tanks
Spills, leaks	Maintenance; proper housekeeping

Hazardous Properties and Nomenclature of Organic Materials

Most compounds in which carbon is the key element are classified as organic. Various relevant classes of organic materials are reviewed in terms of chemical behavior and physical properties. A few basic definitions will be presented first.

Covalent: refers to a chemical bond in which there is an equal/even sharing of bonding electron pairs between atoms. This is typical of the bonding between carbon atoms and between carbon and hydrogen atoms in organic compounds.

Hydrocarbons: chemical compounds consisting primarily of carbon and hydrogen.

Aliphatic: organic compound with the carbon backbone arranged in branched or straight chains (e.g., propane).

Aromatic: organic molecular structure having the benzene ring (C_6H_6) as the basic unit (e.g., toluene, xylene).

Saturated: the condition of an organic compound in which each constituent carbon is covalently linked to four different atoms. This is generally a stable configuration (e.g., $CH_3CH_2CH_3$, propane).

Isomers: different structural arrangements with the same chemical formula, (e.g., n-butane and t-butane).

Unsaturated: an organic compound containing double or triple bonds between carbons (e.g., ethylene [$CH_2=CH_2$]). Multiple bonds tend to be sites of reactivity.

Functional Group: an atom or group of atoms, other than hydrogen, bonded to the chain or ring of carbon atoms (e.g., the -OH group of alcohols, the -COOH group of carboxylic acids, the -O- group of ethers). Functional groups determine the behavior of molecules. Consequently, the unique hazards of an organic compound are often determined by its functional group(s).

Most organic compounds are flammable. They tend to melt and boil at lower temperatures than most inorganic substances. Because many organic compounds volatilize easily at room temperature and possess relatively low specific heats and ignition temperatures, they tend to burn easily. Moreover, organic vapors often have high heats of combustion which, upon ignition, facilitate the ignition of surrounding chemicals, thus compounding the severity of the hazard. In addition, many organic compounds are less stable than inorganics. However, the presence of one or more halogen atoms (F,Cl,Br,I) in the molecular structure of an organic compound increases its stability and inertness to combustion. Thus, partially halogenated hydrocarbons burn with less ease than their nonhalogenated analogs. Fully halogenated derivatives, such as carbon tetrachloride (CCl_4) and certain polychlorinated biphenyls (PCBs) are almost noncombustible.

Most organic compounds are water-insoluble. Notable exceptions are the lower molecular weight alcohols, aldehydes, and ketones, all known to be "polar" molecules. This characteristic is of importance to firefighting because the specific gravity of the compound will then be a major determinant of the suitability of water for the suppression of fires involving the chemical.

Except for alkanes and organic acids, organic compounds tend to react easily with oxidizing agents such as hydrogen peroxide or potassium dichromate. Moreover, a mixture of an oxidizing agent and organic matter is usually susceptible to spontaneous ignition. Notably, except for flammability and oxidation, organic compounds tend to react slowly with other chemicals.

The basic system of aliphatic organic nomenclature is shown in Table 4. The prefix for the name is based on the number of carbons involved and remains the same for each type of compound described. The suffix is determined by the type of compound and is independent of the number of carbons in the molecule. Thus, methane, methanol, methanol (formaldehyde), and methanoic (formic) acid represent an alkane, an alcohol, an aldehyde, and a carboxylic acid, respectively, each with one carbon per molecule. In contrast, methanol, ethanol, and propanol are all alcohols, but with one, two, and three carbons per molecule, respectively.

The boiling points provided in Table 4 show the systematic trends in chemical properties as the number of carbons per molecule increases within a given chemical group, and as the various chemical groups are compared for a specific number of carbons per molecule. Thus, in general, within any group, the larger molecules are less volatile than the smaller ones. Also, alkanes tend to be more volatile than alde-hydes. Systematic trends can also be observed for other properties, such as water solubility. It should be noted than the boiling points provided in Table 4 are for the straight-chain isomers of the molecules. If the values for branched chain molecules are included, the comparisons become complicated.

Alkenes and alkynes are similar in structure to the alkanes except the alkenes contain a carbon-to-carbon double bond (C=C) and the alkynes contain a carbon-to-carbon triple bond (C≡C). The name prefixes are exactly the same as for the alkanes with the same number of carbons, but the endings are -one for compounds with double bonds and their derivatives and -yne for compounds with triple bonds and their derivatives. Ethene (ethylene) and propene (propylene) are alkenes. Ethyne (acetylene) is an alkyne. Aromatics are molecules based on single or triple benzene rings. Some of the more common aromatics include benzene, toluene, xylene, and phenol.

Table 4. Nomenclature for Aliphatic Compounds

No. Carbons	Prefix	Boiling Point, °C			Acids	
		Alkanes	Alcohols	Aldehydes	Ending	b.p.°C
1	Meth	-150	65		anoic (formic)	100
2	Eth	-90	78	20	(acetic)	120
3	Prop	-40	95	50	(propionic)	140
4	But	0	120	75	(butyric)	160
5	Pent	35	140	105	(valeric)	185
6	Hex	70	160	130		205
7	Hept	100	175	155		225
8	Oct	125	195	170		240
9	Non	150	215	185		255
10	Dec	175	230	210		270
11	Undec	195				

As previously mentioned, benzene is a 6-carbon ring with the formula C_6H_6. The ring has alternating double and single bonds, and is quite stable. The substitution of a methyl group (-CH$_3$) for one of the hydrogens gives methyl benzene or toluene. The substitution of another methyl group gives dimethyl benzene or xylene. Substitution of a hydroxyl (-OH) for a hydrogen on the benzene ring gives hydroxy benzene or phenol. Aromatics can also be named more specifically based on a system of assigning names or numbers to various positions on the benzene ring. By using the numbering system for the carbons on single or multiple benzene rings in combination with the names of the relevant substituents, any aromatic compound can be assigned a unique name. The following are important properties of individual functional groups.

Alkanes: Presented as (C$_n$H$_{2n+2}$), these are saturated hydrocarbons. The lower molecular weight alkanes (ethane through butane) are gases at standard temperature and pressure. The remainder are water-insoluble liquids, that are lighter than water and thus form films or oil slicks on the surface of water. Hence, water is not used to suppress fires involving materials, such as gasoline, that include substantial proportions of liquid alkanes. Alkanes are relatively unreactive with most acids, bases, and mild oxidizing agents. However, with addition of sufficient heat, alkanes will react and burn in air or oxygen when ignited. In fact, low molecular weight alkanes (LPG, butane, gasoline) are commonly used as fuels. Consequently, the biggest hazard from alkanes is flammability.

Organic Carboxylic Acids: (RCOOH) are usually weak acids but can be very corrosive to skin. However, The substitution of Cl atoms on the carbon next to the carboxylic carbon, produces a stronger acid. Thus, trichloracetic acid is almost a strong acid whereas acetic acid is a weak one.

Organic Sulfonic Acids: (RSO$_2$H) are generally stronger acids than organic carboxylic acids.

Organic Bases: (such as amines, RNH$_2$) are weak bases but can be corrosive to skin or other tissue.

Alcohols: (ROH) are not very reactive. The lower molecular weight alcohols (methanol, ethanol, propanol) are completely miscible with water, but the heavier alcohols tend to be less soluble. Most common alcohols are flammable. Aromatic alcohols like phenol are not as flammable (flashpoint = 79°C) and are fairly water soluble (~9 g/L).

Alkenes: Also known as olefins, and denoted as C$_n$H$_{2n}$ the compounds are unsaturated hydrocarbons with a single carbon-to-carbon double bond per molecule. The alkenes are very similar to the alkanes in boiling point, specific gravity, and other physical characteristics. Like alkanes, alkenes are at most only weakly polar. Alkenes are insoluble in water but quite soluble in nonpolar solvents like benzene. Because alkenes are mostly insoluble liquids that are lighter than water and flammable as well, water is not used to suppress fires involving these materials. Because of the double bond, alkenes are more reactive than alkanes.

Esters: These are not very reactive. Only the lowest molecular weight esters have appreciable solubility in water (e.g., ethyl acetate, 8 percent). Methyl and ethyl esters are more volatile than the corresponding unesterified acids. Most common esters are flammable. Esters have sweet to pungent odors.

Ethers: (R-O-R) are low on the scale of chemical reactivity. Aliphatic ethers are generally volatile, flammable liquids with low boiling points and low flashpoints. Well known hazardous ethers include diethyl ether, dimethylether, tetrahydrofuran. Beyond their flammability, ethers present an additional hazard, they react with atmospheric oxygen in the presence of light to form organic peroxides.

Organic Peroxides: (R-O-O-R) are very hazardous. Most of the compounds are so sensitive to friction, heat, and shock that they cannot be handled without dilution. As a result, organic peroxides present a serious fire and explosion hazard. Commonly encountered organic peroxides include benzoyl peroxide, peracetic acid, and methyl ethyl ketone peroxide.

Aldehydes and Ketones: These share many chemical properties because they possess the carbonyl (C=O) group as a common feature of their structure. Aldehydes and ketones have lower boiling points and higher vapor pressures than their alcohol counterparts. Aldehydes and ketones through C_4 are soluble in water and have pronounced odors. Ketones are relatively inert while aldehydes are easily oxidized to their counterpart organic acids.

Two important hazardous characteristics are flammability and toxicity. Flammability, the tendency of a material to burn, can be subjectively defined. Many materials that we normally do not consider flammable will burn, given high enough temperatures. Flammability cannot be gauged by the heat content of materials. Fuel oil has a higher heat content than many materials considered more flammable because of their lower flashpoint. In fact, flashpoint has become the standard for gauging flammability. The most common systems for designating flammability are the Department of Transportation (DOT) definitions, the National Fire Protection Association's (NFPA) system, and the Environmental Protection Agency's (EPA) Resource Conservation and Recovery Act's (RCRA) definition of ignitable wastes, all of which use flashpoint in their schemes. The NFPA diamond, which comprises the backbone of the NFPA Hazard Signal System, uses a four-quadrant diamond to display the hazards of a material. The top quadrant (red quadrant) contains flammability information in the form of numbers ranging from zero to four. Materials designated as zero will not burn. Materials designated as four rapidly or completely vaporize at atmospheric pressure and ambient temperature, and will burn readily (flashpoint $<73°F$ and boiling point $<100°F$).

The NFPA defines a flammable liquid as one having a flashpoint of 200°F or lower, and divides these liquids into five categories:

1.   Class IA: liquids with flashpoints below 73°F and boiling points below 100°F. An example of a Class IA flammable liquid is n-pentane (NFPA Diamond: 4).

2.   Class IB: liquids with flashpoints below 73°F and boiling points at or above 100°F. Examples of Class IB flammable liquids are benzene, gasoline, and acetone (NFPA Diamond: 3).

3.   Class IC: liquids with flashpoints at or above 73°F and below 100°F. Examples of Class 1C flammable liquids are turpentine and n-butyl acetate (NFPA Diamond: 3).

4.   Class II: liquids with flashpoints at or above 100°F but below 140°F. Examples of Class II flammable liquids are kerosene and camphor oil (NFPA Diamond: 2).

5.   Class III: liquids with flashpoints at or above 140°F but below 200°F. Examples of Class III liquids are creosote oils, phenol, and naphthalene. Liquids in this category are generally termed combustible rather than flammable (NFPA Diamond: 2).The DOT system designates those materials with a flashpoint of 100°F or less as flammable, those between 100°F and 200°F as combustible, and those with a flashpoint of greater than 200°F as nonflammable. EPA designates those wastes with a flashpoint of less than 140°F as ignitable hazardous wastes. These designations serve as useful guides in storage, transport, and spill response. However, they do have limitations, particularly from a safety standpoint. Since these designations are somewhat arbitrary, it is useful to understand the basic concepts of flammability.

The elements required for combustion are few--a substrate, oxygen, and a source of ignition. The substrate, or flammable material, occurs in many classes of compounds, but most often is organic. Generally, compounds within a given class exhibit increasing heat contents with increasing molecular weights (see Table 5). Other properties specific to the substrate that are important in determining flammable hazards are the auto-ignition temperature, boiling point, vapor pressure, vapor density. Auto-ignition temperature (the temperature at which a material will spontaneously ignite) is more important in preventing fire from spreading (e.g., knowing what fire protection is needed to keep temperatures below the ignition point) but can also be important in spill or material handling situations. For example, gasoline has been known to spontaneously ignite when spilled onto an overheated engine or manifold. The boiling point and vapor pressure of a material are important not only because vapors are more easily ignited than liquids, but also because vapors are more readily transportable than liquids (they may disperse, or when heavier than air, flow to a source of ignition and flash back). Vapors with densities greater than one do not tend to disperse but rather to settle into sumps, basements, depressions in the ground, or other low areas, thus representing active explosion hazards. Oxygen, the second requirement for combustion, is generally not limiting. Oxygen in the air is sufficient to support combustion of most materials within certain limits. These limitations are compound specific and are called the explosive limits in air. The upper and lower explosive limits (UEL and LEL) of several common materials are given in Table 6.

Table 5. Heat Content/Increasing Weight Relationships

Compound	MW	Heat Content Kg Calories/gm.MW
methane	16	210.8
ethane	30	368.4
propane	44	526.3
methanol	32	170.9
ethanol	46	327.6
propanol	60	480.7

Table 6. Explosive Limits of Hazardous Materials

Compound	LEL %	UEL %	FP °F	Vapor Density
Acetone	2.15	13	-4	2.0
Acetylene	2.50	100	Gas	0.9
Ammonia	16	25	Gas	0.6
Benzene	1.30	7.1	12	7.8
Carbon	12.4	74	Gas	1.0
Gasoline	1.4	7.6	-45	3-4
Hexane	1.1	7.5	-7	3.0
Toluene	1.2	7.1	40	3.1
Vinyl chloride	3.6	33	Gas	2.2
p-Xylene	1.0	6.0	90	3.7

The source of ignition may be physical (such as a spark, electrical arc, small flame, cigarette, welding operation, or a hot piece of equipment), or it may be chemical in nature, such an an exothermic reaction. In any case, when working with or storing flammables, controlling the source of ignition is often the easiest and safest way to avoid fires or explosions.

Once a fire has started, control of the fire can be accomplished in several ways: through water systems (by reducing the temperature), carbon dioxide or foam systems (by limiting oxygen), or through removal of the substrate (by shutting off valves or other controls).

Toxicology: Toxicology is the science that studies the harmful effects chemicals can have on the body. All chemicals affect man to some degree, depending on the time of exposure, concentration, and human susceptibility. One chemical may only cause a slight rash or dizziness while another may result in cancer or death. It is the degree of exposure and toxicity that are of practical concern.

The means by which chemicals enter the body are inhalation (breathing), ingestion (swallowing), and absorption (skin or living tissue contact). Once in the system these chemicals may produce such symptoms as tissue irritation, rash, dizziness, anxiety, narcosis, headaches, pain, fever, tremors, shortness of breath, birth defects, paralysis, cancer, and death, to mention a few. The amount of chemical that enters the body is called the "dose." The relationship that defines the body response to the dose given is called the "dose-response curve". The lowest dose causing a detectable response is the "threshold limit." The "limit" is dependent on factors such as particle size of contaminant, solubility, breathing rate, residence time in the system, and human susceptibility.

To accomplish meaningful studies, measurements of various parameters are essential. Dose is one of them, and in inhalation studies dose is proportional to the air concentration of the contaminant multiplied by the length of time it is breathed. The units of concentration are ppm (a volume/volume description of concentration-- parts of air contaminant per one million parts of the air mixture) for gases and vapors, and mg/m^3 (a weight/volume description—milligrams of air contaminant per cubic meter of air mixture). Other concentration units exist, such as fibers per cubic centimeter (f/cc) for asbestos, and "rems" for radiation. Dose for oral or skin applications is measured by weight or volume in assigned units such as grams or cubic centimeters. Toxicity data are presented in the literature by such terms as "LD_{50}" and "LC_{50}", that lethal dose per kilogram of body weight or lethal concentration that can kill 50 percent of an animal population. Such data are found, for example, in the Registry of Toxic Effects of Chemical Substances (RTECS). With data such as these obtained from animals closely resembling the human in biochemistry, relative toxicities can be established to characterize chemicals. These data in conjunction with air contaminant threshold limit values (TLV) or permissible exposure limits (PEL), set by law for short periods of exposure or eight-hour, time-weighted average exposure, have produced safe working exposure limits for the worker. Many of these values are contained in the OSHA Standards and the American Conference of Governmental Industrial Hygienist's (ACGIH) in their publications on *Threshold Limit Values and Biological Exposure Indices.*

Human response to chemicals may be described by two types of biological effects--acute and chronic. An acute effect generally results after a single significant exposure, with severe symptoms developing rapidly and coming quickly to a crisis. An example of an acute effect is a few minutes exposure to carbon monoxide of various concentrations that cause headache, dizziness, or death.

The chronic effect results from a repeated dose or exposure to a substance over a relatively prolonged period of time. Examples of chronic effects are possible reduction in life span, increased susceptibility to other diseases, and cancer as a result of smoking. Some materials, such as lead, can bioaccumulate (be stored in the body) and cause continuing effects, or reach a threshold value where an effect on the body occurs after a prolonged period of time, or "latency" period. An example of such a chemical is asbestos, which may produce asbestosis twenty years after the initial exposure.

An effect which exists but has not been widely studied because of its immensity and related problems is "synergism." Synergism occurs when the effect of two chemicals is greater than or less than either chemical alone. Inhalation of isopropyl alcohol and carbon tetrachloride can be well below safe concentration limits separately, but together, produce severe effects including renal failure. Toxicology and epidemiology, the sciences that study diseases in a general population, are closely related.

Most of the present occupational concentration limits for hazardous material have resulted from illnesses and deaths of workers, and from use of both disciplines. Some materials cause genetic changes that can cause cancer (carcinogen), mutation (mutagens), and birth defects (teratogens). These effects are often hard to document due to latency periods and synergisms.

The U.S. OSHA Hazard Communication Standard, 29 CFR 1910.1200, has categorized certain target organ effects, including examples of signs and symptoms and chemicals which have been found to cause such effects. These examples are presented to illustrate the range and diversity of effects and hazards found in the workplace, and the broad scope employers must consider in this area, but they are not intended to be all-inclusive. These are summarized for the reader in Table 7.

Table 7. Target Organ Effects Categorized Under the Hazard Communication Act

---

Hepatotoxins: Chemicals which produce liver damage; *Signs and Symptoms*: Jaundice; liver enlargement; *Chemicals*: Carbon tetrachloride; nitrosamines.

Nephrotoxins: Chemicals which produce kidney damage; *Signs and Symptoms*: Edema; proteinuria; *Chemicals*: Halogenated hydrocarbons; uranium.

Neurotoxins: Chemicals which produce their primary toxic effects on the nervous system; *Signs and Symptoms*: Narcosis; Behavioral changes; decrease in motor functions; *Chemicals*: Mercury; carbon disulfide.

Agents which act on the blood or hematopoietic system: Decreases hemoglobin function; deprive body tissues of oxygen; *Signs and Symptoms*: Cyanosis; loss of consciousness; *Chemicals*: Carbon monoxide; cyanides.

Agents which damage the lung: Chemicals which irritate or damage the pulmonary tissue; *Signs and Symptoms*: Cough; tightness in chest; shortness of breath; *Chemicals*: Silica; asbestos.

Reproductive toxins: Chemicals which affect the reproductive capabilities including chromosomal damage (mutations) and effects on fetuses (teratogenesis); *Signs and Symptoms*: Birth defects; sterility; *Chemicals*: Lead: KEPONE.

Cutaneous hazards: Chemical which affect the dermal layer of the body; *Signs and Symptoms*: Defatting of the skin; rashes; irritation; *Chemicals*: Ketones; chlorinated compounds.

Eye hazards: Chemicals which effect the eye or visual capacity; *Signs and Symptoms*: Conjunctivitis; corneal damage; *Chemicals*: Organic solvents; acids.

Flammability of Hydrocarbons

Hydrocarbons are derivatives from petroleum or crude, but within the context of our immediate discussions, we shall use the terms petroleum liquids and hydrocarbon liquids as being interchangeable. From a fire standpoint, there are only two categories of petroleum liquids, namely *flammable liquids* and *combustible liquids*. Both categories of materials will burn; however, it is into which of these two categories that a liquid belongs that determines its relative fire hazard. Of the two categories, it is the flammables that are considered to be more hazardous, principally because they release ignitable vapors at lower temperatures (a concept consistent with the life cycle theory of fire).

Fire hazard is viewed from the standpoint of safety, to which in the United States, the Occupational Safety and Health Standard (OSH) is often used as the basis for classification of flammables versus a combustible material. Additionally, the U.S. Department of Transportation also has very specific definitions regarding classification of fire hazards based on safe transport of materials. For initial discussions we will adhere to the OSHA definitions, and later refer to distinctions in U.S. federal definitions which are legal standards.

For flammable liquids, the OSHA standard defines a flammable material based upon the liquid's flash point temperature. Any liquid having a flash point below 100°F is classified as being flammable. The definition for flammable liquids given by the National Fire protection Association (NFPA) includes the additional criteria that the liquid's vapor cannot exceed 40 psi at a liquid temperature of 100°F.

From a practical standpoint, these criteria refer simply to the fact that any material with a flash point temperature of 100°F or less is capable of releasing vapors at a rate sufficient to be ignitable, and hence represents the greatest danger from a fire standpoint due to the possibility of spontaneous combustion. It is important to note that there are many materials that are capable of vaporizing at extremely low temperatures. A common example is gasoline whose flash point is -40 °F.

The combustible liquid category are thus those liquids whose flash points are above 100°F. The category for petroleum liquids covers a range from the 100 °F flash point of kerosene to the flash point of 450 °F of some motor oils. Although these materials are less hazardous than flammable liquids, they still represent fire hazards and under certain conditions are as dangerous as flammables. Some typical examples of the two categories of fire hazards for petroleum liquids are given in Table 8.

Table 8. Common Examples of Flammable and Combustible Petroleum Liquids

Flammables	Flash point, °F	Combustible	Flash point, °F
Gasoline	-40	Kerosene	100
Ethers	-30	Fuel oils	100-140
Acetone	-4	Diesel oil	130
Methanol	52	Lubricating oil	300
Crude Oil/Naphtha	20-90	Asphalt	400

The term *flash point* is sometimes confusing and its definition should be carefully considered. The term basically refers to the temperature that a liquid must be at before it will provide the fuel in vapor form necessary for the condition of spontaneous combustion to occur. Perhaps more accurate a definition is that it refers to the lowest temperature a liquid may be and still have the ability to liberate flammable vapor at a sufficient rate that, when mixed with the proper amounts of air, the air-fuel mixture will flash in the presence of a source of energy or ignition source. This provides a more pragmatic viewpoint on how fires occur. In essence, when a material liberates vapor, this vapor represents fuel. When it combines with oxygen in air in the proper amounts, we now have a flammable mixture, and hence all that is needed to complete the fire triangle is a source of energy. It is the vapor (air-fuel mixture) that burns, and not the material itself. It, therefore, stands to reason that, assuming sufficient amounts of air to be present, the greater the volume of released vapor, the larger and more intense will be the fire.

Another term that is often given attention to is the *fire point*. The fire point temperature refers to the temperature which the liquid must be at before released vapor is in sufficient quantity to continue to burn. With the flash point temperature, the amounts of vapor being released at the exact flash point temperature will not sustain the fire and, after flashing across the liquid surface, the flame extinguishes. For many materials the fire point is only a few degrees above the flash point, but regardless, the flash point is perhaps the more universally accepted basis of classifying a fire hazard largely because a flash fire will generally be sufficient to ignite combustible materials.

Fire is an exothermic (heat-liberating) reaction. There must be a continuous feedback of energy (heat) to keep the reaction going. Also, heat is dissipated from the fire by one or more of the methods of transferring heat: conduction, convection, and radiation. Heat energy is also fed back to the fire by radiation from the flame, and this source of heat keeps the fire going. If we could devise a way to interrupt that feedback of heat to the fuel, the continuity of the fire would be broken, and the fire would go out. Hence, a fire-extinguishing agent is needed that siphons heat energy away from the fire, reduces the temperature of the material burning, and cools the surroundings below the ignition temperature of the fuel, so that there would not be a re-ignition of flammable vapors once the fire was extinguished.

Water is the most common extinguishing agent that performs this task. Water has many disadvantages, however. Some of the drawbacks to the use of water as an extinguishing agent include its propensity to conduct electricity (which, of course, is deadly if the water is applied incorrectly), its low viscosity (which allows it to run off a wall instead of sticking there), and a high surface tension (which prevents it from penetrating tightly arranged materials). Water also allows heat to be radiated through it, freezes at a relatively high temperature, splashes about, and displaces many flammable liquids, causing them to spread rapidly, while burning all the time. This list of problems also includes the fact that water itself will violently react with many of the hazardous materials it is supposed to control.

In addition to the fact that water is relatively inexpensive and is usually available in large quantities, there are two specific properties of water that make it invaluable. Those properties are its latent heat of vaporization and its specific heat. The latent heat of vaporization of a substance is defined as the amount of heat a material must absorb when it changes from a liquid to a vapor or gas. The specific heat of a substance is defined as the ratio between the amount of heat necessary to raise the temperature of a substance and the amount of heat necessary to raise the same weight of water by the same number of degrees.

The specific heat of water is important because it is so high in relation to the specific heat of other materials; this fact means that it takes more energy to raise the temperature of water than just about any other material. Therefore, the temperature of the materials to which water has been applied will drop faster than the temperature of water will rise. The specific heat may be reported as the number of calories needed to raise the temperature of one gram of the material $1\,°C$, or the number of British Thermal Units (Btu) needed to raise one pound of the material, $1\,°F$. Therefore, when water is applied to a fire, it begins absorbing heat from the fire, thereby cooling the fire down while the water heats up. For every Btu absorbed, the temperature of the water will rise $1\,°F$ per pound of water involved. The important thing to remember here is that the rise in temperature of the water is caused by heat energy absorbed from the fire. The water is siphoning the heat away from the burning material. The temperature of the water will continue to rise, as long as the fire is producing heat, until it reaches its boiling point of $212\,°F$. At this time the latent heat of vaporization of water comes into play. At $212\,°F$ the water is still a liquid and will remain a liquid unless more energy is received from the fire. At this time, there is a phase change from liquid to vapor, with no increase in temperature; that is, water as a liquid at $212\,°F$ converts to water vapor at $212\,°F$. It is at this phase change that the latent heat of vaporization of water does its work, for while water will absorb 1 Btu per pound for every increase of $1\,°F$, up to $212\,°F$, at $212\,°F$ when the phase change occurs, 970 Btu are absorbed per pound. That sudden, rapid, and massive withdrawal of heat energy from the fire at this time is what gives water its tremendous fire-extinguishing capabilities, which are so valuable as to overcome the previously mentioned disadvantages. Heat is withdrawn from the burning material so rapidly, and in such large quantities, that the temperature of the burning fuel drops dramatically, usually well below its ignition temperature. When this happens, of course, the fire goes out. The latent heat of vaporization also explains why steam at $212\,°F$ is hotter than boiling water at $212\,°F$. The live steam has 970 Btu's of energy more than the boiling water. This latent heat of vaporization also explains why materials wet with water are difficult, and sometimes impossible, to ignite. If a combustible substance has absorbed enough water to be considered wet, or just damp, this water will act as a barrier to ignition by its evaporation as it is heated. As heat is applied to the wet substance, the water begins to evaporate (go through the phase change from a liquid to a vapor). To make this phase change, the water must absorb 1 Btu for every pound of water present for every 1°F it rises until it reaches 212°F, whereupon it must absorb 970 Btu for every pound of water present.

Before any combustible material that has been wet with water can burn, the water (which has preferentially been absorbing the applied heat and thus keeping the combustible material itself from heating to its ignition temperature) must be driven off. If in the process of driving off the water enough heat energy from the potential ignition source has been used up so that there is not enough left (for example a burnt-out match) to raise the combustible to its ignition temperature, there will be no fire.

Water, of course, does not work with all materials. There is a special class of materials that are water reactive, and hence water becomes an unacceptable extinguishing agent. For these class of materials another approach to eliminating the fire is taken. Specifically, we must remove the oxidizer leg from the fire triangle; i.e. cut off the supply of oxygen which fuels the air to fuel mixture.

From a fire standpoint, the two main categories of petroleum liquids are flammable and combustible, and are determined mainly by the liquid's flash point. Both categories of liquids will burn but it is into which of these two categories the liquid belongs that determines its relative fire hazard. As already noted, flammable liquids are generally considered the more hazardous of the two categories mainly because they release ignitable vapors. Following OSHA definitions, a flammable material is any liquid having a flash point below 100°F. The NFPA expands this definition by including the stipulation that the vapor cannot exceed 40 psi (pounds per square inch) at a liquid temperature of 100°F, with the theory being that such liquids are capable of releasing vapor at a rate sufficient to be ignitable. Since this aspect of the definition relating to vapor pressure has little fire-ground application it is often ignored. However, it is important to note that if the heat from a fire raises the liquid temperature to a temperature above the liquid's flash point, it will automatically increase the vapor pressure inside a closed container. Any other source of sufficient heat will produce the same result.

Within the combustible liquid category are those materials with a flash point above 100 °F. Combustible liquids are considered less hazardous than flammable liquids because of their higher flash points. However, this statement can be misleading since there are circumstances when it is not a valid assumption. It is possible for certain combustible liquids to be at their flash point when a hot summer sun has been striking their metal container for some time. Additionally, during the transportation of some combustible products, the product is either preheated or a heat source is maintained to make the product more fluid than it would be at atmospheric temperatures. One reason this is done is to facilitate transportation or pumping; i.e. to aid with the movement of a material that is very viscous, such as asphalt or tar. Also, some materials classed as combustible solids will be heated to their melting point. Naphthalene is one example of this treatment. Naphthalene might be heated to a temperature above its melting point, which is about 176 °F. Despite its fairly high ignition temperature (almost 980 °F), it would not be unreasonable to surmise that a spill of liquid naphthalene could present a serious fire hazard. Fortunately, with naphthalene, quick action with adequate amounts of water applied as spray streams should cool and solidify it, thus greatly minimizing the fire risk. It is important to note that a combustible liquid at or above its flash point will behave in the same manner that a flammable liquid would in a similar emergency. As an example No.2 fuel oil when heated to a temperature of 150°F can be expected to act or react in the same way gasoline would at 50°F. In most instances, however, to reach this elevated temperature will require the introduction of an external heat source. Some common examples of combustible petroleum liquids are given in Table 9.

It is important to note that the extinguishing techniques, controlling actions, or fire-prevention activities implemented can differ greatly depending upon which of the two categories the liquid falls in.

To have the ability to categorize a liquid correctly when it is not so identified, it is only necessary to know its flash point. By definition, the flash point of a liquid determines whether a liquid is flammable or combustible. The categories of liquids are further subdivided into classes according to the flash point plus the boiling point of certain liquids.

These divisions are summarized in Table 10, which shows that flammable liquids fall into Class 1, and combustible liquids into Classes 2 and 3. The products that are at the low end (100°F) of the Class 2 combustible-liquid group might be thought of as borderline cases. These could act very much like flammable liquids if atmospheric temperatures were in the same range.

Table 9. Examples of Petroleum Liquids that are Combustible

Product	Flash Point (°F)	Product	Flash Point (°F)
Kerosene	100+	Lubricating Oil	300
Fuel Oils/Diesel	100 - 140	Asphalt	400

Table 10. Classes of Flammable and Combustible Liquids

Class	Flash Point (°F)	Boiling Point (°F)
1	Below 100	-
1A	Below 73	Below 100
1B	Below 73	At or above 100
1C	73-99	-
2	100-139	-
3	140 or above	Below 100
3A	140-199	At or above 100
3B	200 or above	-

It is not a common industry practice to identify either stationary or portable (mobile) liquid containers by the class of liquid it contains. The usual practice is to label either **FLAMMABLE** or **COMBUSTIBLE** and include the required U.S. Department of Transportation placard.

Basically, the flash point is the temperature a liquid must be at before it will provide the fuel vapor required for a fire to ignite. A more technical definition for flash point is: The lowest temperature a liquid may be at and still have the capability of liberating flammable vapors at a sufficient rate that, when united with the proper amounts of air, the air-fuel mixture will flash if a source of ignition is presented. The amounts of vapor being released at the exact flash-point temperature will not sustain the fire and, after flashing across the liquid surface, the flame will go out. It must be remembered that at the flash-point temperature, the liquid is releasing vapors and, as with other ordinary burnable materials, it is the vapors that burn. The burning process for both ordinary combustible solids and liquids requires the material to be vaporized. It may also be in the form of a very fine mist, which will be instantly vaporized if a source of heat is introduced. It is not the actual solid or the liquid that is burning, but the vapors being emitted from it. For this reason, when we speak of a fuel we are referring to the liberated vapor. It is an accepted phenomenon, assuming sufficient amounts of air to be present, that the greater the volume of released vapor, the larger the fire will be.

The technical literature sometimes refers to the "fire point," which in most instances is just a few degrees above the flash point temperature, and is the temperature the liquid must be before the released vapor is in sufficient quantity to continue to burn, once ignited. However, because a flash fire will normally ignite any Class "A" combustible present in the path of the flash, it is reasonable to accept the flash point as being the critical liquid temperature in assessing a fire hazard.

Any of the other combustibles ignited by the flash fire, that is, wood, paper, cloth, etc., once burning, could then provide the additional heat necessary to bring the liquid to its fire point.

A crucial objective upon arrival of the first responding fire forces is to determine if the liquid present is a product that is vaporizing at the time or, if it is not, and what condition may be present that is capable of providing the required heat to cause the liquid to reach its flash point. This information would have a direct influence on the selection of control and/or extinguishing activities. An emergency involving a petroleum liquid, which is equal to or above flash point, means that a fuel source consisting of flammable vapors will be present. This, in turn, means the responding fire-fighting forces will be faced with either a highly hazardous vapor cloud condition or with a fire if ignition has occurred before arriving at the scene. Conversely, if it is a liquid at a temperature below its flash point, then fuel would not be immediately available to burn.

As explained earlier by the theories of fire, a source of air or more specifically, oxygen must be present. A reduction in the amount of available air to below ideal quantities causes the fire to diminish. Moreover, reduce the fuel quantity available and the fire will also diminish in size. Almost all extinguishing techniques developed are methods of denying the fire one or both of these requirements. By cooling a material below its flash point, vapor production is halted, thus removing the fuel from the fire. When utilizing a smothering-type extinguishing agent, the principle involved consists of altering the air-fuel mixture. When the vapor is no longer in its explosive range, the fire dies, either due to insufficient fuel or a lack of oxygen. The flash point tells us the conditions under which we can expect the fuel vapor to be created, but it is the *explosive range* which tells us that a certain mixture of fuel vapor and air is required for the vapor to become ignitable. The terms flammable limit and combustible limit are also used to describe the explosive range. These three terms have identical meanings and can be used interchangeably.

The area between the LEL and the UEL is what is known as the explosive range. The figures given for the amount of fuel vapor required to place a substance within its explosive range are given as a percentage of the total air-fuel mixture. To compute how much air is required to achieve this mixture, subtract the listed percentage from 100 percent: the remainder will be the amount of air needed. Even though it is only the oxygen contained in the air that the fire consumes, flammable ranges are shown as air-fuel ratios because it is the air that is so readily available. Any air-fuel mixture in which the vapor is above the UEL, or any air-fuel mixture in which the vapor is below the LEL, will not burn.

Using gasoline as an example, the explosive range can be computed as follows:

	LEL(%)	UEL (%)
Gasoline vapor	1.5	7.6
Air	98.5	92.4
Total volume	100	100

This example helps to illustrate that large volumes of air are required to burn gasoline vapors. The explosive ranges for the different grades of gasoline, or even those of most other petroleum liquids, are such that average explosive-range figures that are suitable for use by the fire fighter would be the LEL at 1 percent and the UEL at about 7 percent. The vapor content of a contaminated atmosphere may be determined through the use of a combustible gas-detecting instrument, referred to as an explosimeter. If a fire involving a petroleum liquid does occur, an extinguishing technique that may be appropriate is the altering of the air-fuel mixture. One technique utilized will necessitate the use of an extinguishing agent such as a foam with the capability of restricting the air from uniting with the vapor. Another technique is to prevent the liquid from having the ability to generate vapor. Usually this is a cooling action and is accomplished with water spray streams. In both cases, extinguishment is accomplished as a result of altering the air-fuel mixture to a point below the LEL for the specific liquid.

We will now devote attention to the so-called *ignition temperature*. Consider the emergency situation where there is a spill of gasoline. We may immediately conclude that two of the requirements for a fire exist. First, the gasoline, which would be at a temperature above its flash point, will be releasing flammable vapors; thus a source of fuel will be present. Moreover, there is ample air available to unite with the fuel thus there is the potential for the mixture to be in its explosive range. The only remaining requirement needed to have a fire is a source of heat at or above the ignition temperature of gasoline. Technically speaking, all flammable vapors have an exact minimum temperature that has the capability of igniting the specific air-vapor mixture in question. This characteristic is referred to as the ignition temperature and could range from as low as 300 °F for the vapor from certain naphthas to over 900 °F for asphaltic material vapor. Gasoline vapor is about halfway between; at around 600 °F. A rule of thumb for the ignition temperature of petroleum-liquid vapors is 500 °F. This figure may appear low for several of the hydrocarbon vapors, but it is higher than that of most ordinary combustibles, and is close enough to the actual ignition temperatures of the products most frequently present at emergency scenes to give a suitable margin of safety.

In emergency situations, it is best to take conservative approaches by assuming that all heat sources are of a temperature above the ignition temperature of whatever liquid may be present. This approach is not an overreaction when it is realized that almost all the normally encountered spark or heat sources are well above the ignition temperature of whatever petroleum liquid might be present. Among the more common   sources of ignition would be smoking materials of any kind (cigarettes, cigars, etc.), motor vehicles, and equipment powered by internal combustion engines: also electrically operated tools or equipment, as well as open-flame devices such as torches and flares.

The removal of any and all potential ignition sources from the area must be instituted immediately and methodically. The operation of any motor vehicle, including diesel-powered vehicles, must not be permitted within the immediate vicinity of either a leak or spill of a flammable liquid. The probability of a spark from one of the many possible sources on a motor vehicle is always present. Also, under no circumstances should motor vehicles be allowed to drive through a spill of a petroleum product.

Ignition sources are not necessarily an external source of heat; it could be the temperature of the liquid itself. Refineries and chemical plants frequently operate processing equipment that contains a liquid above its respective ignition temperature. Under normal operating conditions, when the involved liquid is totally contained within the equipment, no problems are presented because the container or piping is completely filled with either liquid or vapor. If full and totally enclosed, it means there can be no air present; thus an explosive or ignitable mixture cannot be formed. If the enclosed liquid which in certain stages of its processing may be above the required ignition temperature should be released to the atmosphere, there is a possibility that a vapor-air mixture could be formed and hence, ignition could occur. This type of ignition is referred to as *auto-ignition*. Auto-ignition is defined as the self-ignition of the vapors emitted by a liquid heated above its ignition temperature and that, when escaping into the atmosphere, enter into their explosive range. Some typical ignition temperatures for various petroleum liquids are 600 °F for gasoline, 550 °F for naphtha and petroleum ethers, 410 °F for kerosine, and 725 °F for methanol. From the above discussions, the important elements that are responsible for a fire are:

- Fuel in the form of a vapor that is emitted when a liquid is at or above its flash point temperature.

- Air that must combine with the vapor in the correct amounts to place the mixture in the explosive range.

- Heat, which must be at least as hot as the ignition temperature, must then be introduced.

In addition to fuel, oxygen, and energy, the tetrahedron of fire theory identifies the chemical chain reaction of the flame as a requirement for a fire to sustain itself. The fourth side of the tetrahedron is the chain reaction, however, from a practical standpoint it does not appear to have a significant influence on normal fire-control practices. It is known that when using a dry chemical, extinguishment is achieved by the interruption of the chain reaction propagating the flame rather than by a smothering action, however this knowledge doesn't really alter the practical application of this technique to fire fighting.

Petroleum liquids have certain characteristics that can exert an influence on the behavior of the liquid and/or vapor that is causing the problem. For this reason, these features may have a bearing on the choice of control practices or extinguishing agents under consideration. These characteristics include the weight of the vapor, the weight of the liquid, and whether the liquid will mix readily with water. The specific properties of importance are vapor density, specific gravity, and water solubility. Before discussing these important physical properties, let's first examine the data in Table 11 which lists the flammability limits of some common gases and liquids.

Table 11. Limits of Flammability of Gases and Vapors, % in Air

Gas or Vapor	LEL	UEL	Gas or Vapor	LEL	UEL
Hydrogen	4.00	75.0	Ethylene	3.1	32.0
Carbon monoxide	12.50	74.0	Propylene	2.4	10.3
Ammonia	15.50	26.60	Butadiene	2.00	11.50
Hydrogen sulfide	4.30	45.50	Butylene	1.98	9.65
Carbon disulfide	1.25	44.0	Amylene	1.65	7.70
Methane	5.30	14.0	Acetylene	2.50	81.00
Ethane	3.00	12.5	Benzene	1.4	7.1
Propane	2.20	9.5	Toluene	1.27	6.75
Butane	1.90	8.5	Styrene	1.10	6.10
Iso-butane	1.80	8.4	o-Xylene	1.00	6.00
Pentane	1.50	7.80	Naphthalene	0.90	...
Iso-pentane	1.40	7.6	Anthracene	0.63	...
Hexane	1.20	7.5	Cyclo-propane	2.40	10.4
Heptane	1.20	6.7	Cyclo-hexene	1.22	4.81
Octane	1.00	3.20	Cyclo-hexane	1.30	8.0
Nonane	0.83	2.90	Methyl cyclohexane	1.20	...
Decane	0.67	2.60	Gasoline-regular	1.40	7.50
Dodecane	0.60	...	Gasoline-92 Oct	1.50	7.60
Tetradecane	0.50	...	Naphtha	1.10	6.00

Two general conclusions can be drawn from the data in Table 11. First, the lower the material's LEL, obviously the more hazardous. Also note that there are some materials that have wide explosive ranges. This aspect is also significant from a fire standpoint. As an example, comparing hydrogen sulfide to benzene, although the LEL for H_2S is more than 3 times greater, its explosive range is 7 times wider. This would suggest that H_2S is an extremely hazardous material even though its LEL is relatively high. In fact, H_2S fires are generally so dangerous that the usual practice is to contain and allow burning to go to completion rather than to fight the fire. With the above general introduction, we now focus attention on some of the engineering controls used to limit emissions. In particular, attention is given in the next section to pressure relieving systems.

The proper design and operation of these systems for emergency pressure let-downs and equipment turnaround events are crucial to the safe operation of a refinery or chemical plant operation. Designs are often very specific to the refinery operation, however there are still many commonalities between systems, that enable a number of generalizations to be made. A significant part of the discussion focuses on flares.

PRESSURE RELIEVING SYSTEMS

There are large volumes of hydrocarbon gases that are produced and handled in refinery and petrochemical plants. Generally, these gases are used as fuel or as raw material for further processing. In the past, however, large quantities of these gases were considered waste gases, and along with waste liquids, were dumped to open pits and burned, producing large volumes of black smoke. With modernization of processing units, this method of waste-gas disposal, even for emergency gas releases, has been eliminated. Nevertheless, petroleum refineries are still faced with the problem of safe disposal of volatile liquids and gases resulting from scheduled shutdowns and sudden or unexpected upsets in process units. Emergencies that can cause the sudden venting of excessive amounts of gases and vapors include fires, compressor failures, overpressures in process vessels, line breaks, leaks, and power failures. Uncontrolled releases of large volumes of gases also constitute a serious safety hazard to personnel and equipment. A system for disposal of emergency and waste refinery gases consists of a manifolded pressure-relieving or blowdown system, and a blowdown recovery system or a system of flares for the combustion of the excess gases, or both. Many older refineries, however, do not operate blowdown recovery systems. In addition to disposing of emergency and excess gas flows, these systems are used in the evacuation of units during shutdowns and turnarounds. Normally a unit is shut down by depressuring into a fuel gas or vapor recovery system with further depressuring to essentially atmospheric pressure by venting to a low-pressure flare system. Thus, overall emissions of refinery hydrocarbons are substantially reduced.

Refinery pressure-relieving systems, commonly called blowdown systems, are used primarily to ensure the safety of personnel and protect equipment in the event of emergencies such as process upset, equipment failure, and fire. In addition, a properly designed pressure relief system permits substantial reduction of hydrocarbon emissions to the atmosphere.

The equipment in a refinery can operate at pressures ranging from less than atmospheric to 1,000 psig and higher. This equipment must be designed to permit safe disposal of excess gases and liquids in case operational difficulties or fires occur. These materials are usually removed from the process area by automatic safety and relief valves, as well as by manually controlled valves, manifolded to a header that conducts the material away from the unit involved. One of the preferred methods of disposing of the waste gases that cannot be recovered in a blowdown recovery system is by burning in a smokeless flare. Liquid blowdowns are usually conducted to appropriately designed holding vessels and reclaimed. A blowdown or pressure-relieving system consists of relief valves, safety valves, manual bypass valves, blowdown headers, knockout vessels, and holding tanks. A blowdown recovery system also includes compressors and vapor surge vessels such as gas holders or vapor spheres. Flares are usually considered as part of the blowdown system in a modern refinery. The pressure-relieving system can be used for liquids or vapors or both. For reasons of economy and safety, vessels and equipment discharging to blowdown systems are usually segregated according to their operating pressure. In other words, there is a high-pressure blowdown system for equipment working, for example, above 100 psig, and low-pressure systems for those vessels with working pressures below 100 psig. Butane and propane are usually discharged to a separate blowdown drum, which is operated above atmospheric pressure to increase recovery of liquids. Usually a direct-contact type of condenser is used to permit recovery of as much hydrocarbon liquid as possible from the blowdown vapors. The non-condensables are burned in a flare system. A typical pressure-relieving system for flaring operations used not only as a safety measure but also as a means of reducing the emission of hydrocarbons to the atmosphere. A typical installation includes four separate collecting systems as follows: (1) a low-pressure blowdown system for vapors from equipment with working pressure below 100 psig, (2) a high-pressure blowdown system for vapors from equipment with working pressures above 100 psig, (3) a liquid blowdown system for liquids at all pressures, and (4) a light-ends blowdown for butanes and lighter hydrocarbon blowdown products. The liquid portion of light hydrocarbon products released through the light-ends blowdown system is recovered in a drum near the flare. A backpressure of 50 psig is maintained on the drum, which minimizes the amount of vapor that vents through a backpressure regulator to the high-pressure blowdown line. The high-pressure, low-pressure, and liquid-blowdown systems all discharge into the main blowdown vessel. Any entrained liquid is dropped out and pumped to a storage tank for recovery. Offgas from this blowdown drum flows to a vertical vessel with baffle trays in which the gases are contacted directly with water, which condenses some of the hydrocarbons and permits their recovery. The overhead vapors from the sump tank flow to the flare system manifold for disposal by burning in a smokeless flare system.

The design of a pressure relief system is one of the most important problems in the planning of a refinery or petrochemical plant. The safety of personnel and equipment depends upon the proper design and functioning of this type of system. The consequences of poor design can be disastrous. A pressure relief system can consist of one relief valve, safety valve, or rupture disc, or of several relief devices manifolded to a common header. Usually the systems are segregated according to the type of material handled, that is, liquid or vapor, as well as to the operating pressures involved.

The several factors that must be considered in designing a pressure relief system are (1) the governing code, such as that of ASME (American Society of Mechanical Engineers); (2) characteristics of the pressure relief devices; (3) the design pressure of the equipment protected by the pressure relief devices; (4) line sizes and lengths, and (5) physical properties of the material to-be relieved to the system. In discussing pressure relief systems, the following terms are commonly used.

Relief Valve: A relief valve is an automatic pressure relieving device actuated by the static pressure upstream of the valve. It opens further with increase of pressure over the set pressure. It is used primarily for liquid service.

Safety Valve: A safety valve is an automatic relieving device actuated by the static pressure upstream of the valve and characterized by full opening or pop action upon opening. It is used for gas or vapor service.

Rupture Disc: A rupture disc consists of a thin metal diaphragm held between flanges.

Maximum Allowable Working Pressure: The maximum allowable working pressure (that is, design pressure), as defined in the construction codes for unfired pressure vessels, depends upon the type of material, its thickness, and the service condition set as the basis for design. The vessel may not be operated above this pressure or its equivalent at any metal temperature higher than that used in its design; consequently, for that metal temperature, it is the highest pressure at which the primary safety or relief valve may be set to open.

Operating Pressure: The operating pressure of a vessel is the pressure, in psig, to which the vessel is usually subjected in service. A processing vessel is usually designed to a maximum allowable working pressure, in psig, that will provide a suitable margin above the operating pressure in order to prevent any undesirable operation of the relief valves. It is suggested that this margin be approximately 10 percent higher, or 25 psi, whichever is greater.

Set Pressure: The set pressure, in psig, is the inlet pressure at which the safety or relief valve is adjusted to open.

Accumulation: Accumulation is the pressure increase over the maximum allowable working pressure of the vessel during discharge to the safety or relief valve expressed as a percent of that pressure or pounds per square inch.

Over Pressure: Over pressure is the pressure increase over the set pressure of the primary relieving device. It is the same as accumulation when the relieving device is set at the maximum allowable working pressure of the vessel. When the set pressure of the first safety or relief valve to open is less than the maximum allowable working pressure of the vessel the over pressure may be greater than 10 percent of the set pressure of the first safety or relief valve.

Blowdown: Blowdown is the difference between the set pressure and the reseating pressure of a safety or relief valve, expressed as a percent of a set pressure or pounds per square inch.

Lift: Lift is the rise of the disc in a safety or relief valve.

Backpressure: Backpressure is the pressure developed on the discharge side of the safety valves. Superimposed backpressure is the pressure in the discharge header before the safety valve opens (discharged from other valves).

Built-up Pressure: Built-up backpressure is the pressure in the discharge header after the safety valve opens.

Safety Valves

Nozzle-type safety valves are available in the conventional or balanced-bellows configurations. Backpressure in the piping downstream of the standard-type valve affects its set pressure, but theoretically, this backpressure does not affect the set pressure of the balanced-type valve. Owing, however, to imperfections in manufacture and limitations of practical design, the balanced valves available vary in relieving pressure when the backpressure reaches approximately 40 percent of the set pressure. The actual accumulation depends upon the manufacturer.

Until the advent of balanced valves, the general practice in the industry was to select safety valves that start relieving at the design pressure of the vessel and reach full capacity at 3 to 10 percent above the design pressure. This overpressure was defined as accumulation. With the balanced safety valves, the allowable accumulation can be retained with smaller pipe size. Each safety valve installation is an individual problem. The required capacity of the valve depends upon the condition producing the overpressure.

Rupture Discs

A rupture disc is an emergency relief device consisting of a thin metal diaphragm carefully designed to, rupture at a predetermined pressure. The obvious difference between a relief or safety valve and a rupture disc is that the valve reseats and the disc does not. Rupture discs may be installed in parallel or series with a relief valve. To prevent an incorrect pressure differential from existing, the space between the disc and the valve must be maintained at atmospheric pressure. A rupture disc is usually designed to relieve at 1.5 times the maximum allowable working pressure of the vessel. In determining the size of a disc, three important effects that must be evaluated are low rupture pressure, elevated temperatures, and corrosion. Manufacturers can supply discs that are guaranteed to burst at plus or minus 5 percent of their rated pressures. The corrosive effects of a system determine the type of material used in a disc. Even a slight amount of corrosion can drastically shorten disc life. Discs are available with plastic linings, or they can be made from pure carbon materials.

The discharge piping for relief and safety valves and rupture discs should have a minimum of fittings and bends. There should be minimum loading on the valve, and piping should be used with adequate supports and expansion joints. Suitable drains should be used to prevent liquid accumulation in the piping and valves.

Flares

Smoke is the result of incomplete combustion. Smokeless combustion can be achieved by: (1) adequate heat values to obtain the minimum theoretical combustion temperatures, (2) adequate combustion air, and (3) adequate mixing of the air and fuel. An insufficient supply of air results in a smoky flame. Combustion begins around the periphery of the gas stream where the air and fuel mix, and within this flame envelope the supply of air is limited. Hydrocarbon side reactions occur with the production of smoke. In this reducing atmosphere, hydrocarbons crack to elemental hydrogen and carbon, or polymerize to form hydrocarbons. Since the carbon particles are difficult to burn, large volumes of carbon particles appear as smoke upon cooling.

Side reactions become more pronounced as molecular weight and unsaturation of the fuel gas increase. Olefins, diolefins, and aromatics characteristically burn with smoky, sooty flames as compared with paraffins and naphthenes. A smokeless flame can be obtained when an adequate amount of combustion air is mixed sufficiently with the fuel so that it burns completely and rapidly before any side reactions can take place. Combustion of hydrocarbons in the steam-inspirated-type elevated flare appears to be complete. The air pollution problem associated with the uncontrolled disposal of waste gases is the venting of large volumes of hydrocarbons and other odorous gases and aerosols. The preferred control method for excess gases and vapors is to recover them in a blowdown recovery system and, failing that, to incinerate them in an elevated type flare. Such flares introduce the possibility of smoke and other objectionable gases such as carbon monoxide, sulfur dioxide, and nitrogen oxides. Flares have been further developed to ensure that this combustion is smokeless and in some cases nonluminous. Luminosity does attract attention to the refinery operation and in certain cases can cause bad public relations. There is also the consideration of military security in which nonluminous emergency gas flares would be desirable. It is important to note that the hydrocarbon and carbon monoxide emissions from a flare can be much greater than those from a properly operated refinery boiler or furnace. Other combustion contaminants from a flare include nitrogen oxides. The importance of these compounds to the total air pollution problem depends upon the particular conditions in a particular locality.

Other air contaminants that can be emitted from flares depend upon the composition of the gases burned. The most commonly detected emission is sulfur dioxide, resulting from the combustion of various sulfur compounds (usually hydrogen sulfide) in the flared gas. Toxicity, combined with low odor threshold, make venting of hydrogen sulfide to a flare an unsuitable and sometimes dangerous method of disposal. In addition, burning relatively small amounts of hydrogen sulfide can create enough sulfur dioxide to cause crop damage or local nuisance. Materials that tend to cause health hazards or nuisances should not be disposed of in flares. Compounds such as mercaptans or chlorinated hydrocarbons require special combustion devices with chemical treatment of the gas or its products of combustion.

The ideal refinery flare is a simple device for safe and inconspicuous disposal of waste gases by combustion. Hence, the ideal flare is a combustion device that burns waste gases completely and smokelessly. There are, in general, three types of flares: elevated flares, ground-level flares, and burning pits. The burning pits are reserved for extremely large gas flows caused by catastrophic emergencies in which the capacity of the primary smokeless flares is exceeded. Ordinarily, the main gas header to the flare system has a water seal bypass to a burning pit. Excessive pressure in the header blows the water seal and permits the vapors and gases to vent a burning pit where combustion occurs. This is rarely practiced today, except for parts of South America and Eastern Europe.

The essential parts of a flare are the burner, stack, seal, liquid trap, controls, pilot burner, and ignition system. In some cases, vented gases flow through chemical solutions to receive treatment before combustion. As an example, gases vented from an isomerization unit that may contain small amounts of hydrochloric acid are scrubbed with caustic before being vented to the flare.

Elevated flares are the most commonly used system. Smokeless combustion can be obtained in an elevated flare by the injection of an inert gas to the combustion zone to provide turbulence and inspirate air. A mechanical air-mixing system would be ideal but is not economical in view of the large volume of gases typically handled. The most commonly encountered air-inspirating material for an elevated flare is steam. Three main types of steam injected elevated flares are in use. These types vary in the manner in which the steam is injected into the combustion zone. In the first type, there is a commercially available multiple nozzle, as shown in Figure 1, which consists of an alloy steel tip mounted on the top of an elevated stack. Steam injection is accomplished by several small jets placed concentrically around the flare tip. These jets are installed at an angle, causing the steam to discharge in a converging pattern immediately above the flare tip. A second type of elevated flare has a flare tip with no obstruction to flow, that is, the flare tip is the same diameter as the stack. The steam is injected by a single nozzle located concentrically within the burner tip. In this type of flare, the steam is premixed with the gas before ignition and discharge.

A third type of elevated flare is equipped with a flare tip constructed to cause the gases to flow through several tangential openings to promote turbulence. A steam ring at the top of the stack has numerous equally spaced holes about 1/8 inch in diameter for discharging steam into the gas stream.

The injection of steam in this latter flare may be automatically or manually controlled. In most cases, the steam is proportioned automatically to the rate of gas flow; however, in some installations, the steam is automatically supplied at maximum rates, and manual throttling of a steam valve is required for adjusting the steam flow to the particular gas flow rate. There are many variations of instrumentation among various flares, some designs being more desirable than others.

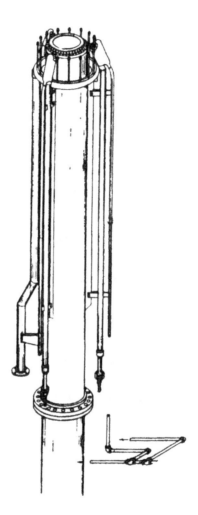

Figure 1. Smokeless flare burner.

For economic reasons all designs attempt to proportion steam flow to the gas flow rate. Steam injection is generally believed to result in the following benefits: (1) Energy available at relatively low cost can be used to inspirate air and provide turbulence within the flame, (2) steam reacts with the fuel to form oxygenated compounds that burn readily at relatively low temperatures, (3) water-gas reactions also occur with this same end result, and (4) steam reduces the partial pressure of the fuel and retards polymerization. Inert gases such as nitrogen have also been found effective for this purpose; however, the expense of providing a diluent such as this is prohibitive.

There are four principal types of ground level flare: Horizontal venturi, water injection, multijet, and vertical venturi. In a typical horizontal, venturi-type ground flare system, the refinery flare header discharges to a knockout drum where any entrained liquid is separated and pumped to storage. The gas flows to the burner header, which is connected to three separate banks of standard gas burners through automatic valves of the snap-action type that open at predetermined pressures. If any or all of the pressure valves fail, a bypass line with a liquid seal is provided (with no valves in the circuit), which discharges to the largest bank of burners. The automatic-valve operation schedule is determined by the quantity of gas most likely to be relieved to the system.

The allowable back-pressure in the refinery flare header determines the minimum pressure for the control valve and the No. 1 burner bank. On the assumption that the first valve was set at 3 psig, then the second valve for the No. 2 burner bank would be set for some higher pressure, say 5 psig. The quantity of gas most likely to be released then determines the size and the number of burners for this section. Again, the third most likely quantity of gas determines the pressure setting and the size of the third control valve. Together, the burner capacity should equal the maximum expected flow rate. The valve-operating schedule for the system is set up as follows:(1) When the relief header pressure reaches 3 psig, the first control valve opens and the four small venturi burners go into operation. The controller setting keeps the valve open until the pressure decreases to about 1-1/2 psig; (2) When the header pressure reaches 5 psig, the second valve opens and remains open until the pressure drops to about 3 psig; (3) When the pressure reaches 6 psig, the third valve opens and remains open until the pressure decreases to 4 psig; (4) At about 7 psig, the gas blows the liquid seal.

Another common type of ground flare used in petroleum refineries has a water spray to inspirate air and provide water vapor for the smokeless combustion of gases. This flare requires an adequate supply of water and a reasonable amount of open space. The structure of the flare consists of three concentric stacks. The combustion chamber contains the burner, the pilot burner, the end of the ignitor tube, and the water spray distribitor ring. The primary purpose of the intermediate stack is to combine the water spray that it will be mixed intimately with burning gases. The outer stack confines the flame and directs it upward. Water sprays in elevated flares are not too practical for several resons. Difficulty is experienced in keeping the water spray in the flame zone, and the scale formed in the waterline tends to plug the nozzles. Water main pressure dictates the height to which water can be injected without the use of a booster pump. For a 100- to 250-foot stack, a booster pump would undoubtedly be required. Rain created by the spray from the flare stack is objectionable from the standpoint of corrosion of nearby structures and other equipment. Water is not as effective as steam for controlling smoke with high gas flow rates, unsaturated materials, or wet gases. The water spray flare is economical when venting rates are not too high and slight smoking can be tolerated.

The miltijet-type ground flare is designed to burn excess hydrocarbons without smoke, noise, or visible flame. These generally tend to be less expensive than the steam-injected type, on the assumption that new steam facilities must be installed to serve a steam-injected flare unit. Where the steam can be diverted from noncritical operations such as tank heating, the cost of the multijet flare and the steam inspirating elevated flare may be similar. These flares use two sets of burners; the smaller group handles normal gas leakage and small gas releases, while both burner groups are used at higher flaring rates. This sequential operation is controlled by two water-sealed drums set to release at different pressures. In extreme emergencies, the multijet burners are bypassed by means of a water seal that directs the gases to the center of the stack. This seal blows at flaring rates higher than the design capacity of the flare. At such an excessive rate, the combustion is both luminous and smoky, but the unit is usually sized so that an overcapacity flow would be a rare occurrence. The overcapacity line may also be designed to discharge through a water seal to a nearby elevated flare rather than to the center of a multijet stack. Similar staging could be accomplished with automatic valves or backpressure regulators; however, in this case, the water seal drums are used because of reliability and ease of maintenance. The staging system is balanced by adjusting the hand control butterfly valve leading to the first-stage drum. After its initial setting, this valve is locked into position. The vertical, venturi-type ground flare is a design based upon the use of commercial-type venturi burners. This type of flare has been used to handle vapors from gas-blanketed tanks, and vapors displaced from the depressuring of butane and propane tank trucks. Since the commercial venturi burner requires a certain minimum pressure to operate efficiently, a gas blower must be provided. A compressor takes vapors from tankage and discharges them through a water seal tank and a flame arrestor to the flare. This type of arrangement can readily be modified to handle different volumes of vapors by the installation of the necessary number of burners. This type of flare is suitable for relatively small flows of gas of a constant rate. Its main application is in situations where other means of disposing of gases and vapors are not available.

Effect of Steam Injection: A flare installation that does not inspirate an adequate amount of air or does not mix the air and hydrocarbons properly emits dense, black clouds of smoke that obscure the flame. The injection of steam into the zone of combustion causes a gradual decrease in the amount of smoke, and the flame becomes more visible.

When trailing smoke has been eliminated, the flame is very luminous and orange with a few wisps of black smoke around the periphery. The minimum amount of steam required produces a yellowish- orange, luminous flame with no smoke. Increasing the amount of steam injection further decreases the luminosity of the flame. As the steam rate increases, the flame becomes colorless and finally invisible during the day. At night this flame appears blue. An injection of an excessive amount of steam causes the flame to disappear completely and be replaced with a steam plume. An excessive amount of steam may extinguish the burning gases and permit unburned hydrocarbons to discharge to the atmosphere. When the flame is out, there is a change in the sound of the flare because a steam hiss replaces the roar of combustion.

The commercially available pilot burners are usually not extinguished by excessive amounts of steam, and the flame reappears as the steam injection rate is reduced. As the use of automatic instrumentation becomes more prevalent in flare installations, the use of excessive amounts of steam and the emission of unburned hydrocarbons decrease and greater steam economies can be achieved. In evaluating flare installations, controlling the volume of steam is important. Too little steam results in black smoke, which, obviously, is objectionable. Conversely, excessive use of steam produces a white steam plume and an invisible emission of unburned hydrocarbons. A condition such as this can also be a serious air pollution problem.

The venturi-type ground flare, as previously discussed, consists of burners, pilots, ignitors, and control valves. The total pressure drop permitted in a given installation depends upon the characteristics of the particular blowdown system. In general, the allowable pressure drop through the relief valve headers, liquid traps, burners, and so forth, must not exceed one-half the internal unit's relieving pressure. The burner cut-in schedule is based upon a knowledge of the source, frequency, and quantity of the release gases. Pressure downstream of the control valves must be adequate to provide stable burner operation.

Flare installations designed for relatively small gas flows can use clusters of commercially avail able venturi burners. For large gas releases, special venturi burners must be constructed. The venturi (air-inspirating) burners are installed in clusters with a small venturi-type pilot burner in the center. This burner should be connected to an independent gas source. The burners may be mounted vertically or horizontally.

The burners should fire through a refractory wall to provide protection for personnel and equipment. Controls can be installed to give remote indication of the pilot burner's operation. For large-capacity venturi burners, field tests are necessary to obtain the proper throat-to-orifice ratio and the minimum pressure for stable burner operation.

INHALATION HAZARDS FROM TANKER OPERATIONS

When a compartment of a tank vehicle or tanker is filled through an open overhead hatch or bottom connection, the incoming liquid displaces the vapors in the compartment to the atmosphere. Except in rare instances, where a tank vehicle or tanker is free of hydrocarbon vapor, as when being used for the first time, the displaced vapors consist of a mixture of air and hydrocarbon concentration, depending upon the product being loaded, the temperature of the product and of the tank compartment, and the type of loading. Ordinarily, but not always, when gasoline is loaded, the hydrocarbon concentration of the vapors is from 30 to 50 percent by volume and consists of gasoline fractions ranging from methane through hexane. The volume of vapors produced during the loading operation, as well as their composition, is greatly influenced by the type of loading or filling employed. The types in use throughout the industry may be classified under two general headings, overhead loading and bottom loading. Overhead loading may be further divided into splash and submerged filling. In splash filling, the outlet of the delivery tube is above the liquid surface during all or most of the loading. In submerged filling the outlet of the delivery tube is extended to within 6 inches of the bottom and is submerged beneath the liquid during most of the loading. Splash filling generates more turbulence and therefore more hydrocarbon vapors than submerged filling does, other conditions being equal. On the basis of a typical 50 percent splash filling operation, vapor losses from the overhead filling of tank vehicles with gasoline have been determined empirically to amount to 0.1 to 0.3 percent of the volume loaded.

The equipment required for bottom loading is simpler than that used for overhead loading. Loading by this method is accomplished by connecting a swing-type loading armor hose at ground level to a matching fitting on the underside of the tank vehicles. Aircraft type, quick-coupling valves are used to ensure a fast, positive shutoff and prevent liquid spills. All the loading is submerged and under a slight pressure; thus, turbulence and resultant production of vapors are minimized. The method employed for loading marine tankers is essentially a bottom-loading operation. Liquid is delivered to the various compartments through lines that discharge at the bottom of each compartment. The vapors displaced during loading are vented through a manifold line to the top of the ship's mast fox dischargers the atmosphere. In addition to the emissions resulting from the displacement of hydrocarbon vapors from the tank vehicles, additional emissions during loading result from evaporation of spillage, drainage, and leakage of product.

An effective system for control of vapor emissions from loading must include a device to collect the vapors at the tank vehicle hatch and a means for disposal of these vapors. Four types of vapor collectors or closures, fitting the loading tube, have been developed for use during overhead-loading operations of trucks. All are essentially plug shaped devices that fit-into the hatch openings and have a central channel through which gasoline can flow into the tank vehicle compartment. This central channel, actually a section of the loading tube, is surrounded by an annular vapor space. Entry into this vapor space is achieved through openings on the bottom of the closure that are below the point of contact of the external closure surface with the sides of the hatch opening. Thus, vapors are prevented from passing around the closure and out of the hatch, and must flow instead into the annular space, which in turn, is connected to a hose or pipe leading to a vapor disposal system.

Vapors displaced from tank vehicles during the bottom-loading operation are more easily collected than those that result from overhead loading. The filling line and the vapor collection line are independent of each other. The vapor collection line is usually similar to the loading line, consisting of a flexible hose or swing-type arm connected to a quick-acting valve fitting on the dome of the vehicle. This fitting could be placed at ground level simplify the operation further. A check valve must be installed on the vapor collection line to prevent backflow of vapor to the atmosphere when the connection to the tank vehicle is broken.

In designing for complete vapor pickup at the tank vehicle hatch, several factors, including tank settling, liquid drainage, and topping off must be considered. The settling of a tank vehicle due to the weight of product being added requires that provision be made for vertical travel of the leading arm to follow the motion of the vehicle so that the vapor collector remains sealed in the tank hatch during the entire loading cycle. Two solutions to the problem of settling have been used. The first, applicable to pneumatically operated arms includes the continuous application of air pressure to the piston in the air cylinder acting on the arm. The arm is thus forced to follow the motion of the vehicle without need for clamping or fastening the vapor collector to the tank vehicle. The second solution, employed on counterweighted and torsion spring loading arms, provides for locking the vapor collector tank vehicle hatch. The arm then necessarily follows the motion of the vehicle. The second solution is also applicable to vapor collection, arms or hoses that are connected to the top of a tank vehicle during bottom loading. The second problem, that of preventing considerable liquid drainage from a loading arm as it is withdrawn after completion of filling operations, has been adequately solved. The air valve that operates the air cylinder of pneumatically operated loading arms may be modified by addition of an orifice on the discharge side of the valve. The orifice allows 30 to 45 seconds to elapse before the loading assembly clears the hatch compartment. This time interval is sufficient to permit complete draining of liquid into tank compartments from arms fitted with loading valves located in an outboard position. Loading arms with inboard valves require additional drainage time and present the problem of gasoline retention in the horizontal section of the arm. The third factor to be considered in the design of an effective vapor collection system is topping off. Topping off is the term applied to the loading operation during which the liquid level is adjusted to the capacity marker inside the tank vehicle compartment. Since the loading arm is out of the compartment hatch during the topping operation, vapor pickup by the collector is nil. Metering the desired volumes during loading is one solution to the problem. Metered loading must, however, be restricted to empty trucks or to trucks prechecked for loading volume available. Accuracy of certain totalizing meters or preset stop meters is satisfactory for loading without the need for subsequent open topping. An interlock device for the pneumatictype loading arms, consisting of pneumatic control or mechanical linkage, prevents opening of the loading valve unless the air cylinder valve is in the down position. Topping off is not a problem when bottom loading is employed. Metered loading, or installation of a sensing device in the vehicle compartments that actuates a shutoff valve located either on the truck or the loading island, eliminates the need for topping off.

The methods of disposing of vapors collected during loading operations may be considered under three headings: Using the vapors as fuel, processing the vapors for recovery of hydrocarbons, or effecting a vapor balance system in conjunction with submerged loading. The first method of disposal, using the vapors directly as fuel, may be employed when the loading facilities are located in or near a facility that includes fired heaters or boilers.

In a typical disposal system, the displaced vapors flow through a drip pot to a small vapor holder that is gas blanketed to prevent forming of explosive mixtures. The vapors are drawn from the holder by a compressor and are discharged to the fuel gas system.

The second method of disposal uses equipment designed to recover the hydrocarbon vapors. Vapors have been successfully absorbed in a liquid such as gasoline or kerosine. If the loading facility is located near a refinery or gas absorption plant, the vapor line can be connected from the loading facility to an existing vapor recovery system through a regulator valve. Vapors are recovered from loading installations distant from existing processing facilities by use of package units.

Explosive mixtures must be prevented from existing in this unit. This is accomplished by passing the vapors displaced at the loading, rack through a saturator countercurrently to gasoline pumped from storage. The saturated vapors then flow to the vaporsphere. Position of the diaphragm in the vaporsphere automatically actuates a compressor that draws the vapors from the sphere and injects them at about 200 psig into the absorber. Countercurrent flow of stripped gasoline from the saturator or of fresh gasoline from storage is used to absorb the hydrocarbon vapors. Gasoline from the absorber bottoms is returned to storage while the tail gases, essentially air, are released to the atmosphere through a backpressure regulator. Some difficulty has been experienced with air entrained or dissolved in the sponge gasoline returning to storage. Any air released in the storage tank is discharged to the atmosphere saturated with hydrocarbon vapors. A considerable portion of the air can be removed by flashing the liquid gasoline from the absorber in one or more additional vessels operating at successively lower pressures.

Another type of package unit adsorbs the hydrocarbon vapors on activated carbon. The vapors displaced during bottom filling are minimal. A volume displacement ratio of vapor to liquid of nearly 1:1 is usually achieved. A closed system can then be employed by returning all the displaced vapors to a storage tank. The storage tank should be connected to a vapor recovery system.

OIL-WATER EFFLUENT SYSTEMS

Oil-water effluent systems are found in the three phases of the petroleum industry; namely, production, refining, and marketing. The systems vary in size and complexity though their basic function remains the same, that is, to collect and separate wastes, to recover valuable oils, and to remove undesirable contaminants before discharge. In the production of crude oil, wastes such as oily brine, drilling muds, tank bottoms, and free oil are generated. Of these, the oilfield brines present the most difficult disposal problem because of the large volume encountered. Community disposal facilities capable of processing the brines to meet local water pollution standards are often set up to handle the treatment of brines. Among the traditional methods of disposal of brines has been injection into underground formations. A typical collection system associated with the crude-oil production phase of the industry usually includes a number of small gathering lines or channels transmitting waste water from wash tanks, leaky equipment, and treaters to an earthen pit, a concrete-lined sump, or a steel waste-water tank. A pump decants waste water from these containers to water-treating facilities before injection into underground formations or disposal to sewer systems. Any oil accumulating on the surface of the water is skimmed off to storage tanks.

The effluent disposal systems found in refineries are larger and more elaborate than those in the production phase. A typical modern refinery gathering system usually includes gathering lines, drain seals, junction boxes, and channels of vitrified clay or concrete for transmitting waste water from processing units to large basins or ponds used as oil-water separators. These basins are sized to receive all effluent water, sometimes even including rain runoff, and may be earthen pits, concrete-lined basins, or steel tanks. Liquid wastes discharging to these systems originate at a wide variety of sources such as pump glands, accumulators, spills, cleanouts, sampling lines, relief valves, and many others The types of liquid wastes may be classified as waste water with: oil present as free oil, emulsified oil, or as oil coating on suspended matter; chemicals present as suspensoids, emulsoids, or solutes. These chemicals include acids, alkalies, phenols, sulfur compounds, clay, and others. Emissions from these varied liquid wastes can best be controlled by properly maintaining, isolating, and treating the wastes at their source; by using efficient oil-water separators; and by minimizing the formation of emulsions.

The waste water from the process facilities and treating units just discussed flows to the oil water separator for recovery of free oil and settleable solids. Factors affecting the efficiency of separation include temperature of water, particle size, density, and amounts and characteristics of suspended matter. Stable emulsions are not affected by gravity-type separators and must be treated separately. The oil-water separator design must provide for efficient inlet and outlet construction, sediment collection mechanisms, and oil skimmers. Reinforced concrete construction has been found most desirable for reasons of economy, maintenance, and efficiency.

The effluent water from the oil-water separator may require further treatment before final discharge to municipal sewer systems, channels, rivers, or streams. The type and extent of treatment depend upon the nature of the contaminants present, and on the local water pollution ordinances governing the concentration and amounts of contaminants to be discharged in refinery effluent waters. The methods of final-effluent clarification to be briefly discussed here include (1) filtration, (2) chemical flocculation, and (3) biological treatment. Several different types of filters may be used to clarify the separator effluent. Hay-type filters, sand filters, and vacuum precoat filters are the most common. The selection of any one type depends upon the properties of the effluent stream and upon economic considerations. Methods of treatment are either by sedimentation or flotation. In sedimentation processes, chemicals such as copper sulfate, activated silica, alum, and lime are added to the waste-water stream before it is fed to the clarifiers. The chemicals cause the suspended particles to agglomerate and settle out. Sediment is removed from the bottom of the clarifiers by mechanical scrapers. Effectiveness of the sedimentation techniques in the treatment of separator effluents is limited by the small oil particles contained in the waste water. These particles, being lighter than water, do not settle out easily. They may also become attached to particles of suspended solids and thereby increase in buoyancy. In the flotation process a colloidal floc and air under pressure are injected into the waste water. The stream is then fed to a clarifier through a backpressure valve that reduces the pressure to atmospheric. The dissolved air is suddenly released in the form of tiny bubbles that carry the particles of oil and coalesced solids to the surface where they are skimmed off by mechanical flight scrapers. Of the two, the flotation process has the potential to become the more efficient and economical. Biological treating units such as trickling filters activated sludge basins, and stabilization basins have been incorporated into modern refinery waste disposal systems. By combining adsorption and oxidation, these units are capable of reducing oil, biological oxygen demand, and phenolic content from effluent water streams. To prevent the release of air pollutants to the atmosphere, certain pieces of equipment, such as clarifiers, digesters, and filters, used in biological treatment should be covered and vented to recovery facilities or incinerated.

From an air pollution standpoint the most objectionable contaminants emitted from liquid waste streams are hydrocarbons, sulfur compounds, and other malodorous materials. The effect of hydrocarbons in smog-producing reactions is well known, and sulfur compounds such as mercaptans and sulfides produce very objectionable odors, even in high dilution. These contaminants can escape to the atmosphere from openings in the sewer system, open channels, open vessels, and open oil-water separators. The large exposed surface, area of these separators requires that effective means of control be instituted to minimize hydrocarbon losses to the atmosphere from this source. The most effective means of control of hydrocarbon emissions from oil-water separators has been the covering of forebays or primary separator sections. Either fixed roofs or floating roofs are considered acceptable covers. Separation and skimming of over 80 percent of the floatable oil layer is effected in the covered sections. Thus, only a minimum of oil is contained in the effluent water, which flows under concrete curtains to the open afterbays or secondary separator sections. The explosion hazard associated with fixed roofs is not present in a floating-roof installation. These roofs are similar to those developed for storage tanks. The floating covers are built to fit into bays with about 1 inch of clearance around the perimeter. Fabric or rubber may be used to seal the gap between the roof edge and the container wall. The roofs are fitted with access manholes, skimmers, gage hatches, and supporting legs. In operation, skimmed oil flows through lines from the skimmers to a covered tank (floating roof or connected to vapor recovery) or sump and then is pumped to deemulsifying processing facilities. Effluent water from the oil-water separator is handled in the manner described previously.

In addition to covering the separator, open sewer lines that may carry volatile products are converted to closed, underground lines with waterseal-type vents. Junction boxes are vented to vapor recovery facilities, and steam is used to blanket the sewer lines to inhibit formation of explosive mixtures. Accurate calculation of the hydrocarbon losses from separators fitted with fixed roofs is difficult because of the many variables of weather and refinery operations involved.

Isolation of certain odor- and chemical-bearing liquid wastes at their source for treatment before discharge of the water to the refinery waste water-gathering system has been found to be the most effective and economical means of minimizing odor and chemicals problems. The unit that is the source of wastes must be studied for possible changes in the operating process to reduce wastes. In some cases the wastes from one process may be used to treat the wastes from another. Among the principal streams that are treated separately are oil-in-water emulsions, sulfur bearing waters, acid sludge, and spent caustic wastes.

Oil-in-water emulsions are wastes that can be treated at their source. An oil-in-water emulsion is a suspension of oil particles in water that cannot be divided effectively by means of gravity alone. Gravity-type oil-water separators are generally, ineffective in breaking the emulsions, and means are provided for separate treatment where the problem is serious. Oil-in-water emulsions are objectionable in the drainage system since the separation of otherwise recoverable oil may be impaired by their presence. Moreover, when emulsions of this type are discharged into large bodies of water, the oil is released by the effect of dilution, and serious pollution of the water may result. Formation of emulsions may be minimized by proper design of process equipment and piping. Both physical and chemical methods are available for use in breaking emulsions. Physical methods of separation include direct application of heat, distillation, centrifuging, filtration, and use of an electric field. The effectiveness of any one method depends upon the type of emulsion to be treated.

Sulfur-Bearing Waters: Sulfides and mercaptans are removed from wastewater streams by various methods. Some refineries strip the wastewater in a column with live steam. The overhead vapors from the column are condensed and collected in an accumulator from which the noncondensables flow to sulfur recovery facilities or are incinerated. Flue gas has also been used as the stripping medium. Bottoms water from steam stripping towers, being essentially sulfide free, can then be drained to the refinery's sewer system. Oxidation of sulfides in waste water is also an effective means of treatment. Air and heat are used to convert sulfides and mercaptans to thiosulfates, which are water soluble and not objectionable. Chlorine is also used as an oxidizing agent for sulfides. It is added in stoichiometric quantities proportional to the waste water. This method is limited by the high cost of chlorine. Water containing dissolved sulfur dioxide has been used to reduce sulfide concentration in waste waters. For removing small amounts of hydrogen sulfide, copper sulfate and zinc chloride have been used to react and precipitate the sulfur as copper and zinc sulfides. Hydrogen sulfide may be released, however, only if the water treated with these compounds contacts an acid stream.

Acid Sludge: The acid sludge produced from treating operations varies with the stock treated and the conditions of treatment. The sludge may vary from a low-viscosity liquid to a solid. Methods of disposal of this sludge are many and varied. Basically, they may be considered under the following general headings: Disposal by burning as fuel, or dumping in the ground; processing to produce byproducts such as ammonium sulfate, metallic sulfates, oils, tars, and other materials; processing for recovery of acid. The burning of sludge results in discharge to the atmosphere of excessive amounts of sulfur dioxide and sulfur trioxide from furnace stacks. If sludge is solid or semisolid it may be buried in specially constructed pits. This method of disposal, however, creates the problem of acid leaching out to adjacent waters. Recovery of sulfuric acid from sludge is accomplished essentially by either hydrolysis or thermal decomposition processes. Sulfuric acid sludge is hydrolyzed by heating it with live steam in the presence of water. The resulting product separates into two distinct phases. One phase consists of diluted sulfuric acid with a small amount of suspended carbonaceous material, and the second phase, of a viscous acid-oil layer. The dilute sulfuric acid may be (1) neutralized by alkaline wastes, (2) reacted chemically with ammonia-water solution to produce ammonium sulfate for fertilizer, or (3) concentrated by heating.

AIR EMISSIONS FROM VALVES

Valves are employed in every phase of the petroleum industry where petroleum or petroleum product is transferred by piping from one point to another. There is a great variety of valve designs, but, generally, valves may be classified by their application as flow control or pressure relief. Manual and automatic flow control valves are used to regulate the flow of fluids through a system. Included under this classification are the gate, globe, angle, plug, and other common types of valves. These valves are subject to product leakage from the valve stem as a result of the action of vibration, heat, pressure, corrosion, or improper maintenance of valve stem packing. Pressure relief and safety valves are used to prevent excessive pressures from developing in process vessels and lines. The relief valve designates liquid flow while the safety valve designates vapor or gas flow. These valves may develop leaks because of the corrosive action of the product or because of failure of the valve to reseat properly after blowoff. Rupture discs are sometimes used in place of pressure relief valves. Their use is restricted to equipment in batch-type processes.

The maintenance and operational difficulties caused by the inaccessibility of many pressure relief valves may allow leakage to become substantial. Emissions vary over a wide range. Liquid leakage results in emissions from evaporation of liquid

minute per square foot). The required tower size is dependent upon: (1) cooling range (hot water minus cold water temperature); (2) approach (cold water minus wet bulb temperature); (3) amount of liquid to be cooled; (4) wet bulb temperature; (5) air velocity through cell; and (6) tower height. Cooling towers used in conjunction with equipment processing hydrocarbons and their derivatives are potential sources of air pollution, and hence inhalation hazards, because of possible contamination of the water. The cooling water may be contaminated by leaks from the process side of heat-exchange equipment, direct and intentional contact with process streams, or improper process unit operation. As this water is passed over a cooling tower, volatile hydrocarbons and other materials accumulated in the water readily evaporate into the atmosphere. When odorous materials are contained in the water a nuisance is easily created. Inhibitors or additives used in the cooling tower to combat corrosion or algae growth should not cause any significant air pollution emissions, nor should the water-softening facilities common to many cooling towers be a problem.

The control of hydrocarbon discharges or of release of odoriferous compounds at the cooling tower is not practical. Instead, the control must be at the point where the contaminant enters the cooling water. Hence, systems of detection of contamination in water, proper maintenance, speedy repair of leakage from process equipment and piping, and good housekeeping programs in general are necessary to minimize the air pollution occurring at the cooling tower. Water that has been used in contact with process streams, as in direct-contact or barometric-type condensers, should be eliminated from the cooling tower if this air pollution source is to be completely controlled. Greater use of fin-fan coolers can also control the emissions indirectly by reducing or eliminating the volume of cooling water to be aerated in a cooling tower.

In certain refining operations, air is blown through heavier petroleum fractions for the purpose of removing moisture or agitating the product. The exhaust air is saturated with hydrocarbon vapors or aerosols, and, if discharged directly to the atmosphere, is a source of air pollution. The extent of airblowing operations and the magnitude of emissions from the equipment vary widely among refineries. Emissions from airblowing for removal of moisture, or for agitation of products may be minimized by replacing the airblowing equipment with mechanical agitators and incinerating the exhaust vapors.

Refinery operations frequently require that a pipeline be used for more than one product. To prevent leakage and contamination of a particular product, other products connecting and product feeding lines are customarily "blinded off. "Blinding a line" is the term commonly used for the inserting of a flat, solid plate between two flanges of a pipe connection. Blinds are normally used instead of valves to isolate pipelines because a more positive shutoff can be secured and because of generally lower costs. In opening, or breaking, the flanged connection to insert the blind, spillage of product in that portion of the pipeline can occur. The magnitude of emissions to the atmosphere from this spillage is a function of the vapor pressure of the product, type of ground surface beneath the blind, distance to the nearest drain, and amount of liquid holdup in the pipeline. Emissions to the atmosphere from the changing of blinds can be minimized by pumping out the pipeline and then flushing the line with water before breaking the flange. In the case of highly volatile hydrocarbons, a slight vacuum may be maintained in the line. Spillage resulting from blind changing can also be minimized by use of "line" blinds in place of the common "slip" blinds. Line blinds do not require a complete break of the flange connection during the changing operation. These blinds use a gear mechanism to release the spectacle plate without actually breaking the line. Combinations of this device in conjunction with gate valves are available to allow changing of the line blind while the line is under pressure from either direction.

MISCELLANEOUS AIR EMISSIONS

Turnaraounds: Periodic maintenance and repair of process equipment are essential to refinery operations. A major phase of the maintenance program is the shutting down and starting up of the various units, usually called a turnaround. The procedure for shutting down a unit varies from refinery to refinery and between units in a refinery. In general, shutdowns are effected by first shutting off the heat supply to the unit and circulating the feed stock through the unit as it cools. Gas oil may be blended into the feedstock to prevent its solidification as the temperature drops. The cooled liquid is then pumped out to storage facilities, leaving hydrocarbon vapors in the unit. The pressure of the hydrocarbon vapors in the unit is reduced by evacuating the various items of equipment to a disposal facility such as a fuel gas system, a vapor recovery system, a flare, or in some cases, to the atmosphere. Discharging vapors to the atmosphere is undesirable from the standpoint of air pollution control since as much as several thousand pounds of hydrocarbons or other objectionable vapors or odors can be released during a shutdown. The residual hydrocarbons remaining in the unit after depressuring are purged out with steam, nitrogen, or water. Any purged gases should be discharged to the aforementioned disposal facilities. Condensed steam and water effluent that may be contaminated with hydrocarbons or malodorous compounds during purging should be handled by closed water-treating systems.

Tank Cleaning: Storage tanks in a refinery require periodic cleaning and repair. For this purpose, the contents of a tank are removed and residual vapors are purged until the tank is considered safe for entry by maintenance crews. Purging can result in the release of hydrocarbon or odorous material in the form of vapors to the atmosphere. These vapors should be discharged to a vapor recovery system or flare. When the vapors in the tank are released to a recovery or disposal system before the tank is opened for maintenance, the emissions are considered negligible. When the stored liquid is transferred to another tank, and the emptied vessel is then opened for maintenance without purging to a recovery or disposal system, in which case the emission to the atmosphere can be considered to be equal to the weight of hydrocarbon vapor occupying the total volume of the tank at the reported pressure.

Steam cleaning of railroad tank cars used for transporting petroleum products can similarly be a source of emissions if the injected steam and entrained hydrocarbons are vented directly to the atmosphere. Although no quantitative data are available to determine the magnitude of these emissions, the main objection to this type of operation is its nuisance-causing potential. Some measure of control of these emissions maybe effected by condensing the effluent steam and vapors. The condensate can then be separated into hydrocarbon and water phases for recovery. Noncondensable vapors should be incinerated.

Use of Vacuum Jets: Certain refinery processes are conducted under vacuum conditions. The most practical way to create and maintain the necessary vacuum is to use steam-actuated vacuum jets, singly or in series. Barometric condensers are often used after each vacuum jet to remove steam and condensable hydrocarbons. The effluent stream from the last stage of the vacuum jet system should be controlled by condensing as much of the effluent as is practical and incinerating the noncondensables in an afterburner or heater firebox. Condensate should be handled by a closed treating system for recovery of hydrocarbons. The hot well that receives water from the barometric condensers may also have to be enclosed and any off gases incinerated.

Compressor Engine Exhausts: Refining operations require the use of various types of gas compressors. These machines are often driven by internal combustion engines that exhaust air contaminants to the atmosphere. Although these engines are normally fired with natural gas and operate at essentially constant loads, some unburned fuel passes through the engine. Oxides of nitrogen are also found in the exhaust gases as a result of nitrogen fixation in the combustion cylinders.

Chapter 4
INDOOR AIR QUALITY MANAGEMENT

INTRODUCTION

The quality of the indoor environment in any building is a result of the interaction between the site, climate, building system (original design and later modifications in the structure and mechanical systems), construction techniques, contaminant sources (building materials and furnishings, moisture, processes and activities within the building and outdoor sources), and the activities of the building occupants. To generalize, the following factors are involved in the development of indoor air quality problems:

Source: there is a source of contamination or discomfort indoors, outdoors, or within the mechanical systems of the building.

HVAC: the HVAC system is not able to control existing air contaminants and ensuring thermal comfort (temperature and humidity conditions that are comfortable for most occupants).

Pathways: one or more pollutant pathways connect the pollutant source to the occupants and a driving force exists to move pollutants along the pathway(s).

Occupants: building occupants are present.

It is important to understand the role that each of these factors may play in order to prevent, investigate, and resolve and manage indoor air quality problems.

Indoor air contaminants can originate within the building or be drawn in from outdoors. If contaminant sources are not controlled, problems can arise, even if the HVAC system is properly designed and well-maintained. Sources can be from outside the building; from operating equipment, from human activities, and other or miscellaneous sources. Sources outside a building include contaminated outdoor air, emissions from nearby sources, soil gas, or moisture or standing water.

Contaminated outdoor air can include pollen, dust, fungal spores, industrial pollutants, general vehicle exhaust. Emissions from nearby sources include exhaust from vehicles on nearby roads or in parking lots or garages, loading docks, odors from dumpsters, re-entrained (drawn back into the building) exhaust from the building itself or from neighboring buildings, unsanitary debris near the outdoor air intake. Soil gas generally refers to radon, leakage from underground fuel tanks, contaminants from previous uses of the site (e.g., landfills), pesticides. Moisture or standing water promoting excess microbial growth; rooftops after rainfall, crawlspace can also be a major source of indoor air quality problems.

Equipment sources can be of two types; namely, HVAC system and non-HVAC system equipment. In an HVAC system, the sources of contamination may be dust or dirt in ductwork or other components, microbiological growth in drip pans, humidifiers, ductwork, coils, improper use of biocides, sealants, and/ or cleaning compounds, improper venting of combustion products, refrigerant leakage. From non-HVAC equipment, the emissions can be from office equipment (volatile organic compounds, ozone), supplies (solvents, toners, ammonia), emissions from shops, labs, cleaning processes, elevator motors and other mechanical systems.

The human or personal activities that can contribute to poor indoor air quality are actions such as smoking, cooking, body odor, cosmetic odors. Personal activities sources can also be related to housekeeping activities, such as cleaning materials and procedures emissions from stored supplies or trash, use of deodorizers and fragrances, airborne dust or dirt (e.g., circulated by sweeping and vacuuming). Another source may be maintenance activities - microorganisms in mist from improperly maintained cooling towers airborne dust or dirt volatile organic compounds from use of paint, caulk, adhesives, and other products pesticides from pest control activities emissions from stored supplies.

Building components and furnishings may also be a contributing factor or source. These can be locations that produce or collect dust or fibers; including, textured surfaces such as carpeting, curtains, and other textiles, open shelving, old or deteriorated furnishings, or materials containing damaged asbestos.

Unsanitary conditions and water damage can harbor microbiological growth on or in soiled or water-damaged furnishings, microbiological growth in areas of surface condensation, standing water from clogged or poorly designed drains, dry traps that allow the passage of sewer gas. In addition, chemicals released from building components or furnishings such as volatile organic compounds or, inorganic compounds can create problems.

Other or miscellaneous sources can be accidental releases, such as spills of water or other liquids, microbiological growth due to flooding or to leaks from roofs, piping, fire damage (soot, PCBs from electrical equipment, odors). Special use areas and mixed use buildings, although intended to isolate problems, can be a source of contamination in the common-use areas of a building. These can be smoking lounges, laboratories, print shops, art rooms, exercise rooms, beauty salons, and food preparation areas. Other potential sources are redecorating/remodeling/repair activities, emissions from new furnishings, dust and fibers from demolition resulting in odors and volatile organic and inorganic compounds from paint, caulk, adhesives, and microbiologicals released from demolition, or remodeling activities.

Indoor air often contains a variety of contaminants at concentrations that are far below any standards or guidelines for occupational exposure. Given our present knowledge, it is difficult to relate complaints of specific health effects to exposures to specific pollutant concentrations, especially since the significant exposures may be to low levels of pollutant mixtures. This chapter provides an overview of indoor air quality issues and management practices. Proper indoor air quality management is an integral part of any program dealing with safe industry practices.

HVAC SYSTEMS

An HVAC system includes all heating, cooling, and ventilation equipment that services a building: that is, it includes furnaces or boilers, chillers, cooling towers, air handling units, exhaust fans, ductwork, filters, steam (or heating water) piping. Discussions that follow apply both to central HVAC systems and to individual components used as stand-alone units.

A proper HVAC system provides thermal comfort, distributes adequate amounts of outdoor air to meet ventilation needs of all building occupants, isolates and removes odors and contaminants through pressure control, filtration, and exhaust fans. One of the important roles of any HVAC system is to provide thermal comfort.

A number of variables interact to determine whether people are comfortable with the temperature of the indoor air. The activity level, age, and physiology of each person affect the thermal comfort requirements of that individual. The American Society of Heating, Refrigerating, and Air Conditioning Engineers (ASHRAE) Standard 55-1981 describes the temperature and humidity ranges that are comfortable for most people engaged in largely sedentary activities. The ASHRAE standard assumes "normal" indoor clothing. Added layers of clothing reduce the rate of heat loss. Uniformity of temperature is important to comfort. When the heating and cooling needs of rooms within a single zone change at different rates, rooms that are served by a single thermostat may be at different temperatures. Temperature stratification is a common problem caused by convection, which is the tendency of light, warm air to rise and heavier, cooler air to sink, thus causing a circulation of air patterns. If the air is not properly mixed by the ventilation system, the temperature near the ceiling can be several degrees warmer than at floor level. Even if air is properly mixed, insulated floors over unheated spaces can create discomfort in some climate zones. Large fluctuations of indoor temperature can also occur when controls have a wide "dead band" (a temperature range within which neither heating nor cooling takes place).

Radiant heat transfer may cause people located near very hot or very cold surfaces to be uncomfortable even though the thermostat setting and the measured air temperature are within the comfort range. Buildings with large window areas sometimes have acute problems of discomfort due to radiant heat gains and losses, with the locations of complaints shifting during the day as the sun angle changes. Large vertical surfaces can also produce a significant flow of naturally-convecting air, resulting in drafty conditions. Adding insulation to walls helps to moderate the temperature of interior wall surfaces. Closing curtains reduces heating from direct sunlight and isolates building occupants from exposure to window surfaces (which, lacking insulation, are likely to be much hotter or colder than the walls).

Humidity is an important factor in achieving thermal comfort. Raising relative humidity reduces the ability to lose heat through perspiration and evaporation, so that the effect is similar to raising the temperature. Humidity extremes can also create other IAQ (Indoor Air Quality) problems. Excessively high or low relative humidities can produce discomfort, while high relative humidities can promote the growth of mold and mildew.

Most air handling units distribute a blend of outdoor air and recirculated indoor air. HVAC designs may also include units that introduce 100% outdoor air or that simply transfer air within the building. Uncontrolled quantities of outdoor air enter buildings by infiltration through windows, doors, and gaps in the exterior construction of the building. Thermal comfort and ventilation needs are met by supplying "conditioned" air (a blend of outdoor and recirculated air that has been filtered, heated or cooled, and sometimes humidified or dehumidified).

Large buildings often have exterior ("core") spaces in which constant cooling is required to compensate for heat generated by occupants, equipment, and lighting, while perimeter rooms may require heating or cooling depending on outdoor conditions. Two of the most common HVAC designs used in modem public and commercial buildings are **constant volume** and **variable air volume** systems. Constant volume systems are designed to provide a constant airflow and to vary the air temperature to meet heating and cooling needs. The percentage of outdoor air may be held constant, but is often controlled either manually or automatically to vary with outdoor temperature and humidity. Controls may include a mini setting that should allow the system to meet ventilation guidelines for outdoor air quantities under design conditions.

In contrast, variable air volume (VAV) systems condition supply air to a constant temperature and ensure thermal comfort by varying the airflow to occupied spaces. Most early VAV systems did not allow control of the outdoor air quantity, so that a decreasing amount of outdoor air was provided as the flow of supply air was reduced. More recent designs ensure a minimum supply of outdoor air with static pressure devices in the outdoor air stream. Additional energy-conserving features such as economizer control or heat recovery are also found in some buildings. Quality design, installation, and testing and balancing are essential to the proper operation of all types of HVAC systems, especially VAV systems, as are regular inspections, and maintenance.

The amount of outdoor air considered adequate for proper ventilation has vaned substantially over time. The current guideline issued by ASHRAE is ASHRAE Standard 62-1989. The building code that was in force for older systems may well have established a lower amount of ventilation (in cubic feet of outdoor air per minute per person) than is currently recommended.

Controlling Odors and Air Pollutants

One technique for controlling odors and contaminants is to dilute them with outdoor air. In other words - ' dilution is the solution to pollution'. Dilution can work only if there is a consistent and appropriate flow of supply air that mixes effectively with room air. The term "ventilation efficiency" is used to describe the ability of the ventilation system to distribute supply air and remove internally generated pollutants. Current research focuses on ways to measure ventilation efficiency and interpret the results of those measurements.

Another technique for isolating odors and contaminants is to design and operate the HVAC system so that pressure relationships between rooms are controlled. This type of control technique is accomplished by adjusting the air quantities that are supplied to and removed from each room. If more air is supplied to a room than is exhausted, the excess air leaks out of the space and the room is said to be under **positive pressure.** If less air is supplied than is exhausted, air is pulled into the space and the room is said to be under **negative pressure.**

Control of pressure relationships is critically important in mixed use buildings or buildings with special use areas. Lobbies and buildings in general are often designed to operate under positive pressure to minimize the infiltration of unconditioned air, with its potential to cause drafts and introduce dust, dirt, and thermal discomfort. Without proper operation and maintenance, these pressure differences are not likely to remain as originally designed.

A third technique is to use local exhaust systems (sometimes known as dedicated exhaust ventilation systems) to isolate and remove contaminants by maintaining negative pressure in the area around the contaminant source. Local exhaust can be linked to the operation of a particular piece of equipment used to treat an entire room. Air should be exhausted to the outdoors, not recirculated, from locations which produce significant odors and high concentrations of contaminants. Spaces where local exhaust is used must be provided with make-up air and the local exhaust must function in coordination with the rest of the ventilation system. Under some circumstances, it may be acceptable to transfer conditioned air from relatively clean parts of a building to comparatively dirty areas and use it as make-up air for a local exhaust system. The advantage of such a transfer is that it can achieve significant energy savings.

Air cleaning and filtration devices designed to control contaminants are found as components of HVAC systems (for example, filter boxes in ductwork) and can also be installed as independent units. The effectiveness of air cleaning depends upon proper equipment selection, installation operation, and maintenance.

Contaminant Pathways

Airflow patterns in buildings result from the combined action of mechanical ventilation systems, human activity, and natural forces. Pressure differentials created by these forces move airborne contaminants from areas of relatively higher pressure to areas of relatively lower pressure through any available openings. The HVAC system is generally the predominant pathway and driving force for air movement in buildings. However, all of a building's components (walls, ceilings, floors, penetrations, HVAC equipment, and occupants) interact to affect the distribution of contaminants. For example, as air moves from supply registers or diffusers to return air grilles, it is diverted or obstructed by partitions, walls, and furnishings, and redirected by openings that provide pathways for air movement. On a localized basis, the movement of people has a major impact on the movement of pollutants. Some of the pathways change as doors and windows open and close. It is useful to think of the entire building - the rooms and the connections (e.g., chases, corridors, stairways, elevator shafts) between them as part of the air distribution system. Additionally, natural forces exert an important influence on air movement between zones and between the building's interior and exterior.

Both the stack effect and wind can overpower a building's mechanical system and disrupt air circulation and ventilation, especially if the building envelope is leaky. Stack effect is defined as the pressure driven flow produced by convection (the tendency of warm air to rise). The stack effect exists whenever there is an indoor-outdoor temperature difference and becomes stronger as the temperature difference increases. As heated air escapes from upper levels of the building, indoor air moves from lower to upper floors, and replacement outdoor air is drawn into openings at the lower levels of buildings. Stack effect airflow can transport contaminants between floors by way of stairwells, elevator shafts, utility chases, or other openings. Wind effects are transient, creating local areas of high pressure (on the windward side) and low pressure (on the leeward side) of buildings.

Depending on the leakage openings in the building exterior wind can affect the pressure relationships within and between rooms. The basic principle of air movement from areas of relatively higher pressure to areas of relatively lower pressure can produce many patterns of contaminant distribution, including: local circulation in the room containing the pollutant source, air movement into adjacent spaces that are under lower pressure, recirculation of air within the zone containing the pollutant source or in adjacent zones where return systems overlap, movement from lower to upper levels of the building, air movement into the building through either infiltration of outdoor air or reentry of exhaust air. Air moves from areas of higher pressure to areas of lower pressure through any available openings. A small crack or hole can admit significant amounts of air if the pressure differentials are high enough (which may be very difficult to assess).

Even when the building as a whole is maintained under positive pressure, there is always some location (for example, the outdoor air intake) that is under negative pressure relative to the outdoors. Entry of contaminants may be intermittent, occurring only when the wind blows from the direction of the pollutant source. The interaction between pollutant pathways and intermittent or variable driving forces can lead to a single source causing IAQ complaints in areas of the building that are distant from each other and from the source.

BUILDING OCCUPANTS AND IAQ ISSUES

The term "building occupants" refers to people who spend extended time periods (e.g., a full workday) in the building. Visitors are also occupants; they may have different tolerances and expectations from those who spend their entire workdays in the building, and are likely to be more sensitive to odors. Groups that may be particularly susceptible to effects of indoor air contaminants include allergic or asthmatic individuals, people with respiratory disease, people whose immune systems are suppressed due to chemotherapy, radiation therapy, disease, or other causes, contact lens wearers. Some other groups are particularly vulnerable to exposures of certain pollutants. For example, people with heart disease may be more affected by exposure at lower levels of carbon monoxide than healthy individuals. Children exposed to environmental tobacco smoke have been shown to be at higher risk of respiratory illnesses and those exposed to nitrogen dioxide have been shown to be at higher risk from respiratory infections. Because of varying sensitivity among people, one individual may react to a particular IAQ problem while surrounding occupants have no ill effects. Symptoms that are limited to a single person can also occur when only one work station receives the majority of the pollutant dose. In other cases, complaints may be widespread. A single indoor air pollutant or problem can trigger different reactions in different people. Some may not be affected at all. Information about the types of symptoms can sometimes lead directly to solutions. However, symptom information is more likely to be useful for identifying the timing and conditions under which problems occur.

The effects of IAQ problems are often nonspecific symptoms rather than clearly defined illnesses. Symptoms commonly attributed to IAQ problems include: headache, fatigue, shortness of breath, sinus congestion, coughing, sneezing, eye, nose, and throat irritation, skin irritation, dizziness, nausea. All of these symptoms, however, may also be caused by other factors, and are not necessarily due to air quality deficiencies. Health and comfort are used to describe a spectrum of physical sensations. For example, when the air in a room is slightly too warm for a person's activity level, that person may experience mild discomfort. If the temperature continues to rise, discomfort increases and symptoms such as fatigue, stuffiness, and headaches can appear. Some complaints are often related to the discomfort end of the spectrum. One of the most common IAQ complaints is related to the presence of a strange odor. Odors are often associated with a perception of poor air quality, whether or not they cause symptoms. Environmental stressors such as improper lighting, noise, vibration, overcrowding, ergonomic stressors, and job-related psychosocial problems, (such as job stress) can produce symptoms that are similar to those associated with poor air quality.

The term **sick building syndrome (SBS)** is used to describe cases in which building occupants experience acute health and comfort effects that are linked to the time they spend in the building, but in which no specific illness or cause can be identified. The complaints may be localized in a particular room or zone or may be widespread throughout the building. Many different symptoms have been associated with SBS, including respiratory complaints, irritation, and fatigue. Analysis of air samples often fails to detect high concentrations of specific contaminants. The problem may be caused by any or all of the following: the combined effects of multiple pollutants at low concentrations, other environmental stressors (e.g., overheating, poor lighting, noise), ergonomic stressors, job-related psychosocial stressors (e.g., overcrowding, labor-management problems).

Building-related illness (BRI) is a term referring to illness brought on by exposure to the building air, where symptoms of diagnosable illness are identified (e.g., certain allergies or infections) and can be directly attributed to environmental agents in the air. Legionnaire's disease and hypersensitivity pneumonitis are examples of BRI that can have serious, even life threatening consequences.

A small percentage of the population may be sensitive to a number of chemicals in indoor air, each of which may occur at very low concentrations. The existence of this condition, which is known as **multiple chemical sensitivity (MCS)**, is a matter of considerable controversy. MCS is not currently recognized by the major medical organizations. Medical opinion is divided, and further research is needed. The applicability of access for the disabled and worker's compensation regulations to people who believe they are chemically sensitive is becoming a concern for facility managers.

Sometimes a number of building occupants experience serious health problems (e.g., cancer, miscarriages, Lou Gehrig's disease) over a relatively short time period. These **clusters** of health problems are occasionally blamed on indoor air quality, and can produce anxiety among building occupants. Establishing a communication system that can help prevent indoor air quality problems and resolve problems if they do arise is a critical issue in the proper management of IAQ problems. Effective communication can encourage building occupants to improve their work environment through positive contributions. Many indoor air quality problems can be prevented if staff and building occupants understand how their activities affect IAQ. If a company already has a health and safety committee functioning to promote good working conditions, it is easy to add indoor air quality to their list of concerns. It is important to define the responsibilities of building management, staff, and occupants in relation to indoor air quality. These responsibilities can be formalized by incorporating them into documents such as employee manuals or lease agreements. Educate occupants about the permitted uses and maximum occupancy of different areas within the building and make sure that appropriate ventilation is provided for the activities that are permitted. Indoor air quality complaints often arise in mixed-use buildings. For example, kitchen staff expect food odors as part of their work, but nearby office workers may find cooking odors distracting and unpleasant. Inform occupants about the importance of keeping the building management informed about significant changes in the number of people regularly using particular areas of the building. The ventilation systems in buildings are designed and operated to supply air to projected ranges of occupants. If the occupancy rate becomes a problem, it may be helpful to refer to a standard reference such as ASHRAE Standard 62-1989 to show occupants that keeping occupancy within the ventilation capacity serves the goal of providing a quality work environment and is not an arbitrary decision by building management. Management should review plans that may involve increases in the number of occupants, relocation of walls or partitions, installation of new equipment, or changes in the use of space. Building owners, facility managers, and occupants share responsibility for monitoring new equipment installation and changes in the use of space. The review process allows potential indoor air quality problems to be identified so that the HVAC system can be modified as needed. Only authorized maintenance personnel should adjust air supply or exhaust vents; however, if occupants are expected to follow such a "hands-off' policy, facility management must respond promptly to IAQ complaints.

Many organizations have established procedures for responding to occupant complaints that can be modified to include indoor air quality concerns. To avoid frustrating delays, building occupants need to know how to express their complaints about IAQ Most importantly, they need to know how to locate responsible staff and where to obtain complaint forms. This information can be posted on bulletin boards, circulated in memos or newsletters, or publicized by some other means. Complaints should be handled promptly, with every incident given serious attention. It is advisable to establish a recordkeeping system that cross-references documentation on complaints with records of equipment operation and maintenance. The recordkeeping system can help to resolve complaints by collecting information in a form that highlights patterns of problems (for example, complaints that occur at a regular time of day or in the same area of the building). In many cases, building managers may be alerted to potential indoor air quality problems by complaints from occupants. The complaints can be vague, to the effect that one or more people feel "sick" or "uncomfortable" or that someone has noticed an unusual odor. They may be specific, blaming a particular material as the cause of discomfort or health problems. People are usually reacting to a real problem, so their complaints should be taken seriously. However, they may attribute their symptoms to the wrong cause, so their theories about the problem should be heard respectfully but weighed cautiously. Indoor air quality problems can sometimes be identified and resolved quickly. On other occasions, complaints originate from the interaction of several variables, and detailed investigation may be necessary in order to resolve the problem.

Listening and responding to building occupants is critical to achieving a successful resolution of indoor air quality complaints. IAQ complaints may be grounded in poor indoor air quality, thermal conditions, noise, glare, or even job stress. However, it is in the building manager's best interest to respond to all complaints about the indoor environment promptly and seriously and to establish credibility through open communication with building occupants. The biggest mistake that building managers can make in the face of an IAQ complaint is to underestimate the problems that can result if building occupants believe that no action is being taken or that important information is being withheld. Without open communication, any IAQ problem can become complicated by anxiety, frustration, and distrust, delaying its resolution. Paying attention to communication, as well as problem-solving, helps to ensure proper support and cooperation of building occupants as the complaint is investigated and resolved. The messages to convey are that management believes it is important to provide a healthy and safe building, that good indoor air quality is an essential component of a healthful indoor environment, and that complaints about indoor air quality are taken seriously. Make certain that occupants know how to contact the responsible personnel who can receive and respond to IAQ complaints. Tenants may also have an internal system for channeling complaints, for example through a health and safety representative, supervisor, or company doctor. Indoor air quality complaints that can be resolved quickly and that involve small numbers of people (e.g., annoying but harmless odors from an easily-identified source) can be handled matter-of-factly like other minor problems without risking confusion and bad feeling among other building occupants. Communication becomes a more critical issue when there are delays in identifying and resolving the problem and when serious health concerns are involved. If the problem seems to be widespread or potentially serious, it is advisable to work with your health and safety committee. If you do not have a health and safety committee, consider forming one, or establishing a joint management-tenant IAQ task force. Productive relations will be enhanced if occupants are given basic information during the process of investigation and mitigation. Potential critics can become allies if they are invited to be part of the problem-solving process and become better educated about IAQ and building operations. Building managers may be understandably reluctant to share test results or consultants' reports with their tenants or employees, but secrecy in such matters can backfire if information leaks out at a later time. Building management staff can be encouraged to talk directly with occupants both at the time a complaint occurs and later during a diagnostic investigation. Their observations about patterns of symptoms or building conditions may provide helpful information.

Confidentiality of records can be important to occupants, especially if they are concerned that IAQ complaints will lead to negative reactions from their employers. There may be legal penalties for violating confidentiality of medical records. By reassuring occupants that privacy will be respected, investigators are more likely to obtain honest and complete information. It is advisable to explain the nature of investigative activities, so that rumors and suspicions can be countered with factual information.

Notices or memoranda can be delivered directly to selected occupants or posted in general use areas. Newsletter articles or other established communication channels can also be used to keep building occupants up-to-date. Problems can arise from saying either too little or too much. Premature release of information when data-gathering is still incomplete can produce confusion, frustration, and mistrust at a later date. Similar problems can result from incorrect representation of risk - assuming the worst case (or the best). However, if progress reports are not given, people may think nothing (or something terrible) is happening. It is good practice to clear each piece of information with the facility manager, building owner, or legal counsel. Vague discomfort, intermittent symptoms, and complex interactions of job stress with environmental factors, which make IAQ problems difficult to investigate, can also obscure the effects of mitigation efforts. Even after the proper mitigation strategy is in place, it may take days or weeks for contaminants to dissipate and symptoms to disappear. If building occupants are informed that their symptoms may persist for some time after mitigation, the inability to bring instant relief is less likely to be seen as a failure.

AUDITING PRACTICES AND THE IAQ PROFILE

An IAQ profile provides a description of the features of the building structure, function, and occupancy that impact indoor air quality. Upon completion of the IAQ profile, one should have an understanding of the current status of air quality in the building and baseline information on the factors that have a potential for causing problems in the future. The IAQ profile can help building management to identify potential problem areas and prioritize budgets for maintenance and future modifications. Combined with information on lighting, security, and other important systems, it can become an owner's manual that is building-specific and will serve as a reference in a variety of situations. The key questions to answer while developing the IAQ profile are: How was this building originally intended to function? Consider the building components and furnishings, mechanical equipment (HVAC and nonHVAQ), and the occupant population and associated activities. Is the building functioning as designed? Find out whether it was commissioned. Compare the information from the commissioning to its current condition. What changes in building layout and use have occurred since the original design and construction? Find out if the HVAC system has been reset and retested to reflect current usage. What changes may be needed to prevent IAQ problems from developing in the future? Consider potential changes in future uses of the building.

The process of developing an IAQ profile should require only a modest effort, from a few days to a few weeks of staff time, depending on the complexity of the building and the amount of detailed information collected. The work can be done in pieces over a longer period, if necessary, to fit into a building manager's busy schedule. Over time, it is desirable to make some actual measurements of airflow, temperature, relative humidity, carbon dioxide (CO), and/or pressure differentials (e.g., in each of the air handling zones or other subareas of the building). These measurements provide far better information on current conditions than can be obtained from the plans and specifications, even if as-built records are available. Also, few buildings have been adequately commissioned, so the system may never have delivered the airflows shown on the design drawings. In the event of litigation around future IAQ complaints, the value of the IAQ profile as a resource document will be enhanced by real-world measurements. Refer to the ASHRAE standard on commissioning. The EPA document on designing for good indoor air quality also contains information on the process of commissioning buildings.

Many of the resources necessary for the IAQ profile should already be on hand. Additional information can be collected by the staff person or persons who have the following skills: basic understanding of HVAC system operating principles, ability to read architectural and mechanical plans and understand manufacturer's catalog data on equipment ability to identify items of office equipment, ability to work cooperatively with building occupants and gather information about space usage ability to collect information about HVAC system operation, equipment condition, and maintenance schedules, authority to collect information from subcontractors about work schedules and materials used (particularly cleaning and pest control activities), ability to understand the practical meaning of the information contained in the Material Safety Data Sheets (MSDSs). If direct measurements are to be included in the IAQ profile, the staff should have the tools and training to make the following measurements: air volumes at supply diffusers and exhaust grilles, CO concentration, temperature, relative humidity at different pressure differentials, assessment of thermal and ventilation load requirements. The type of information needed for an IAQ profile is similar to that which is collected when solving indoor air quality problems, but includes the entire building rather than focusing on areas that may have caused an identified problem. The IAQ profile should be an organized body of records that can be referred to in planning for renovations, negotiating leases and contracts, or responding to future complaints.

The process of gathering information for the IAQ profile can be divided into three major stages: (1) Collect and review existing records. (2) Conduct a walkthrough inspection of the building. (3) Collect detailed information on the HVAC system, pollutant pathways, pollutant sources, and building occupancy. These three steps constitute an indoor air quality audit. The first two stages should be carried out as quickly as possible, but the third stage can be handled as time allows so that it does not interfere with other staff responsibilities.

Recordkeeping Practices

Initial efforts should be devoted to collecting any available documents that describe the construction and operation of the building: architectural and mechanical plans, specifications, submittals, sheet metal drawings, commissioning reports, adjusting and balancing reports, inspection records, and operating manuals. Many buildings may lack some or all of these documents. If there are no commissioning reports or balancing reports, actual ventilation quantities may be different from those indicated on mechanical design drawings. If there are no operating or maintenance manuals for HVAC equipment, it is difficult for staff to carry out an adequate preventive maintenance program.

An examination of the original architectural and mechanical design should be made to gain an understanding of the building's layout and intended functions. Identify and note locations in which changes in equipment or room usage create a potential for indoor air quality problems and give them special attention during the walkthrough inspection. Items of interest and the questions they suggest could include the following:

- Commissioning reports: Was the building properly commissioned when it was first constructed, including testing and balancing of the RVAC system?

- Operating manuals: Do staff members understand how the HVAC equipment is intended to operate?

- Remodeled areas: Has the HVAC system layout been changed to accommodate new walls, rearranged partitions, or similar architectural modifications?

- Addition, removal, or replacement of HVAC equipment: Where the original equipment has been replaced, do the newer units have the same capacity as the originals? Has new equipment been properly installed and tested? Where equipment has been removed, is it no longer needed?

- Changes in room use: Is there a need for additional ventilation (supply and/or exhaust) due to increased occupant population or new activities within any area of the building? Have new items of equipment (non-HVAQ been provided with local exhaust where needed? Look for unusual types or quantities of equipment such as copy machines or computer terminals.

Check HVAC maintenance records against equipment lists. Collect existing maintenance and calibration records and check them against the construction documents (e.g., equipment lists and mechanical plans). Equipment that has been installed in inaccessible or out-of-the-way locations is frequently overlooked during routine maintenance.

This is particularly true of items such as filter boxes and small capacity exhaust fans. If there is an organized record of past occupant complaints about the building environment, review those complaints to identify building areas that deserve particular attention.

Walkthrough Inspection of the Building

The intent of the walkthrough inspection is to acquire a good overview of occupant activities and building functions and to look for IAQ problem indicators. No specific forms are suggested for this stage of IAQ profile development. However, the investigator should have a sketch plan of the building, such as a small floor plan showing fire exits, so that his or her notes can be referenced to specific locations.

Detailed measurements of temperature, humidity, airflow, or other parameters are more appropriate to a later stage of profile development. However, chemical smoke can be used to observe airflow patterns and pressure relationships between special use areas or other identified pollutant sources and surrounding rooms. Odors in inappropriate locations may indicate that ventilation system components require adjustment or repair.

The value of IAQ ventilation measurement tools to your operation will grow as you become more familiar with handling indoor air quality concerns. For example, if you do not own a direct-reading carbon dioxide monitor, it is not necessary to acquire one for the IAQ profile. Those who already have access to this type of instrument can take readings during the walkthrough as a way to obtain baseline information about normal operating conditions or identify problem locations. If you begin to suspect that underventilation is a consistent problem, you may decide that it would be helpful to obtain more ventilation monitoring equipment.

A walkthrough inspection provides an opportunity to introduce facility staff and other building occupants to the topic of indoor air quality and to understand current staff (and contractor) responsibilities in relation to housekeeping and maintenance activities. Advance notice of the inspection will make it seem less intrusive and may encourage staff and other occupants to remember important information. Discussion of routine activities in the building will help to clarify elements that should be included in the IAQ management plan. Ask staff members about their job responsibilities, training and experience. It will be helpful to meet with responsible staff and contractors to discuss facility operation and maintenance issues (e.g., HVAC, plumbing, electric, interior maintenance). HVAC maintenance schedule (e.g., filter changes, drain pan maintenance) use and storage of chemicals schedule of shipping and receiving, handling of vehicles at loading dock scheduling and other procedures for isolating odors, dust, and emissions from painting, roof repair, and other contaminant-producing activities should be examined. Others areas to examine include budgeting (e.g., including staff influence on budget decisions), housekeeping, cleaning schedules, trash storage and schedule of refuse removal, use and storage of chemicals, pest control, schedule and location of pesticide applications, use and storage of chemicals, pest control activities other than use of pesticides.

The walkthrough inspection can be used to identify areas with a high potential for IAQ problems. The following are general indicators of IAQ problems: odors, dirty or unsanitary conditions (e.g., excessive dust), visible fungal growth or moldy odors (often associated with problem of too much moisture), sanitary conditions in equipment such as drain pans and cooling towers poorly-maintained filters signs of mold or moisture damage at walls (e.g., below windows, at columns, at exterior corners), ceilings, and floors staining and discoloration.

The walkthrough should focus on uneven temperatures, persistent odors, drafts, sensations of stuffiness. You may find that occupants are attempting to compensate for an HVAC system that doesn't meet their needs. Look for propped-open corridor doors, blocked or taped-up diffusers, popped-up ceiling tiles, people using individual fans/heaters or wearing heavier (or lighter) clothing than normal.

Overcrowding issues should be noted. Future occupant density is estimated when the ventilation system for a building is designed. When the actual number of occupants approaches or exceeds this occupant design capacity, managers may find that IAQ complaints increase. At that point, the outdoor air ventilation rate will have to be increased. However, the ventilation and cooling systems may not have sufficient capacity to handle the increased loads from the current use of the space.

Check for underventilation caused by obstructed vents, faulty dampers or other HVAC system malfunctions, or from problems within the occupied space. Furniture, papers, or other materials can interfere with air movement around thermostats or block airflow from wall or floor-mounted registers. If office cubicles are used, a small space (i.e., two to four inches) between the bottom of the partitions and the floor may improve air circulation.

Lift a ceiling tile and examine the plenum for potential problems. Walls or full-height partitions that extend to the floor above can obstruct or divert air movement in ceiling plenums unless transfer grilles have been provided. If fire dampers have been installed to allow air circulation through walls or partitions, confirm that the dampers are open. Construction debris and damaged or loose material in the plenum area may become covered with dust and can release particles and fibers.

Be aware of areas that contain unusual types or quantities of equipment such as copy machines or computer terminals. Also look for instances of over-illumination. High concentrations of electrical fixtures and equipment can overwhelm the ventilation and cooling systems.

Confirm that the HVAC system maintains appropriate pressure relationships to isolate and contain odors and contaminants in mixed use buildings and around special use areas. Examples of special use areas include attached parking garages, loading docks, print shops, smoking lounges, janitorial closets, storage areas, and kitchens.

Check the outdoor air intakes to see whether they are located near contaminant sources (e.g., plumbing vents, exhaust outlets, dumpsters, loading docks, or other locations where vehicles idle). See if the space containing the HVAC system is clean and dry. Examples of problems include: cleaning or other maintenance supplies stored in mechanical room; dust and dirt buildup on floors and equipment; moisture in mechanical room because of inadequate insulation, lack of conditioned air, or failure to provide for air movement. Unsanitary conditions in the mechanical room are particularly a problem if unducted return air is dumped into and circulated through the mechanical room.

The collection of detailed information for the IAQ profile can be handled as time is available. Areas that have been identified as presenting potential IAQ problems should be given the highest priority.

Use maintenance records to inspect HVAC equipment and make sure that it is in good operating condition. Identify items of equipment that need to be repaired, adjusted, or replaced. Record control settings and operating schedules for HVAC equipment for comparison to occupancy schedules and current uses of space.

Using a sketch plan of the building that was begun during the walkthrough inspection, indicate architectural connections (e.g., chases) and mechanical connections (e.g., ductwork, temperature control zones). Observe and record airflow between spaces intended to run positive or negative and the areas that surround them (including airflow between perimeter rooms and outdoors). Note that hidden pathways such as chases may travel both vertically and horizontally and transport pollutants over long distances.

Record potential pollutant sources in the building. Note the locations of major sources. Major sources such as large items of equipment can be recorded on the floor plan. Record the names and locations of chemicals or hazardous substances used or stored within the building, such as those that may be contained in cleaning materials, biocides, paints, caulks, and adhesives. Ask your suppliers to provide you with Material Safety Data Sheets. You may be unaware of the potential hazards of some materials that are commonly used in public and commercial buildings. For example: In 1990, EPA eliminated the sale of mercury-containing interior latex paint. (Enamel paints do not contain mercury.) People are urged not to use exterior latex paint indoors, as it may contain mercury. If you have paint in storage that may have been manufactured before August 20, 1990, you may contact the manufacturer, the National Pesticide Telecommunication Network (1-800-858-7378), or your State Health Department for guidance. In 1990, EPA banned the use of chromium chemicals in cooling towers, because the chemicals have been shown to be carcinogenic. Heating system steam should not be used in the HVAC humidification system, as it may contain potentially harmful chemicals such as corrosion inhibitors.

Part of the walkthrough should focus on collecting information on building occupancy. This should include information on the way each area of the building is used, its source of outdoor air, and whether or not it is equipped with local exhaust. If underventilation is suspected, estimate ventilation rates in cubic feet per minute per person or per square foot floor area, for comparison to guidelines such as design documents, applicable building codes, or the recommendations of ASHRAE 621989. Underventilation problems can occur even in areas where ventilation rates apparently meet ASHRAE guidelines; proper distribution and mixing of supply air with room air are also essential for good ventilation.

If the information collected as you develop the IAQ profile indicates that you have one or more IAQ problems, prioritize these problems by considering the apparent seriousness of their consequences. For example, combustion gas odors demand a more rapid response than thermostats that are out of calibration.

MANAGEMENT PLANS

Any IAQ management system will be successful only if it is organized to fit the specific needs of the building. Managing a building for good indoor air quality involves reviewing and amending current practice (and establishing new procedures, if necessary) to:

Operate and maintain HVAC equipment: keep all equipment and controls in proper working order and to keep interior of equipment and ductwork clean and dry.

Oversee activities of staff, tenants, contractors, and other building occupants that impact indoor air quality: smoking , housekeeping , building maintenance , shipping and receiving pest control food preparation and other special uses.

Maintain communications with occupants so that management will be informed of complaints about the indoor environment in a timely way: identify building management and staff with IAQ responsibilities   use health and safety committees.

Educate staff, occupants, and contractors about their responsibilities in relation to indoor air quality: staff training , lease arrangements , contracts.

Identify aspects of planned projects that could affect indoor air quality and manage projects so that good air quality is maintained: this includes redecorating, renovation, or remodeling, relocation of personnel or functions within the building , new construction.

Development of the management plan involves reviewing and revising staff responsibilities so that IAQ considerations become incorporated into routine procedures. Organizations may assign responsibility for operations, recordkeeping, purchasing, communications, planning, and policymaking in many different ways. However, the key elements of good IAQ management remain the same: It is important to understand the fundamental influences that affect indoor air quality. This can be achieved by becoming familiar with literature on IAQ and keeping abreast of new information. Select an IAQ manager with the following attributes: clearly defined responsibilities, adequate authority and resources. Use the IAQ profile and other available information to: evaluate the design, operation, and usage of the building, identify potential IAQ problem locations, identify staff and contractors whose activities affect indoor air quality. Review and revise staff responsibilities to ensure that responsibilities that may affect indoor air quality are clearly assigned. In addition, establish lines of communication for sharing information pertaining to: equipment in need of repair or replacement, plans to remodel, renovate, or redecorate, new uses of building space or increases in occupant population, installation of new equipment.

Review standard procedures and make necessary revisions to promote good indoor air quality, such as: terms of contracts (e.g., pest control, leases), scheduling of activities that produce dust, emissions, odors, scheduling of equipment operation, inspection, and maintenance, specifications for supplies (e.g., cleaning products, construction materials, furnishings), policy regarding tobacco smoking within the building.

Review the existing recordkeeping system and make necessary revisions to: establish a system for logging IAQ related complaints, obtain Material Safety Data Sheets for hazardous materials used and stored in the building. Educate building staff, occupants, and contractors about their influence on indoor air quality by: establishing a health and safety committee, instituting training programs as needed. IAQ problems may occur even in buildings whose owners and managers conscientiously apply the best available information to avoid such problems. Those who can demonstrate their ongoing efforts to provide a safe indoor environment are in a strong legal and ethical position if problems do arise.

IAQ management will be facilitated if one individual is given overall responsibility for IAQ. Whether or not this person is given the title of "IAQ Manager," he or she should have a good understanding of the building's structure and function and should be able to communicate with tenants, facility personnel, and building owners or their representatives about IAQ issues. The IAQ manager's ongoing responsibilities might include: developing the IAQ profile, overseeing the adoption of new procedures, establishing a system for communicating with occupants about IAQ issues, coordinating staff efforts that affect indoor air quality, and making sure that staff have the information (e.g., operating manuals, training) and authority to carry out their responsibilities, reviewing all major projects in the building for their IAQ implications, reviewing contracts and negotiating with contractors (e.g., cleaning services, pest control contractors) whose routine activities in the building could create IAQ problems, periodically inspecting the building for indicators of IAQ problems, managing IAQ-related records, responding to complaints or observations regarding potential IAQ problems, conducting an initial walkthrough investigation of any IAQ complaints. If the IAQ manager was not actively involved in developing the IAQ profile, one of the first tasks will be to review the profile carefully. The manager can start by also identifying building locations with a potential for IAQ problems, staff and contractors whose activities impact indoor air quality, and other building occupants whose activities impact indoor air quality. In addition to information from the IAQ profile, it may be helpful to review lease forms and other contractual agreements for an understanding of the respective legal responsibilities of the building management, tenants, and contractors. Incorporation of IAQ concerns into legal documents helps to ensure the use of proper materials and procedures by contractors and can help to limit the load placed on ventilation equipment by occupant activities. The assignment of responsibilities varies widely between organizations, depending upon the routine activities to be carried out and the capabilities of the available personnel. It would not be appropriate for this document to suggest how IAQ-related responsibilities should be allocated in your organization. For example, issues of access in buildings with tenant-occupied space highlight the need for cooperation between building managers and the tenants' office managers. The building staff may be limited in its access to tenant spaces and tenants may not have access to building operations areas such as mechanical rooms, yet both tenants and building management have responsibilities for maintaining good indoor air quality. Unfortunately, facility personnel are not generally trained to think about IAQ issues as they go about their work. Even though building staff may be observing events and conditions that would indicate potential problems to an experienced IAQ investigator, the staff member's attention may be directed elsewhere. As new practices are introduced to prevent indoor air quality problems, an organized system of recordkeeping will help those practices to become part of routine operations and to "flag" decisions that could affect IAQ. The best results can be achieved by taking time to think about the established channels of communication within your organization, so that new forms can be integrated into decision making with minimum disruption of normal procedures.

Using information from the IAQ profile, the IAQ manager should work with staff and contractors to ensure that building operations and planning processes incorporate a concern for indoor air quality. New procedures, recordkeeping requirements, or staff training programs may be needed. The flow of information between the IAQ manager and staff, occupants, and contractors is particularly important. Good indoor air quality requires prompt attention to changing conditions that could cause IAQ problems, such as installation of new equipment or furnishings, increases in occupant population, or new uses of rooms. Indoor air quality can be affected both by the quality of maintenance and by the materials and procedures used in operating and maintaining the building components including the HVAC system. Facility staff who are familiar with building systems in general and with the features of their building in particular are an important resource in preventing and resolving indoor air quality problems. Facility personnel can best respond to indoor air quality concerns if they understand how their activities affect indoor air quality. It may be necessary to change existing practices or introduce new procedures in relation to the following:

Equipment operating schedules: Confirm that the timing of occupied and unoccupied cycles is compatible with actual occupied periods, and that the building is flushed by the ventilation system before occupants arrive. ASHRAE 62-1989 provides guidance on lead and lag times for HVAC equipment. In hot, humid climates, ventilation may be needed during long unoccupied periods to prevent mold growth.

Control of odors and contaminants: Maintain appropriate pressure relationships between building usage areas. Avoid recirculating air from areas that are strong sources of contaminants (e.g., smoking lounges, chemical storage areas, beauty salons). Provide adequate local exhaust for activities that produce odors, dust, or contaminants, or confine those activities to locations that are maintained under negative pressure (relative to adjacent areas). For example, loading docks are a frequent source of combustion odors. Maintain the rooms surrounding loading docks under positive pressure to prevent vehicle exhaust from being drawn into the building. Make sure that paints, solvents and other chemicals are stored and handled properly, with adequate (direct exhaust) ventilation provided. If local filter traps and adsorbents are used, they require regular maintenance. Have vendors provide Material Safety Data Sheets (MSDSs).

Ventilation quantities: Compare outdoor air quantities to the building design goal and local and State building codes and make adjustments as necessary. It is also informative to see how your ventilation rate compares to ASHRAE 62-1989, because that guideline was developed with the goal of preventing IAQ problems.

HVAC equipment maintenance schedules: Inspect all equipment regularly (per recommended maintenance schedule) to ensure that it is in good condition and is operating as designed (i.e., as close to the design setpoints for controls as possible). Most equipment manufacturers provide recommended maintenance schedules for their products.

Components that are exposed to water (e.g., drainage pans, coils, cooling towers, and humidifiers) require scrupulous maintenance in order to prevent microbiological growth and the entry of undesired microbiologicals or chemicals into the indoor airstream.

HVAC inspections: Be thorough in conducting these inspections. Items such as small exhaust fans may operate independently from the rest of the HVAC system and are often ignored during inspections. As equipment is added, removed, or replaced, document any changes in function, capacity, or operating schedule for future reference.

It may also be helpful to store equipment manuals and records of equipment operation and maintenance in the same location as records of occupant complaints for easy comparison if IAQ problems arise.

Building maintenance schedules: Try to schedule maintenance activities that interfere with HVAC operation or produce odors and emissions (e.g., painting, roofing operations) so that they occur when the building is unoccupied. Inform occupants when such activities are scheduled and, if possible, use local ventilation to ensure that dust and odors are confined to the work area.

Purchasing: Review the general information provided by MSDS and request information from suppliers about the chemical emissions of materials being considered for purchase.

Preventive Maintenance Practices

An HVAC system requires adequate preventive maintenance (PM) and prompt attention to repairs in order to operate correctly and provide suitable comfort conditions and good indoor air quality. The HVAC system operator(s) must have an adequate understanding of the overall system design, its intended function, and its limitations. The preventive maintenance program must be properly budgeted and implemented. A well-implemented PM plan will improve the functioning of the mechanical systems and usually save money when evaluated on a life-cycle basis. However, in some buildings, because of budgetary constraints, maintenance is put off until breakdowns occur or complaints arise, following the "if it isn't broken, don't fix it" philosophy. This type of program represents a false economy and often increases the eventual cost of repairs. Poor filter maintenance is a common example of this phenomenon. Filters that are not changed regularly can become a bed for fungal growth, sometimes allowing particles or microorganisms to be distributed within the building. When filters become clogged, the fans use more energy to operate and move less air. If the filters are an inexpensive, low-efficiency type that becomes clogged and then "blows out," the coils then accumulate dirt, causing another increase in energy consumption. Poor air filter efficiency and poor maintenance may cause dirt to build up in ducts and become contaminated with molds, possibly requiring an expensive duct cleaning operation.

The elements of a PM plan include: periodic inspection, cleaning, and service as warranted, adjustment and calibration of control system components, maintenance equipment and replacement parts that are of good quality and properly selected for the intended function. Critical HVAC system components that require PM in order to maintain comfort and deliver adequate ventilation air include: a outdoor air intake opening, damper controls, air filters, drip pans, cooling and heating coils, fan belts, humidification equipment and controls, distribution systems, exhaust fans.

Maintenance "indicators" are available to help facility staff determine when routine maintenance is required. For example, air filters are often neglected (sometimes due to reasons such as difficult access) and fail to receive maintenance at proper intervals. Installation of an inexpensive manometer, an instrument used to monitor the pressure loss across a filter bank, can give an immediate indication of filter condition without having to open the unit to visually observe the actual filter. Computerized systems are available that can prompt staff to carry out maintenance activities at the proper intervals. Some of these programs can be connected to building equipment so that a signal is transmitted to staff if a piece of equipment malfunctions. Individual areas can be monitored for temperature, air movement, humidity, and carbon dioxide, and new sensors are constantly entering the market. These sensors can be programmed to record data and to control multiple elements of the HVAC system.

Indoor air quality complaints can arise from inadequate housekeeping that fails to remove dust and other dirt. On the other hand, cleaning materials themselves produce odors and emit a variety of chemicals. As they work throughout a building, cleaning staff or contractors may be the first to recognize and respond to potential IAQ problems.

Shipping and Receiving

Shipping and receiving areas can create indoor air quality problems regardless of the types of materials being handled. Vehicle exhaust fumes can be minimized by prohibiting idling at the loading dock. This is particularly important if the loading dock is located upwind of outdoor air intake vents. You can also reduce drafts and pollutant entry by pressurizing interior spaces (e.g., corridors) and by keeping doors closed when they are not in use.

Pest Control

Pest control activities that depend upon the use of pesticides involve the storage, handling, and application of materials that can have serious health effects. Common construction, maintenance practices, and occupant activities provide pests with air, moisture, food, warmth, and shelter. Caulking or plastering cracks, crevices, or holes to prevent harborage behind walls can often be more effective than pesticide application at reducing pest populations to a practical minimum. Integrated Pest Management (IPM) is a low-cost approach to pest control based upon knowledge of the biology and behavior of pests. Adoption of an IPM program can significantly reduce the need for pesticides by eliminating conditions that provide attractive habitats for pests. If an outside contractor is used for pest control, it is advisable to review the terms of the contract and include IPM principles where possible. The following items deserve particular attention.

Schedule pesticide applications for unoccupied periods, if possible, so that the affected area can be flushed with ventilation air before occupants return. Pesticides should only be applied in targeted locations, with minimum treatment of exposed surfaces. They should be used in strict conformance with manufacturers' instructions and EPA labels. General periodic spraying may not be necessary. If occupants are to be present, they should be notified prior to the pesticide application. Particularly susceptible individuals could develop serious illness even though they are only minimally exposed. Select pesticides that are species specific and attempt to minimize toxicity for humans and non-target species. Ask contractors or vendors to provide EPA labels and MSDSs. Make sure that pesticides are stored and handled properly consistent with their EPA labels. If only limited areas of the building are being treated, adjust the HVAC system so that it does not distribute contaminated air throughout the rest of the building. Consider using temporary exhaust systems to remove contaminants during the work. It may be necessary to modify HVAC system operation during and after pest control activities (e.g., running air handling units on 100% outdoor air for some period of time or running the system for several complete air exchanges before occupants re-enter the treated space).

Smoking

Although there are many potential sources of indoor air pollution, both research and field studies have shown that environmental tobacco smoke (ETS) is one of the most widespread and harmful indoor air pollutants. Environmental tobacco smoke is a combination of sidestream smoke from the burning end of the cigarette, pipe, or cigar and the exhaled mainstream smoke from the smoker. ETS contains over 4,000 chemicals; 43 of these chemicals are known animal or human carcinogens. Many other chemicals in ETS are tumor promoters, tumor initiators co-carcinogens (i.e., chemicals that are able to cause cancer when combined with another substance), or cancer precursors (i.e., compounds that can make it easier form other carcinogenic chemicals). In 1986, *The Health Consequences of Involuntary Smoking: A Report of the Surgeon General on Environmental Tobacco Smoke* concluded that ETS was a cause of lung cancer in healthy nonsmokers and that "the scientific case against involuntary smoking as a public health risk is more than sufficient to justify appropriate remedial action, and the goal of any remedial action must be to protect the nonsmoker from environmental tobacco smoke." In the same year, the National Research Council of the National Academy of Sciences issued a report, *Environmental Tobacco Smoke: Measuring Exposures and Assessing Health Effects,* which also concluded that passive smoking increases the risk of lung cancer in adults. In June 1991, NIOSH issued a *Current Intelligence Bulletin (#54)* on ETS in the workplace that dealt with lung cancer and other health effects. In its *Bulletin,* NIOSH stated that the weight of evidence is sufficient to conclude that ETS can cause lung cancer in non-smokers (i.e., those who inhale ETS). It recommended that the preferable method to protect nonsmokers is the elimination of smoking indoors and that the alternative method is to require that smoking be permitted only in separately ventilated smoking areas. The NIOSH *Bulletin* emphasized that provision of such isolated areas should be viewed as an interim measure until ETS can be completely eliminated indoors. Smoking areas must be separately ventilated, negatively pressurized in relation to surrounding interior spaces, and supplied with much more ventilation than nonsmoking areas. The NIOSH *Bulletin* also recommends that the air from the smoking area should be exhausted directly outdoors and not recirculated within the building or vented with the general exhaust for the building. ASHRAE Standard 621989 recommends that smoking areas be supplied with 60 cubic feet per minute (60 cfm) per occupant of outdoor air; the standard also recognized that using transfer air, which is pulled in from other parts of the building, to meet the standard is common practice. Both EPA and NIOSH advise that building owners or facility managers considering the introduction of smoking cessation programs, which for the most part, is widespread in the United States today.

DIAGNOSING IAQ PROBLEMS

Another name for the indoor air quality audit is a diagnostic building investigation. Remember, the goal is to identify and solve the indoor air quality complaint in a way that prevents it from recurring and that does not create other problems. An IAQ investigator should use only the investigative techniques that are needed. Many indoor air quality complaints can be resolved without using all of the diagnostic tools described in this chapter. For example, it may be easy to identify the source of odors that are annoying nearby office workers and solve the problem by controlling pressure relationships (e.g., installing exhaust fans). The use of in-house personnel builds skills that will be helpful in minimizing and resolving future problems. On the other hand, some jobs may be best handled by contractors who have specialized knowledge and experience. In the same way, diagnosing some indoor air quality problems may require equipment and skills that are complex and more sophisticated.

Any IAQ investigation begins with one or more reasons for concern, such as occupant complaints. Some complaints can be resolved very simply (e.g., by asking a few common sense questions of occupants and facility staff during the walkthrough). At the other extreme, some problems could require detailed testing by an experienced IAQ professional. The major steps are information gathering, hypothesis formation, and hypothesis testing. The goal of the investigation is to understand the IAQ problem well enough so that it can be resolved. Many IAQ problems have more than one cause and may respond to (or require) several corrective actions.

An initial walkthrough of the problem area provides information about all four of the basic factors influencing indoor air quality (occupants, HVAC system, pollutant pathways, and contaminant sources). The initial walkthrough may provide enough information to resolve the problem. At the least, it will direct further investigation. For example, if the complaint concerns an odor from an easily identified source, one may want to study pollutant pathways as a next step, rather than interviewing occupants about their patterns of discomfort. As one develops an understanding of how the building functions, where pollutant sources are located, and how pollutants move within the building, several hypotheses or potential explanations of the IAQ complaint may be formed. Building occupants and operating staff are often a good source of ideas about the causes of the problem. For example, they can describe changes in the building that may have occurred shortly before the IAQ problem was noticed (e.g., relocated partitions, new furniture or equipment). Hypothesis development is a process of identifying and narrowing down possibilities by comparing them with observations. Whenever a hypothesis suggests itself, it is reasonable to pause and consider it. Is the hypothesis consistent with the facts collected so far? One may be able to test the hypothesis by modifying the HVAC system or attempting to control the potential source or pollutant pathway to see whether the symptoms or other conditions in the building can be relieved. If the hypothesis successfully predicts the results of planned manipulations, then corrective action can be considered. Sometimes it is difficult or impossible to manipulate the factors you think are causing the IAQ problem; in that case, you may be able to test the hypothesis by trying to predict how building conditions will change over time (e.g., in response to extreme outdoor temperatures).

If the hypothesis or "model" does not seem to be a good predictor of what is happening in the building, you probably need to collect more information about the occupants, HVAC system, pollutant pathways, or contaminant sources. Under some circumstances, detailed or sophisticated measurements of pollutant concentrations or ventilation quantities may be required. Outside assistance may be needed if repeated efforts fail to produce a successful hypothesis or if the information required calls for instruments and procedures that are not available in-house. Analysis of the information collected during the IAQ investigation could produce any of the following results:

The apparent cause(s) of the complaint(s) is (are) identified: Remedial action and follow-up evaluation will confirm whether the hypothesis is correct.

Other IAQ problems are identified that are not related to the original complaints: These problems (e.g., HVAC malfunctions, strong pollutant sources) should be corrected when appropriate.

A better understanding of potential IAQ problems is needed in order to develop a plan for corrective action: It may be necessary to collect more detailed information and/or to expand the scope of the investigation to include building areas that were previously overlooked. Outside assistance may be needed.

The cause of the original complaint cannot be identified: A thorough investigation has found no deficiencies in HVAC design or operation or in the control of pollutant sources, and there have been no further complaints. In the absence of new complaints, the original complaint may have been due to a single, unrepeated event or to causes not directly related to IAQ.

An investigation may require one or many visits to the complaint area. The amount of preparatory work needed before the initial walkthrough varies with the nature and scope of the complaint and the expertise of the investigator, among other factors. For example, an in-house investigator who is already familiar with the layout and mechanical system in the building may begin responding to a complaint about discomfort by going directly to the complaint area to check the thermostat setting and see whether air is flowing from the supply outlets. If the investigator is not familiar with the building or is responding to complaints that suggest a serious health problem, more preparation may be needed before the initial walkthrough. The activities listed below can be directed at a localized "problem area" or extended to include the entire building:

Collect easily-available information about the history of the building and the complaints.

Identify known HVAC zones and complaint areas: Begin to identify potential sources and pollutants (e.g., special use areas near the complaint location). Having a copy of mechanical and floor plans can be helpful at this stage, especially if they are reasonably up-to-date.

Notify the building occupants of the upcoming investigation: Tell them what it means and what to expect.

Identify key individuals needed for access and information: A person familiar with the HVAC systems in the building should be available to assist the investigator at any time during the onsite phase. Individuals who have complained or who are in charge of potential sources (e.g., housekeeping, non-HVAC equipment) should be aware that their information is important and should be contacted for appointments or telephone interviews if they will not be available during the onsite visit.

The initial walkthrough provides an opportunity to question complainants about the nature and timing of their symptoms and to briefly examine the immediate area of the complaint. The investigator attempts to identify pollutant sources and types and observes the condition and layout of the HVAC system serving the complaint area. Staff can be asked to describe the operating schedule of equipment. Obvious problems (e.g., blocked diffusers, malfunctioning air handlers) can be corrected to see if the complaints disappear. The walkthrough can solve many routine IAQ problems and will suggest directions for a more complex investigation, should one be necessary. Some investigators avoid taking any measurements during the initial walkthrough so that they are not distracted from "getting the big picture." Others find that using smoke sticks, digital thermometers, and direct reading CO_2 meters or detector tubes to take occasional measurements helps them develop a feel for the building. Any instruments that will be used should be inspected to make sure they are in working order and calibrated. IAQ investigations generally include the use of, at a minimum: heatless chemical smoke devices and instruments for measuring temperature and humidity. Carbon dioxide measuring devices (detector tubes with a hand pump or a direct reading meter) are helpful for most investigations. Other instruments may be needed as the investigation progresses.

Collecting information on complaints is an essential part of the investigation. Occupant data falls into two categories: complaints of discomfort or other symptoms (e.g., teary eyes, chills) and perceptions of building conditions (e.g., odors, draftiness). Investigators can gather valuable information about potential indoor air problems by listening to occupants, and use that information for: defining the complaint area within the building; suggesting directions for further investigation, either by identifying other events that seem to happen at the same time as the incidents of symptoms or discomfort, or by identifying possible causes for the types of symptoms or discomfort that are occurring; indicating potential measures to reduce or eliminate the problem. If there is a record of occupant complaints, a review of that record can help to define the location of the IAQ problem and identify people who should be interviewed as part of the investigation. Information about the history of complaints could also stimulate theories about potential causes of the problem. The most obvious way to collect information from building occupants is to talk to them in person. If it is not possible to interview everyone who has complained about building conditions, the investigator should attempt to interview a group of individuals that reflects the concerns of the affected areas. The investigation may also include occupant interviews with building occupants who do not have complaints. Then conditions in the complaint area can be compared to conditions in similar building locations where there are no complaints. Many events occur simultaneously in and around a complex building, and it can be very difficult to judge which of those events might be related to the IAQ complaints. In trying to resolve stubborn problems, professional investigators sometimes ask occupants and facility staff to keep day-by-day records. Occupants are asked to record the date and time of symptoms, where they are when the symptoms appear, and any other information that might be useful. Such information could include observations about the severity and duration of symptoms and comments on weather conditions, events, and activities that are happening at the same time. Facility staff are asked to record the date and time of events such as maintenance work, equipment cycles, or deliveries. If symptoms seem to occur at particular times of day, staff can focus their attentions on recording events that occur before and during those periods. Such records are likely to produce more accurate and detailed information than can be obtained by relying on memory.

The pattern of complaints within the building helps to define the complaint area. The timing of symptoms and the types of symptoms reported may provide clues about the cause of the problem. The investigator should look for symptom patterns, and define the complaint area. Use the spatial pattern (locations) of complaints to define the complaint area. Building locations where symptoms or discomfort occur define the rooms or zones that should be given particular attention during the initial investigation. However, the complaint area may need to be revised as the investigation progresses. Pollutant pathways can cause occupant complaints in parts of the building that are far removed from the source of the problems. The investigator should also look for patterns in the timing of complaints. The timing of symptoms and complaints can indicate potential causes for the complaints and provide directions for further investigation. Review the data for cyclic patterns of symptoms (e.g., worst during periods of minimum ventilation or when specific sources are most active) that may be related to HVAC system operation or to other activities in and around the building. Look for patterns in the types of symptoms or discomfort. IAQ investigations often fail to prove that any particular pollutant or group of pollutants are the cause of the problem. Such causal relationships are extremely difficult to establish. There is little information available about the health effects of many chemicals. Typical indoor levels are much lower than the levels at which toxicology has found specific effects. Therefore, it may be more useful to look for patterns of symptoms than for specific pollutant and health effect relationships. Investigators who are not medically trained cannot make a diagnosis and should not attempt to interpret medical records. Also, confidentiality of medical information is protected by law in some jurisdictions and is a prudent practice everywhere. In general terms, indoor air quality is judged to be worse as temperatures rise above 76 °F, regardless of the actual air quality. There is controversy concerning recommended levels of relative humidity. In general, the range of humidity levels recommended by different organizations seems to be 30% to 60% RH. Relative humidities below this level may produce discomfort from dryness. On the other hand, maintaining relative humidities at the lowest possible level helps to restrict the growth of mold and mildew.

The concerns (comfort for the most part) associated with dry air must be balanced against the risks (enhanced microbiological growth) associated with humidification. If temperatures are maintained at the lower end of the comfort range (68-70 °F) during heating periods, relative humidity in most climates will not fall much below 30% (also within the comfort range) in occupied buildings.

IAQ complaints often arise because the quantity or distribution of outdoor air is inadequate to serve the ventilation needs of building occupants. Problems may also be traced to air distribution systems that are introducing outdoor contaminants or transporting pollutants within the building. The investigation should begin with the components of the HVAC system(s) that serve the complaint area and surrounding rooms, but may need to expand if connections to other areas are discovered. Your goal is to understand the design and operation of the HVAC system well enough to answer the following questions:

- Are the components that serve the immediate complaint area functioning properly?

- Is the HVAC system adequate for the current use of the building?

- Are there ventilation (or thermal comfort) deficiencies?

- Should the definition of the complaint area be expanded based upon the HVAC layout and operating characteristics?

An evaluation of the HVAC system may include limited measurements of temperature, humidity, air flow, as well as smoke tube observations. Complex investigations may require more extensive or sophisticated measurements of the same variables (e.g., repeated CO_2 measurements taken at the same location under different operating conditions, continuous temperature and relative humidity measurements recorded with a data logger).

A detailed engineering study may be needed if the investigation discovers problems such as the following:

- airflows are low

- HVAC controls are not working or are working according to inappropriate strategies

- building operators do not understand (or are unfamiliar with) the HVAC system

A review of existing documentation (e.g., plans, specifications, testing and balancing reports) should provide information about the original design and later modifications, particularly: the type of HVAC system (e.g., constant volume, VAV); locations and capacities of HVAC equipment serving the complaint area; the planned use of each building area; supply, return, and exhaust air quantities;

location of the outdoor air intake and of the supply, return, and exhaust registers, diffusers, and grilles that serve the complaint area.

The most useful way to record this information is to make a floor plan of the complaint area and surrounding rooms. You may be able to copy an existing floor plan from architectural or mechanical drawings, fire evacuation plans, or some other source. If there is no documentation on the mechanical system design, much more onsite inspection will be required to under- stand the HVAC system. The HVAC system may have been installed or modified without being commissioned, so that it may never have performed according to design. In such cases, good observations of airflow and pressure differentials are essential. In addition, load analyses may be required.

It is important to note that IAQ complaints are often intermittent. Discussions with staff may reveal patterns that relate the timing of complaints to the cycles of equipment operation or to other events in the building such as painting, installation of new carpeting, or pest control. These patterns are not necessarily obvious. Keeping a day-to-day record may help to clarify subtle relationships between occupant symptoms, equipment operation, and activities in and around the building. If the building is new or if there is a preventive maintenance program with recent test and balance reports, it is possible that the HVAC system is functioning according to its original design. Otherwise it is probable that one or more features of building usage or system operation have changed in ways that could affect indoor air quality. Elements of the on-site investigation can include the following:

Check temperature and/or humidity to see whether the complaint area is in the comfort range: Take more than one measurement to account for variability over time and from place to place. Check thermostat operation. Check whether the supply air temperature corresponds to the design criteria. Use a hygrothermograph (if available) to log temperature and humidity changes in the complaint area.

Check for indicators of inadequate ventilation: Check supply diffusers to see if air is moving (using chemical smoke). If it is not, confirm that the fan system is operating, and then look for closed dampers, clogged filters, or signs of leaks. Compare design air quantities to building codes for the current occupancy or ventilation guidelines (e.g., ASHRAE 62-1989). If the HVAC system, performing as designed, would not provide enough ventilation air for current needs, then there is good reason to believe that actual ventilation rates are inadequate. Measure carbon dioxide in the complaint area to see whether it indicates ventilation problems. Measure air quantities supplied to and exhausted from the complaint area, including calculation of outdoor air quantities. Be aware of damper settings and equipment cycles when you are measuring (e.g., are you evaluating minimum outdoor air, "normal" conditions, or maximum airflow?). Note that evaluation of variable air volume (VAV) systems requires considerable expertise. Compare the measured air quantities to your mechanical system design specifications and applicable building codes. Also compare ventilation rates to ASHRAE 62-1989.

Check that equipment serving the complaint area (e.g., grilles, diffusers, fans) is operating properly: Confirm the accuracy of reported operating schedules and controls sequences; for example, power outages may have disrupted time clocks, fans reported as "always running" may have been accidentally switched off, and controls can be in need of calibration. Check to see that equipment is properly installed. For example, look for shipping screws that were never removed or fans that were reversed during installation, so that they move air in the wrong direction.

Compare the current system to the original design: Check to see that all equipment called for in the original design was actually installed. See whether original equipment may have been replaced by a different model (i.e., a model with less capacity or different operating characteristics).

See whether the layout of air supplies, returns, and exhausts promotes efficient air distribution to all occupants and isolates or dilutes contaminants: If supplies and returns are close together, heatless chemical smoke can be used to check for short-circuiting (supply air that does not mix properly with air in the breathing zone, but moves directly to the return grille). CO_2 can also be used to evaluate air mixing. Use heatless chemical smoke to observe airflow patterns within the complaint area and between the complaint area and surrounding spaces, including outdoors. Compare airflow directions under vari- ous operating conditions. If the system layout includes ceiling plenums, look above the ceiling for interruptions such as walls or full-height partitions.

Consider whether the HVAC system itself may be a source of contaminants: Check for deterioration or unsanitary conditions (e.g., corrosion, water damage or standing water, mold growth or excessive dust in ductwork, debris or damaged building materials in ceiling plenums). If the mechanical room serves as a mixing plenum (i.e., return and outdoor air are drawn through the room into the air handler), check very carefully for potential contaminants such as stored solvents and deteriorated insulation.

In reviewing the HVAC data, consider whether the system is adequate to serve the use of the building and whether the timing, location, and impact of apparent deficiencies appear related to the IAQ complaint. Deficiencies in HVAC design, operation, or maintenance may exist without producing the complaint under investigation; some defects may not cause any apparent IAQ problems. Strategies for corrective actions should be based on comparisons between the original design and the current system; comparing the original uses of space to current uses, and a consideration of the condition of the HVAC system.

Normal safety precautions observed during routine operation of the building must be followed closely during IAQ inspections. When the IAQ investigator is not familiar with the mechanical equipment in that particular facility, an operator or engineer should be present at all times in equipment areas. Potential safety hazards include:

- electrocution;

- injury from contacting fans, belts, dampers or slamming doors;

- burns from steam or hot water lines;

- falls in ventilation shafts or from ladders or roofs.

Investigators evaluating building IAQ generally do not encounter situations in which specific personal protection measures (e.g., protective garments and respirators) are required. However, safety shoes and eyeglasses are generally recommended for working around mechanical equipment. When severe contamination is present (e.g., microbiological, chemical, or asbestos), IAQ investigators may need additional protection in the vicinity of certain building areas or equipment. Such decisions are site-specific and should be made in consultation with an experienced industrial hygienist. General considerations include the following:

Microbiological: Care must be taken when serious building-related illness (e.g., Legionnaire's disease) is under investigation or when extensive microbiological growth has occurred. Investigators with allergy problems should be especially cautious. The array of potential contaminants makes it difficult to know what sort of personal protection will be effective. At a minimum, investigators should minimize their exposure to air in the interior of ducts or other HVAC equipment unless respiratory protection is used. If there is reason to suspect biological contamination (e.g., visible mold growth), expert advice should be obtained about the kind of respiratory protection to use and how to use it. Possible protective measures against severe microbiological contamination include disposable coveralls and properly fitted respirators.

Chemical: Where severe chemical contamination is suspected, specific precautions must be followed if OSHA action levels are approached. Such instances rarely occur in IAQ investigations. One possible exception might be a pesticide spill in a confined space. In this case, an appropriate respirator and disposable coveralls may be needed.

Asbestos: An IAQ investigation often includes inspection above accessible ceilings, inside shafts, and around mechanical equipment. Where material suspected of containing asbestos is not only present, but also has deposited loose debris, the investigator should take appropriate precautions. This might include disposable coveralls and a properly fitted respirator.

Pollutant Pathways and Driving Forces

Unless the IAQ problem is caused by an obvious contaminant located in the complainant's immediate workspace, you will need to understand the patterns of airflow into and within the complaint area. Correction of IAQ problems often involves controlling pollutant movement through sealing of pollutant pathways or manipulation of the pressure relationships. If the complaints being investigated are limited to a few areas of the building, pollutant pathways can be evaluated so that the complaint area is properly defined before conducting the source inventory. If complaints are spread throughout the building, evaluation of pathways could be a very time-consuming process, and it may be more practical to look for major contaminant sources before trying to discover how the contaminants move within the building.

Architectural and mechanical pathways allow pollutants to enter the complaint area from surrounding spaces, including the outdoors. An examination of architectural and mechanical plans can help in developing a list of connections to surrounding areas. These include: doors, operable windows, stairways, elevator shafts with utility chases, ductwork and plenums, areas served by common HVAC controls (e.g., shared thermostats). Onsite inspection is needed to confirm the existence of these connections and to identify other openings (e.g., accidental openings such as cracks and holes). Fire codes usually require that chases and hidden openings be firestopped. Check for the existence and condition of firestops in chases, especially those that connect both vertically and horizontally.

The airflow quantities shown in mechanical plans or in testing and balancing reports can be used to determine the direction of air movement intended by the designer. Onsite examination is necessary to determine the actual direction of airflow at each available pathway. Chemical smoke tubes can be used to determine airflow directions between the complaint area and surrounding spaces (including the outdoors), and to reveal air circulation patterns within the complaint area. A micromanometer (or equivalent) can measure the magnitude of pressure differences between these areas. It may be necessary to make observations under different conditions, as airflow direction can change depending upon weather conditions, windspeed and direction, equipment operation within the building, traffic through doors, and other factors (e.g., as VAV systems throttle back).

Switching air handlers or exhaust fans on and off, opening and closing doors, and simulating the range of operating conditions in other ways can help to show the different ways that airborne contaminants move within the building. Dust tracking patterns around door frames can reveal the dominant direction of air and pollutant movement.

Some investigators study air movement by releasing a small amount of peppermint oil at the opening to a suspected pathway and asking an assistant to sniff for the "toothpaste" smell. If this technique is used, it is important that the assistant have an acute sense of smell. If the building is in use during the investigation, occupants may also notice the odor and could find it distracting. Some investigators prefer to use methods that release an odor during unoccupied periods. Investigators should note two common causes of false negative results (falsely concluding that no pathway exists):

- The nose quickly becomes tolerant of strong odors, so that the assistant may need to take a long rest (breathing fresh air) between tests.

- If there is substantial airflow through the pathway, the peppermint oil odor could be diluted so that it is imperceptible. Tracer gases such as sulfur hexafluoride (SF_6 can provide qualitative and quantitative information on pollutant pathways and ventilation rates. Use of tracer gases to obtain quantitative results requires considerable technical expertise. If it appears that a sophisticated study of pathways (or ventilation rates) is required, you need to use trained investigators.

Pollutant pathway information helps the investigator to understand airflow patterns in and around the complaint area. The pollutant pathway data may indicate a need to enlarge the complaint area, or may direct attention toward contaminant sources that deserve close study.

Evaluate airflow patterns into and within the complaint area. Because of the complexity and variability of air- flow patterns, investigators cannot be expected to understand how air moves within the building under all potential operating conditions. However, data on pathways and driving forces can help to locate potential pollutant sources and to understand how con- taminants are transported to building occupants. The discovery of unexpected pollutant pathways can show a need to study areas of the building that may be distant from the original complaint area.

Throughout the investigation, the building investigator will try to identify pollutant sources that may be causing the occupant complaints. Any public or commercial building is likely to contain a number of sources that produce odors, contaminants, or both.

The investigator's task is to identify the source(s) that may be responsible for the complaint(s). The area included in the pollutant source inventory should be defined by the investigator's understanding of the building's architectural and mechanical layout, as well as the pollutant pathways. Common sense will help to differentiate unusual sources (e.g., spills, strong odors from new furnishings or equipment, stains, vapors) from those that are normally found within or near the building. It is important to note that few sources of indoor air contaminants are both continuous and constant in volume over time. Pollutant concentrations often vary in strength over time, and may not be evident at the time of the site visit. Some sources are subtle and might only be noticed by a trained investigator. As the investigation progresses, the inventory of pollutant sources may need to be revised by expanding the definition of the complaint area or examining specific locations more closely (e.g., under various operating conditions).

Depending upon the nature of the complaint, the investigator may find some of the following activities to be useful: inventory outdoor sources; inventory equipment sources; review building components and furnishings; inventory other potential sources.

If a strong pollutant source is identified in the immediate vicinity of the complaint, a simple test (e.g., sealing, covering, or removing the source) can sometimes reveal whether or not it is the cause of the IAQ problem. If a number of potential sources have been found in and around the complaint area, other data (e.g., the pattern of symptoms, the HVAC system design and operation, and pollutant pathways) may be needed in order to determine which source(s), if any, may be related to the complaint. Strategies for using source information include: identify patterns linking emissions to complaints; evaluate unrelated sources; look for patterns linking emissions from potential sources to the IAQ complaints.

A detailed study of pollutants and sources may involve an engineering evaluation of equipment that is releasing IAQ contaminants, diagnostic sampling to assess sources in operation, or other measurements. These may require skills or instruments that are not available in-house. Although air sampling might seem to be the logical response to an air quality problem, such an approach may not be required to solve the problem and can even be misleading. Air sampling should not be undertaken until some or all of the other investigative activities mentioned previously have been used to collect considerable information. Before beginning to take air samples, investigators should develop a sampling strategy that is based on a comprehensive understanding of how the building operates, the nature of the complaints, and a plan for interpreting the results. It may be desirable to take certain routine air quality measurements during an investigation to obtain a "snapshot" of current conditions. These tests should be limited to those that are indicative of very common IAQ concerns such as temperature, relative humidity, air movement, or carbon dioxide (CO_2). Unusual readings may or may not indicate a problem, and should always be interpreted in perspective, based upon site-specific conditions.

Measurement of specific chemical or biological contaminants can be very expensive. Before expending time and money to obtain measurements of indoor air pollutants, you must decide: how the results will be used (e.g., comparison to standards or guidelines, comparison to levels in complaint-free areas); what substances(s) should be measured; where to take, samples; when to take samples; what sampling and analysis method to use so that the results provide useful information.

It is often worthwhile for building staff to develop skills in making temperature, humidity, airflow, and CO_2 measurements and assessing patterns of air movement (e.g., using chemical smoke).

Although air sampling will generate numbers, it will not necessarily help resolve the IAQ problem. Many IAQ complaints are resolved without sampling or with inconclusive sampling results. The design of an air sampling strategy should fit the intended use of the measurements. Potential uses of indoor air measurements include:

- Comparing different areas of the building or comparing indoor to outdoor conditions in order to: confirm that a control approach has the desired effect of reducing pollutant concentrations or improving ventilation. Establish baseline conditions so that they can be compared to concentrations at other times or locations, such as concentrations in outdoor air, concentrations in areas where no symptoms are reported, expected "background" range for typical buildings without perceived IAQ problems.

- Test a hypothesis about the source of the problem, such as: checking emissions from a piece, of equipment. Testing for "indicator" compounds associated with particular types of building conditions: Peak carbon dioxide (CO_2) concentrations over 1000 ppm (parts per million) are an indicator of under-ventilation. Carbon monoxide (CO) over several ppm indicates inappropriate presence of combustion by-products (which may also account for high CO_2 readings). Compare any measured concentrations to guidelines or standards.

The following occupational exposure standards and guidelines should be referred to: OSHA PELs (Occupational Safety and Health Administration's Permissible Exposure Limits), NIOSH RELs (National Institute for Occupational Safety and Health's Recommended Exposure Limits), ACGIH TLVs (American Conference of Governmental Industrial Hygienists' Threshold Limit Values). Public health guidelines for specific pollutants can be found in the following: EPA National Ambient Air Quality Standards, the World Health Organization Air Quality Guidelines, and the Canadian Exposure Guidelines for Residential Air Quality.

There are no widely accepted procedures to define whether IAQ test results are acceptable. Caution must be used in comparing contaminant concentrations to existing occupational standards and guidelines. Although a contaminant concentration above those guidelines is a problem indicator, occupants may still experience health and comfort problems at concentrations well within those guidelines. It is extremely rare for occupational standards to be exceeded, or even approached in public and commercial buildings, including those experiencing indoor air quality problems. Where specific exposure problems are suspected, more detailed diagnostic testing may be needed to locate or understand major sources, confirm the exposure, and to develop appropriate remedial actions. For example, the control of microbial or pesticide contamination may involve surface or bulk sampling. Surface sampling involves wiping a measured surface area and analyzing the swab to see what organisms are present, while bulk sampling involves analyzing a sample of suspect material. Specialized skills, experience, and equipment may be needed to obtain, analyze, and interpret such measurements.

Measurement of "indicator" compounds such as CO_2 or CO can be a cost-effective strategy. Such measurements can help the investigator understand the nature of the problem and define the complaint area. Air sampling for specific pollutants works best as an investigative tool when it is combined with other types of information gathering. It is prudent to begin a program of chemical sampling only if symptoms or observations strongly suggest that a specific pollutant or a specific source may be the cause of the complaint and if sampling results are important in determining an appropriate corrective action. The identified problem area is an obvious site for air sampling. Measurements taken outdoors and in a control location (e.g., a complaint-free area of the building) are helpful in interpreting results from the complaint area.

The conditions experienced by building occupants are best simulated by sampling air from the "breathing zone" away from the influence of any particular individual. However, if an individual sits at a desk all day (except for brief periods), samplers placed on the desk when the individual is elsewhere can provide a good estimate of that person's exposure.

There are several ways to locate sampling sites for an IAQ investigation. One approach first divides the building into homogeneous areas based on key factors identified in the building inspection and interviews. Examples of how a building might be divided include: control zones (e.g., individual rooms); types of HVAC zones (e.g., interior vs. perimeter); complaint versus noncomplaint areas that define a relationship to major sources (e.g., spaces directly, indirectly, or not impacted by smoking areas or exhaust fumes from a piece of equipment); complaint types. Test sites can then be selected to represent complaints, controls, and potential sources with a reasonable number of samples.

Samples may be designed to obtain "worst case" conditions, such as measurements during periods of maximum equipment emissions, minimum ventilation, or disturbance of contaminated surfaces. Worst-case sample results can be very helpful in characterizing maximum concentrations to which occupants are exposed and identifying sources for corrective measures. It is also helpful to obtain samples during average or typical conditions as a basis of comparison. It may, however, be difficult to know what conditions are typical. Exposure to some pollutants may vary dramatically as building conditions change. Devices that allow continuous measurements of key variables can be helpful.

Symptoms or odors that only occur occasionally will not generally be seen during the IAQ investigation. Air samples should not be taken if an incident is not occurring, unless the purpose of the sample is to establish a baseline for future comparisons. One approach to intermittent IAQ problems is for the IAQ investigator to ask appropriate building staff or other occupants to document changes over time using day-to-day records such as an Occupant Diary and Log of Activities and System Operation. When an odor episode does occur, the building engineer could inspect the air handler and intake area while another staff member documents the status of several potential sources. Another strategy is to manipulate building conditions to create worst-case conditions during the building investigation. Chemical smoke and tracer gases can be used to assess where emissions may travel under various building conditions. Such strategies should be carried out in ways that minimize occupant exposure. Care should be taken in selecting the appropriate measurement techniques and to provide proper analysis so that the results provide useful information.

CONTROLLING INDOOR AIR PROBLEMS

Many types of mitigation (correction) strategies have been implemented to solve indoor air quality problems. Mitigation of indoor air quality problems may require the involvement of building management and staff representing such areas of responsibility as: facility operation and maintenance, housekeeping, shipping and receiving, purchasing, policymaking, staff training. Successful mitigation of IAQ problems also requires the cooperation of other building occupants, including the employees of building tenants. Occupants must be educated about the cause(s) of the IAQ problems and about actions that must be taken or avoided to prevent a recurrence of the problems. Efforts to control indoor air contaminants change the relationships between these factors. There are many ways that people can intervene in these relationships to prevent or control indoor air contaminant problems. Control strategies can be categorized as: source control, ventilation, air cleaning, and exposure control. Successful mitigation often involves a combination of these strategies.

Source Control

All efforts to prevent or correct IAQ problems should include an effort to identify and control pollutant sources. Source control is generally the most cost effective approach to mitigating IAQ problems in which point sources of contaminants can be identified. In the case of a strong source, source control may be the only solution that will work. The following are categories and examples of source control:

Remove or reduce the source: Prohibit smoking indoors or limit smoking to areas from which air is exhausted, not recirculated (NIOSH regards smoking areas as an interim solution). Relocate contaminant-producing equipment to an unoccupied, better ventilated, or exhaust-only ventilated space. Select products which produce fewer or less potent contaminants while maintaining adequate safety and efficiency. Modify other occupant activities.

Seal or cover the source: Improve storage of materials that produce contaminants. Seal surfaces of building materials that emit VOCs such as formaldehyde.

Modify the environment: After cleaning and disinfecting an area that is contaminated by fungal or bacterial growth, control humidity to make conditions inhospitable for regrowth.

Source removal or reduction can sometimes be accomplished by a one-time effort such as thorough cleaning of a spill. In other cases, it requires an ongoing process, such as establishing and enforcing a non-smoking policy. Sealing or covering the source can be a solution in some cases; application of a barrier over formaldehyde-emitting building materials is an example.

Sealing may also involve educating staff or building occupants about the contaminant- producing features of materials and supplies and inspecting storage areas to ensure that containers are properly covered. In some cases, modification of the environment is necessary for effective mitigation. If the indoor air problem arises from microbiological contaminants, for example, disinfection of the affected area may not eliminate the problem.

Regrowth of microbiologicals could occur unless humidity control or other steps, such as adding insulation to prevent surface condensation, are taken to make the environment inhospitable to microbiologicals.

Ventilation

Ventilation modification is often used to correct or prevent indoor air quality problems. This approach can be effective either where buildings are underventilated or where a specific contaminant source cannot be identified. Ventilation can be used to control indoor air contaminants by:

Diluting contaminants with outdoor air may increase the total quantity of supply air (including outdoor air): Increase the proportion of outdoor air to total air, and also the improve air distribution.

Isolating or removing contaminants by controlling air pressure relationships m install effective local exhaust at the location of the source: Avoid recirculation of air that contains contaminants. Locate occupants near supply diffusers and sources near exhaust registers. Use air-tightening techniques to maintain pressure differentials and eliminate pollutant pathways. Make sure that doors are closed where necessary to separate zones. Diluting contaminants by increasing the flow of outdoor air can be accomplished by increasing the total supply airflow in the complaint area (e.g., opening supply diffusers, adjusting dampers) or at the air handling unit, (e.g., cleaning the filter on the supply fan). An alternative is to increase the proportion of outdoor air (e.g., adjusting the outdoor air intake damper, installing minimum stops on variable air volume (VAV) boxes so that they satisfy the outdoor air requirements of ASHRAE 62-1989). Increasing ventilation rates to meet ASHRAE Standard 62-1989 (e.g., from 5 to 15 or 20 cfm/person) does not necessarily significantly increase the total annual energy consumption. The increase appears to be less than 5% in typical commercial buildings. The cost of ventilation is generally overshadowed by other operating costs, such as lighting. Further, improved maintenance can produce energy savings to balance the costs that might otherwise result from increased ventilation. The cost of modifying an existing HVAC system to condition additional outdoor air can vary widely depending upon the specific situation. In some buildings, HVAC equipment may not have sufficient capacity to allow successful mitigation using this approach. Original equipment is often oversized so that it can be adjusted to handle the increased load, but in some cases additional capacity is required. Most ventilation deficiencies appear to be linked to inadequate quantities of outdoor air. However, inadequate distribution of ventilation air can also produce IAQ problems. Diffusers should be properly selected, located, installed, and maintained so that supply air is evenly distributed and blends thoroughly with room air in the breathing zone. Short- circuiting occurs when clean supply air is drawn into the return air plenum before it has mixed with the dirtier room air and therefore fails to dilute contaminants. Mixing problems can be aggravated by temperature stratification. Stratification can occur, for example, in a space with high ceilings in which ceiling-mounted supply diffusers distribute heated air.

The side effects of increased ventilation include:

- mitigation by increasing the circulation of outdoor air requires good outdoor air quality

- increased supply air at the problem location might mean less supply air in other areas

- increased total air in the system and increased outdoor air will both tend to increase energy consumption and may require increased equipment capacity

- any approach which affects airflow in the building can change pressure differences between rooms (or zones) and between indoors and outdoors, and might lead to increased infiltration of unconditioned outdoor air

- increasing air in a VAV system may overcool an area to the extent that terminal reheat units are needed

Ventilation equipment can be used to isolate or remove contaminants by controlling pressure relationships. If the contaminant source has been identified, this strategy can be more effective than dilution. Techniques for controlling air pressure relationships range from adjustment of dampers to installation of local exhaust.

Using local exhaust confines the spread of contaminants by capturing them near the source and exhausting them to the outdoors. It also dilutes the contaminant by drawing cleaner air from surrounding areas into the exhaust airstream. If there are return grilles in a room equipped with local exhaust, the local exhaust should exert enough suction to prevent recirculation of contaminants. Properly designed and installed local exhaust results in far lower contaminant levels in the building than could be accomplished by a general increase in dilution ventilation, with the added benefit of costing less.

Replacement air must be able to flow freely into the area from which the exhaust air is being drawn. It may be necessary to add door or wall louvers in order to provide a path for the make-up air. (Make sure that this action does not violate fire codes.)

Identification of the pollutant source and installation of the local exhaust is critically important. For example, an improperly designed local exhaust can draw other contaminants through the occupied space and make the problem worse. The physical layout of grilles and diffusers relative to room occupants and pollutant sources can be important. If supply diffusers are all at one end of a room and returns are all at the other end, the people located near the supplies may be provided with relatively clean air while those located near the returns breathe air that has already picked up contaminants from all the sources in the room that are not served by local exhaust.

Elimination of pollutant pathways by air sealing (e.g., caulking cracks, closing holes) is an approach that can increase the effectiveness of other control techniques.

It can be a difficult technique to implement because of hidden pathways (e.g., above drop ceilings, under raised flooring, against brick or block walls). However, it can have other benefits such as energy savings and more effective pest control (by eliminating paths used by vermin).

Air Cleaning

Another IAQ control strategy is to clean the air. Air cleaning is usually most effective when used in conjunction with either source control or ventilation; however, it may be the only approach when the source of pollution is outside of the building. Most air cleaning in large buildings is aimed primarily at preventing contaminant buildup in HVAC equipment and enhancing equipment efficiency. Air cleaning equipment intended to provide better indoor air quality for occupants must be properly selected and designed for the particular pollutants of interest (for example, gaseous contaminants can be removed only by gas sorption). Once installed, the equipment requires regular maintenance in order to ensure good performance; otherwise it may become a major pollutant source in itself. This maintenance requirement should be borne in mind if an air cleaning system involving a large number of units is under consideration for a large building. If room units are used, the installation should be designed for proper air recirculation. There are four technologies that remove contaminants from the air: particulate filtration, electrostatic precipitation, negative ion generation, and gas sorption. The first three approaches are designed to remove particulate matter, while the fourth is designed to remove gases.

Particulate filtration removes suspended liquid or solid materials whose size, shape and mass allow them to remain airborne at the air velocity conditions present. Filters are available in a range of efficiencies, with higher efficiency indicating removal of a greater proportion of particles and of smaller particles. Moving to medium efficiency pleated filters is advisable to improve IAQ and increase protection for equipment. However, the higher the efficiency of the filter, the more it will increase the pressure drop within the air distribution system and reduce total airflow (unless other adjustments are made to compensate). It is important to select an appropriate filter for the specific application and to make sure that the HVAC system will continue to perform as designed. Filters are rated by different standards which measure different aspects of performance.

Electrostatic precipitation is another type of particulate control. It uses the attraction of charged particles to oppositely charged surfaces to collect airborne particulate matter. In this process, the particles are charged by ionizing the air with an electric field. The charged particles are then collected by a strong electric field generated between oppositely-charged electrodes. This provides relatively high efficiency filtration of small respirable particles at low air pressure losses. Electrostatic precipitators may be installed in air distribution equipment or in specific usage areas. As with other filters, they must be serviced regularly. Note, however, that electrostatic precipitators produce some ozone. Because ozone is harmful at elevated levels, EPA has set standards for ozone concentrations in outdoor air, and NIOSH and OSHA have established guidelines and standards, respectively, for ozone in indoor air. The amount of ozone emitted from electrostatic precipitators varies from model to model.

Negative ion generators use static charges to remove particles from the indoor air. When the particles become charged, they are attracted to surfaces such as walls, floors, table tops, draperies, and occupants. Some designs include collectors to attract the charged particles back to the unit. Negative ion generators are not available for installation in ductwork, but are sold as portable or ceiling-mounted units. As with electrostatic precipitators, negative ion generators may produce ozone, either intentionally or as a by-product of use.

Gas sorption is used to control compounds that behave as gases rather than as particles (e.g., gaseous contaminants such as formaldehyde, sulfur dioxide, ozone, and oxides of nitrogen). Gas sorption involves one or more of the following processes with the sorption material (e.g., activated carbon, chemically treated active clays): a chemical reaction between the pollutant and the sorbent; a binding of the pollutant and the sorbent; or diffusion of the contaminant from areas of higher concentration to areas of lower concentration. Gas sorption units are installed as part of the air distribution system. Each type of sorption material performs differently with different gases. Gas sorption is not effective for removing carbon monoxide. There are no standards for rating the performance of gaseous air cleaners, making the design and evaluation of such systems problematic. Operating expenses of these units can be quite high, and the units may not be effective if there is a strong source nearby.

Exposure control is an administrative approach to mitigation that uses behavioral methods, such as:

Scheduling contaminant-producing activities to avoid complaints: Schedule contaminant-producing activities to occur during unoccupied periods. Notify susceptible individuals about upcoming events (e.g., roofing, pesticide application) so that they can avoid contact with the contaminants. Scheduling contaminant-producing activities for unoccupied periods whenever possible is simple common sense. It may be the best way to limit complaints about activities (such as roofing or demolition) which unavoidably produce odors or dust.

Relocating or rotating susceptible individuals: Move susceptible individuals away from the area where they experience symptoms. Controlling exposure by relocating susceptible individuals may be the only practical approach in a limited number of cases, but it is probably the least desirable option and should be used only when all other strategies are ineffective in resolving complaints.

Mitigation efforts should be evaluated at the planning stage by considering the following criteria: permanence, operating principle, degree to which the strategy fits the job, ability to institutionalize the solution, durability, installation and operating costs, conformity with codes. Mitigation efforts that create permanent solutions to indoor air problems are clearly superior to those that provide temporary solutions (unless the problems are also temporary). Opening windows or running air handlers on full outdoor air may be suitable mitigation strategies for a temporary problem such as outgassing of volatile compounds from new furnishings, but would not be good ways to deal with emissions from operating equipment. A permanent solution to microbiological contamination involves not only cleaning and disinfection, but also modification of the environment to prevent regrowth.

The most economical and successful solutions to IAQ problems tend to be those in which the operating principle of the correction strategy makes sense and is suited to the problem. If a specific point source of contaminants has been identified, treatment at the source (e.g., by removal, scaling, or local exhaust) is almost always a more appropriate correction strategy than dilution of the contaminant by increased general ventilation. If the IAQ problem is caused by the introduction of outdoor air that contains contaminants, increased general ventilation will only make the situation worse (unless the outdoor air is cleaned). It is important to make sure that one understands the IAQ problem well enough to select a correction strategy whose size and scope fit the job. If odors from a special use area such as a kitchen are causing complaints in a nearby office, increasing the ventilation rate in the office may not be a successful approach. The mitigation strategy should address the entire area affected.

If mechanical equipment is needed to correct the IAQ problem, it must be powerful enough to accomplish the task. For example, a local exhaust system should be strong enough and close enough to the source so that none of the contaminant is drawn into nearby returns and recirculated. A mitigation strategy will be most successful when it is institutionalized as part of normal building operations. Solutions that do not require exotic equipment are more likely to be successful in the long run than approaches that involve unfamiliar concepts or delicately maintained systems. If maintenance or housekeeping procedures or supplies must change as part of the mitigation, it may be necessary to plan for additional staff training, new inspection checklists, or modified purchasing practices. Operating schedules for HVAC equipment may also require modification.

IAQ mitigation strategies that are durable and low-maintenance are more attractive to owners and budding staff than approaches that require frequent adjustment or specialized maintenance skills. New items of equipment should be quiet, energy- efficient, and durable, so that the operators are encouraged to keep them running.

The approach with the lowest initial cost may not be the least expensive over the long run. Other economic considerations include: energy costs for equipment operation; increased staff time for maintenance; differential cost of alternative materials and supplies; and higher hourly rates if odor-producing activities (e.g., cleaning) must be scheduled for unoccupied periods. Although these costs will almost certainly be less than the cost of letting the problem continue, they are more readily identifiable, so an appropriate presentation to management may be required. Any modification to building components or mechanical systems should be designed and installed in keeping with applicable fire, electrical, and other building codes.

Two kinds of criteria can be used to judge the success of an effort to correct an indoor air problem: reduced complaints, and measurement of properties of the indoor air. Reduction or elimination of complaints appears to be a clear indication of success, but that is not necessarily the case. Occupants who see that their concerns are being heard may temporarily stop reporting discomfort or health symptoms, even if the actual cause of their complaints has not been addressed. Lingering complaints may also continue after successful mitigation if people have become upset over the handling of the problem. Ongoing (but reduced) complaints could also indicate that there were multiple IAQ problems and that one or more problems are still unresolved. However, it can be very difficult to use measurements of contaminant levels as a means of determining whether air quality has improved. Concentrations of indoor air pollutants typically vary greatly over time; further, the specific contaminant measured may not be causing the problem. If air samples are taken, readings taken before and after mitigation should be interpreted cautiously. It is important to keep the "before" and "after" conditions as identical as possible, except for the operation of the control strategy. For example, the same HVAC operation, building occupancy and climatic conditions should apply during both measurement periods. "Worst-case" conditions identified during the investigation should be used.

Measurements of airflows, ventilation rates, and air distribution patterns are the more reliable methods of assessing the results of control efforts. Airflow measurements taken during the building investigation can identify areas with poor ventilation; later they can be used to evaluate attempts to improve the ventilation rate, distribution, or direction of flow. Studying air distribution patterns will show whether a mitigation strategy has successfully prevented a contaminant from being transported by airflow.

Solving an indoor air quality problem is a cyclical process of data collection and hypothesis testing. Deeper and more detailed investigation is needed to suggest new hypotheses after any unsuccessful or partially-successful control attempt. Even the best planned investigations and mitigation actions may not produce a resolution to the problem. You may have made a careful investigation, found one or more apparent causes for the problem, and implemented a control system. Nonetheless, your correction strategy may not have caused a noticeable reduction in the concentration of the contaminant or improvement in ventilation rates or efficiency. Worse, the complaints may persist even though you have been successful at improving ventilation and controlling all of the contaminants you could identify.

When you have pursued source control options and have increased ventilation rates and efficiency to the limits of your expertise, you must decide how important it is to pursue the problem further. If you have made several unsuccessful efforts to control a problem, then it may be advisable to seek outside assistance. The problem is probably fairly complex, and it may occur only intermittently or cross the borders that divide traditional fields of knowledge. It is even possible that poor indoor air quality is not the actual cause of the complaints. Bringing in a new perspective at this point can be very effective.

Many IAQ problems are simple to resolve when facility staff have been educated about the investigation process. In other cases, however, a time comes when outside assistance is needed. Professional help might be necessary or desirable in the following situations, as examples:

- Mistakes or delays could have serious consequences (e.g., health hazards, liability exposure, regulatory sanctions).

- Building management feels that an independent investigation would be better received or more effectively documented than an in-house investigation.

- Investigation and mitigation efforts by facility staff have not relieved the IAQ problem.

- Preliminary findings by staff suggest the need for measurements that require specialized equipment and training beyond in-house capabilities.

Aside from private consultants, local, state, or federal government agencies may be able to provide expert assistance or direction in solving IAQ problems. It is particularly important to contact a local or state health department if one suspects that there is a serious building-related illness potentially linked to biological contamination in a building.

If available government agencies do not have personnel with the appropriate skills to assist in solving your IAQ problem, they may be able to direct you to firms in your area with experience in indoor air quality work. Note that even certified professionals from disciplines closely related to IAQ issues (such as industrial hygienists, ventilation engineers, and toxicologists) may not have the specific expertise needed to investigate and resolve indoor air problems. Individuals or groups that offer services should be questioned closely about their related experience and their proposed approach to the problem. As with any hiring process, the better you know your own needs, the easier it will be to select a firm or individual to service those needs. Firms and individuals working in IAQ may come from a variety of disciplines.

Typically, the skills of HVAC engineers and industrial hygienists are useful for this type of investigation, although input from other disciplines such as chemistry, chemical engineering, architecture, microbiology, or medicine may also be important. If problems other than indoor air quality are involved, experts in lighting, acoustic design, interior design, psychology, or other fields may be helpful in resolving occupant complaints about the indoor environment.

Building owners and managers who suspect that they may have a problem with a specific pollutant (such as radon, asbestos, or lead) may be able to obtain assistance from local and state health departments. Government agencies and affected industries have developed training programs for contractors who diagnose or mitigate problems with these particular contaminants.

Firms should be asked to provide references from clients who have received comparable services. In exploring references, it is useful to ask about long-term follow-up. After the contract was completed, did the contractor remain in contact with the client to ensure that problems did not recur?

Consultants being considered should have familiarity with state and local regulations and codes. For example, in making changes to the HVAC system, it is important to conform to local building codes. Heating, cooling, and humidity control needs are different in different geographic regions, and can affect the selection of an appropriate mitigation approach. Getting assurances that all firms under consideration have this knowledge becomes particularly important if it becomes necessary to seek expertise from outside the local area.

If projected costs jump suddenly during the investigation process, the consultants should be able to justify that added cost. The budget will be influenced by a number of factors, including: the complexity of the problem, the size and complexity of the building and its HVAC system(s), the quality and extent of recordkeeping by building staff and management, the type of report or other product required, the number of meetings required (formal presentations can be quite expensive), air sampling (e.g., use of instruments, laboratory analysis) if required.

IAQ MEASUREMENT TECHNIQUES

The following is a brief introduction to making measurements that might be needed in the course of developing an IAQ profile or investigating an IAQ complaint. Emphasis has been placed on the parameters most commonly of interest in non-research studies, highlighting the more practical methods and noting some inappropriate tests to avoid.

Most of the instruments discussed in this section are relatively inexpensive and readily available from many local safety supply companies. Air contaminants of concern in IAQ can be measured by one or more of the methods described below.

Vacuum Pump

A vacuum pump with a known airflow rate draws air through collection devices, such as a filter (catches airborne particles), a *sorbent tube* (which attracts certain chemical vapors to a powder such as carbon), or an *impinger* (bubbles the contaminants through a solution in a test tube). Tests originated for industrial environments typically need to be adjusted to a lower detection limit for IAQ work.

Labs can be asked to report when trace levels of an identifiable contaminant are present below the limit of quantification and detection.

In adapting an industrial hygiene sorbent tube sampling method for IAQ, the investigator must consider at least two important questions.

First: are the emissions to be measured from a product's end use the same as those of concern during manufacturing?

Second: is it necessary to increase the air volume sampled? Such an increase may be needed to detect the presence of contaminants at the low concentrations usually found in non-industrial settings. For example, an investigator might have to increase sampling time from 30 minutes to 5 hours in order to detect a substance at the low concentrations found during IAQ investigations. In cases where standard sampling methods are changed, qualified industrial hygienists and chemists should be consulted to ensure that accuracy and precision remain acceptable.

Direct-reading Meter

Direct-reading meters estimate air concentrations through one of several detection principles. These may report specific chemicals (e.g., CO_2 by infrared light), chemical groups (e.g., certain volatile organics by photoionization potential), or broad pollutant categories (e.g., all respirable particles by scattered light). Detection limits and averaging time developed for industrial use may or may not be appropriate for IAQ.

Detector Tube Kit

Detector tube kits generally include a hand pump that draws a known volume of air through a chemically treated tube intended to react with certain contaminants. The length of color stain resulting in the tube correlates to chemical concentration.

Personal Monitoring Devices

Personal monitoring devices (sometimes referred to as "dosimeters") are carried or worn by individuals and are used to measure that individual's exposure to particular chemical(s). Devices that include a pump are called "active" monitors; devices that do not include a pump are called "passive" monitors. Such devices are currently used for research purposes. It is possible that sometime in the future they may also be helpful in IAQ investigations in public and commercial buildings.

Personal Sampling

The following are OSHA recommended guidelines for air sampling programs normally implemented under the guidance of an industrial hygienist. Unnecessary air sampling can tie up laboratory resources and produce delays in reporting results of necessary sampling. One must evaluate the potential for employee overexposure through observation and screening samples before any partial or full-shift air sampling is conducted. Do not overexpose the employee to gather a sample. Screening with portable monitors, gravimetric sampling, or detector tubes can be used to evaluate the following:

a.   Processes, such as electronic soldering,

b.   Exposures to substances with exceptionally high PELs (Permissible Exposure Limits) in relatively dust-free atmospheres, e.g., ferric oxide and aluminum oxide,

c.   Intermittent processes with substances without STELs (Short Term Exposure Limits),

d.   Engineering controls, work practices, or

e.   To assess the need for personal protection.

The objective of a proper sampling program is to take a sufficient number of samples to obtain a representative estimate of exposure. Contaminant concentrations vary seasonally, with weather, with production levels, and in a single location or job class. The number of samples taken depends on the error of measurement and differences in results. It is important also that if the employer has conducted air sampling and monitoring in the past, a thorough review of the records should be made.

Bulk Samples are often required to assist the laboratory in the proper analysis of field samples. The following are some general sampling procedures:

1.   Screen the sampling area using detector tubes, if appropriate. Determine the appropriate sampling technique. Prepare and calibrate the equipment and prepare the filter media.

2.   Select the employee to be sampled and discuss the purpose of the sampling. Inform the employee when and where the equipment will be removed. Stress the importance of not removing or tampering with the sampling equipment. Turn off or remove sampling pumps before an employee leaves a potentially contaminated area (such as when he/she goes to lunch or on a break).

3.   Instruct the employee to notify the supervisor if the sampler requires temporary removal.

4.   Place the sampling equipment on the employee so that it does not interfere with work performance.

5.   Attach the collection device (filter cassette, charcoal tube, etc.) to the shirt collar or as close as practical to the nose and mouth of the employee). Employee exposure is that exposure which would occur if the employee were not using a respirator. The inlet should always be in

a downward vertical position to avoid gross contamination. Position the excess tubing so as not to interfere with the work of the employee.

6. Turn on the pump and record the starting time.

7. Observe the pump operation for a short time after starting to make sure it is operating correctly.

8. Record the information required.

9. Check pump status every two hours. More frequent checks may be necessary with heavy filter loading. Ensure that the sampler is still assembled properly and that the hose has not become pinched or detached from the cassette or the pump. For filters, observe for symmetrical deposition, fingerprints, or large particles, etc. Record the flow rate, if possible.

10. Periodically monitor the employee throughout the work day to ensure that sample integrity is maintained and cyclical activities and work practices are identified.

11. Take photographs, as appropriate, and detailed notes concerning visible airborne contaminants, work practices, potential interferences, movements, and other conditions to assist in determining appropriate engineering controls.

12. Prepare a blank(s) during the sample period for each type of sample collected. For any given analysis, one blank will suffice for up to 20 samples collected. These blanks may include opened but unused charcoal tubes, and so forth.

13. Before removing the pump at the end of the sample period, check the flow rate to ensure that the rotameter bail is still at the calibrated mark (if there is a pump rotameter). If the ball is no longer at the mark, record the pump rotameter reading.

14. Turn off the pump and record the ending time.

15. Remove the collection device from the pump and seal it as soon as possible. The seal should be attached across sample inlet and outlet so that tampering is not possible.

16. Prepare the samples for transport to laboratory for analysis.

17. Recalibrate pumps after each day of sampling (before charging).

The following are recommended procedures from the OSHA Technical Manual, issued February 5, 1990.

1. Detector Tubes

a.   Each pump should be leak-tested before use. Calibrate the detector tube pump for proper volume at least quarterly or after 100 tubes.

2. Total Dust and Metal Fume

a.   Collect total dust on a pre-weighed, low-ash polyvinyl chloride fitter at a flow rate of about 2 liters per minute (lpm), depending on the rate required to prevent overloading.

b.   Collect metal fumes on a 0.8 micron mixed cellulose ester filter at a flow rate of approximately 1.5 lpm, not to exceed 2.0 lpm. Do not collect metal fumes on a low-ash polyvinyl chloride filter.

c.   Avoid any overloading of the filter, as evidenced by any loose particulate.

d.   Calibrate personal sampling pumps before and after each day of sampling, using a bubble meter method (electronic or mechanical) or the precision rotameter method (that has been calibrated against a bubble meter).

e.   Weigh filters before and after taking the sample.

3. Respirable Dust

a.   Collect respirable silica dust using a clean cyclone equipped with a pre-weighed low-ash polyvinyl chloride filter.

b.   Collect silica only as a respirable dust. A bulk sample should be submitted to the laboratory.

c.   All filters used should be pre-weighed and post-weighed.

d.   Calibration Procedures:(1) Do the calibration at the pressure and temperature where the sampling is to be conducted. (2) Replace the filter with a 1-liter jar containing the cassette holder assembly and cyclone or the open face filter cassette. (3) Connect the tubing from the electronic bubble meter to the inlet of the jar. (4) Connect the tubing from the outlet of the cyclone holder assembly or from the filter cassette to the outlet of the jar and then to the sampling pump. (5) Calibrate the pump. The calibration readings must be within 5% of each other.

e. Cyclone cleaning: (1) Clean the cyclone thoroughly after each use to prevent excess wear or damage and to prevent contamination of a sample. Inspect the cyclone after cleaning for signs of wear or damage, such as scoring. Replace the unit if it appears damaged. (2) Gently clean the interior, avoid scoring the interior surfaces. Never insert anything into the cyclone during cleaning. Refer to Figure 1. (3) Leak test the cyclone at least once a month with regular usage.

4. Organic Vapors and Gases

a. Organic vapors and gases may be collected on activated charcoal, silica gel, or other adsorption tubes using low flow pumps.

b. Immediately before sampling, break off the ends of the charcoal tube so as to provide an opening approximately one-half the internal diameter of the tube. Wear eye protection when breaking ends. Use tube holders, if available, to minimize the hazards of broken glass. Do not use the charging inlet or the exhaust outlet of the pump to break the ends of the charcoal tubes.

c. Use the smaller section of the charcoal tube as a back-up and position it near the sampling pump. The charcoal tube shall be held or attached in an approximately vertical position with the inlet either up or down during sampling.

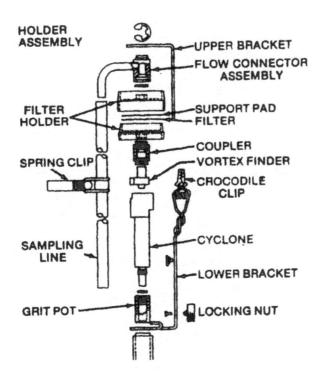

Figure 1. Cyclone (chamber).

d. Draw the air to be sampled directly into the inlet of the charcoal tube and do not allow it to pass through any hose or tubing before entering the pump tubing.

e. Draw the air to be sampled directly into the inlet of the charcoal tube and do not allow it to pass through any hose or tubing before entering the pump tubing.

f. Cap the charcoal tube with the supplied plastic caps immediately after sampling and seal as soon as possible.

g. For other adsorption tubes, follow the same procedures as those for the charcoal tube, with the following exceptions:

(1) Set up the calibration apparatus as shown in Figure 2 replacing the cassette with the solid sorbent tube to be used in the sampling (e.g., charcoal, silica gel, etc.). If a sampling protocol requires the use of two charcoal tubes, then the calibration train must include two charcoal tubes. The air flow must be in the direction of the arrow on the tube.

(2) Calibrate the pump.

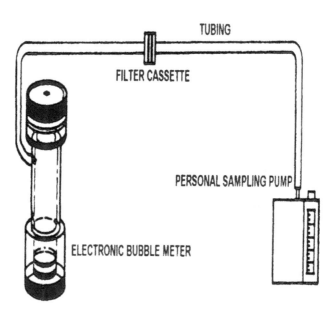

Figure 2. For calibration, the cassette is attached to an electronic bubble meter as shown in the illustration.

5. Midget Impingers/Bubblers

a. Method:

1) Take care in preparing bubblers and impingers to see that frits or tips are not damaged and that joints can be securely tightened.

2) Rinse the impinger/bubbler with the appropriate reagent. Then, add the specified amount of this reagent to the impinger flask. If flasks containing the reagent are transported, caps must be placed on the impinger stem and side arm. To prevent overflowing, do not add over 10 milliliters of liquid to the midget impingers.

3) Collect contaminants in an impinger at a maximum flow rate of 1.0 lpm.

4) The impinger may either be hand-held by the industrial hygienist or attached to the employee's clothing using an impinger holster, in either case, it is very important that the impinger does not tiit, causing the reagent to flow down the side arm to the hose and into the pump.

5) In some instances, it will be necessary to add additional reagent during the sampling period to prevent the amount of reagent from dropping below one-half of the original amount.

6) After sampling, remove the glass stopper and stem from the impinger flask.

7) Rinse the absorbing solution adhering to the outside and inside of the stem directly into the impinger flask with a small amount (1 or 2 ml.) of the sampling reagent. Stopper the flask tightly with the plastic cap provided or pour the contents of the flask into a 20 cc. glass bottle. Rinse the flask with a small amount of the reagent and pour the rinse solution into the bottle. Tape the cap shut to prevent it from coming loose due to vibration. If electrial tape is used, do not "stretch" the tape since it will contract and loosen the cap.

b. Calibration Procedure:

1) Set up the calibration apparatus, replacing the cassette with the impinger/bubbler lied with the amount of liquid reagent specified in the sampling method.

2) Connect the tubing from the electronic bubble meter to the inset of the impinger/bubbler.

3) Connect the outlet of the impinger/bubbler to the tubing to the pump.

4) Calibrate the pump at a maximum flow rate of 1.0 lpm.

6. Vapor Badges

a. Passive diffusion sorbent badges are useful for screening and monitoring certain chemical exposures, especially vapors and gases.

b. Badges are available to detect mercury, nitrogen oxides, ethylene oxide, formaldehyde, etc.

c. Interfering substances should be noted.

For asbestos sampling:

a. Collect asbestos on a special , 0.8 micrometer pore size, 25 mm diameter mixed cellulose ester filter, using a back-up pad.

b. Use fully conductive cassette with conductive extension cowl.

c. Sample open face in worker's breathing zone.

d. Assure that the bottom point (between the extension and the conical black piece) of the cassette is sealed tightly with a shrink band of electrical tape. Point the open end of the cassette down to minimize contamination.

e. Use a flow rate in the range of 0.5 to 2.5 liters per minute. One liter per minute is suggested for general sampling. Office environments allow flow rates of up to 2.5 lpm. Calibrate pump before and after sampling.

f. Sample for as long a time as possible without overloading (obscuring) the filter.

g. Submit at the most, 10 blanks, with a minimum in all cases of 2 blanks. Where possible, collect and submit to the laboratory a bulk sample of the material suspected to be in the air.

h. Mail bulks and air samples separately to avoid cross-contamination. Pack the samples securely to avoid any rattle or shock damage (do not use expanded polystyrene "packing"). Use bubble sheeting as packing. Put identifying paperwork in every package. Do not send samples in plastic bags or in envelopes.

i. Instruct the employee to avoid knocking the cassette and to avoid using a compressed air source that might dislodge the sample.

Equipment Preparation and Calibration

1. Replace alkaline batteries frequently (once a month). Also carry fresh replacement batteries with the equipment.

2. Check the rechargeable Ni-Cad batteries under load (e.g., turn pump on and check voltage at charging jack) before use.

3. Calibrate personal sampling pumps before and after each day of sampling, using either the electronic bubble meter method or the precision rotameter method (that has been calibrated against a bubble meter).

4. Electronic Flow Calibrators:

a. These units are high accuracy electronic bubble flowmeters that provide instantaneous air flow readings and a cumulative averaging of multiple samples. These calibrators measure the flow rate of gases and present the results as volume per unit of time.

b. These calibrators should be used to calibrate all air sampling pumps.

5. When a sampling train requires an unusual combination of sampling media (e.g., glass fiber filter proceeding impinger), the same media/devices should be in line during calibration.

The electronic bubble meter method consists of the following:

(1) Allow the pump to run 5 minutes prior to voltage check.

(2) Assemble the polystyrene cassette filter holder, using the appropriate filter for the sampling method. Compress cassette by using a mechanical press or other means of applying pressure. Use shrink tape around cassette to cover joints and prevent leakage. If a cassette adapter is used, care should be taken to ensure that it does not come in contact with the back-up pad. When calibrating with a bubble meter, the use of cassette adapters can cause moderate to severe pressure drop at high flow rates in the sampling train, which will affect the calibration result, If adapters are used for sampling, then they should be used when calibrating. Nylon adapters can restrict air flow due to plugging over time. Stainless steel adapters are preferred.

(3) Connect the collection device, tubing, pump and calibration apparatus .

(4) A visual inspection should be made of all Tygon tubing connections.

(5) Wet the inside of the electronic flow cell with the supplied soap solution by pushing on the button several times.

(6) Turn on the pump and adjust the pump rotameter, if available, to the appropriate flow rate setting.

(7) Press the button on the electronic bubble meter. Visually capture a single bubble and electronically time the bubble. The accompanying printer will automatically record the calibration reading in liters per minute.

(8) Repeat step 7 until two readings are within 5%.

(9) While the pump is still running, adjust the pump, if necessary.

(10) Repeat the procedures described above for all pumps to be used for sampling. The same cassette and filter may be used for all calibrations involving the same sampling method.

The precision rotameter is a secondary calibration device. If it is to be used in place of a primary device such as a bubble meter, care must be taken to ensure that any introduced error will be minimal and noted. The precision rotameter may be used for calibrating the personal sampling pump in lieu of a bubble meter provided it is: (a) Calibrated with an electronic bubble meter or a bubble meter. (b) Disassembled, cleaned as necessary, and recalibrated. It should be used with care to avoid dirt and dust contamination which may affect the flow. (c) Not used at substantially different temperature and/or pressure from those conditions present when the rotameter was calibrated against the primary source. (d) Used such that pressure drop across it is minimal. If altitude or temperature at the sampling site are substantially different from the calibration site, it is necessary to calibrate the precision rotameter at the sampling site where the same conditions are present.

Filter Weighing Procedure: The step-by-step procedure for weighing filters depends on the make and model of the balance. Consult the manufacturer's instruction book for directions. In addition, follow these guidelines:

1. There shall be no smoking or eating in the weighing area. All filters will be handled with tongs or tweezers. Do not handle the filters with bare hands.
2. Dessicate all filters at least 24 hours before weighing and sampling. Change deSsicant before it completely changes color (e.g., before blue dessicant turns all pink). Evacuate dessicator with a sampling or vacuum pump.
3. Zero the balance prior to use.
4. Calibrate the balance prior to use and after every 10 samples.
5. Immediately prior to placement on the balance, pass all filters over an ionization unit after 12 months of use to the distributor for disposal.
6. Weigh all fitters at least twice.
    a. If there is more than 0.005 milligram difference in the two weighings, repeat the zero and calibration and reweigh the filter.
    b. If there is less than 0.005 milligram difference in the two weighings, average the weights for the final weight.
7. Record all the appropriate weighing information in the weighing log.
8. In reassembling the cassette assembly, remember to add the unweighed backup pad.
9. When weighing the filter after sampling, include any loose material from an overloaded filter and cassette. At all times take care not to exert downward pressure on the weighing pans. Such action may damage the weighing mechanism.

Detector Tubes/Pumps: Detector tube pumps are portable equipment which, when used with a variety of commercially available detector tubes, are capable of measuring the concentrations of a wide variety of compounds in industrial atmospheres. Operation consists of using the pump to draw a known volume of air through a detector tube designed to measure the concentration of the substance of interest. The concentration is determined by a colorimetric change of an indicator which is present in the tube contents.

Detector tubes/pumps are screening instruments which may be used to measure hundreds of organic and inorganic gases and vapors or for leak detection. Some aerosols can also be determined. Detector tubes of a given brand are to be used only with a pump of the same brand.

The tubes are calibrated specifically for the same brand of pump and may give erroneous results if used with a pump of another brand. A limitation of many detector tubes is the lack of specificity. Many indicators are not highly selective and can cross-react with other compounds. Manufacturers' manuals describe the effects of interfering contaminants. Another important consideration is sampling time. Detector tubes give only an instantaneous interpretation of environmental hazards. This may be beneficial in potentially dangerous situations or when celling exposure determinations are sufficient. When long-term assessment of occupational environments is necessary, short-term detector tube measurements may not reflect time-weighted average levels of the hazardous substances present. Detector tubes normally have a shelf-life at 25°C of 1 to 2 years. Refrigeration during storage lengthens the shelf-life. Outdated detector tubes (i.e., beyond the printed expiration date) should never be used.

Detectable concentration ranges are tube-dependent and can be anywhere from one-hundredth to several thousand ppm. The limits of detection depend on the particular detector tube. Accuracy ranges vary with each detector tube. The pump may be handheld during operation (weighing from 8 to 11 ounces), or it may be an automatic type (weighing about 4 pounds) which collects a sample using a preset number of pump strokes.

A full pump stroke for either type of short-term pump has a volume of about 100 cc. In most cases where only one pump stroke is required, sampling time is about one minute. Determinations for which more pump strokes are required take proportionately longer.

Each day prior to use, perform a pump leakage test by inserting an unopened detector tube into the pump and attempt to draw in 100 ml of air. After a few minutes, check for pump leakage by examining pump compression for bellows-type pumps or return to resting position for piston-type pumps. Automatic pumps should be tested according to the manufacturer's instructions. In the event of leakage which cannot be repaired in the field, send the pump to the manufacturer. Record that the leakage test data.

Calibrate the detector tube pump for proper volume measurement at least quarterly. Simply connect the pump directly to the bubble meter with a detector tube in-line. Use a detector tube and pump from the same manufacturer. Wet the inside of the 100 cc bubble meter with soap solution. For volume calibration, experiment to get the soap bubble even with the zero ml mark of the buret. For piston-type pumps, pull the pump handle all the way out (full pump stroke) and note where the soap bubble stops; for bellows-type pumps, compress the bellows fully; for automatic pumps, program the pump to take a full pump stroke.

For either type pump, the bubble should stop between the 95 cc and 105 cc marks. Allow 4 minutes for the pump to draw the full amount of air. Also check the volume for 50 cc (1/2 pump stroke) and 25 cc (1/4 pump stroke) if pertinent. A +5 percent error is permissible. If error is greater than +8 percent, send the pump for repair and recalibration. Record the calibration information required on a calibration log. It may be necessary to clean or replace the rubber bung or tube holder if a large number of tubes have been taken with the pump.

Draeger, Model 31 (bellows): When checking the pump for leaks with an unopened tube. the bellows should not be completely expanded after 10 minutes. Draeger, Quantimeter 1000. Model 1 (automatic): A battery pack is an integral part of this pump. The pack must be charged prior to initial use. One charge is good for 1000 pump strokes. During heavy use, it should be recharged daily. If a "U" (undervoltage) message is continuously displayed in the readout window of this pump, the battery pack should be immediately recharged. Matheson-Kitagawa, Model 8014-400A (piston): When checking the pump for leaks with an unopened tube, the pump handle should be pulled back to the 100-ml mark and locked. After 2 minutes, the handle should be released carefully. After taking 100 to 200 samples, the pump should be cleaned and relubricated. This involves removing the piston from the cylinder, removing the inlet and pressure-relief valve from the front end of the pump, cleaning, and relubricating. Mine Safety Appliances, Samplair Pump, Model A, Part No. 463998 (piston): The pump contains a flow-rate control orifice protected by a plastic filter which periodically needs to be cleaned or replaced. To check the flow rate, the pump is connected to a buret and the piston is withdrawn to the 100 ml position with no tube in the tube holder. After 24-26 seconds, 80 ml of air should be admitted to the pump. Every 6 months the piston should be relubricated with the oil provided. Sensidyne-Gastec, Model 800, Part No. 7010657-1 (piston): This pump can be checked for leaks as mentioned for the Kitagawa pump; however, the handle should be released after 1 minute. Periodic relubrication of the pump head, the piston gasket, and the piston check valve is needed and is use-dependent.

Detector tubes should be refrigerated when not in use to prolong shelf life. They should not be used when cold. They should be kept at room temperature or in a shirt pocket for one hour prior to use. Lubrication of the piston pump may be required if volume error is greater than 5 percent.

Electronic Flow Calibrators: These units are high accuracy electronic bubble flowmeters that provide instantaneous air flow readings and a cumulative averaging of multiple samples. These calibrators measure the flow rate of gases and report volume per unit of time. The timer is capable of detecting a soap film at 80 microsecond intervals. This speed allows under steady flow conditions an accuracy of +/- 0.5% of any display reading. Repeatability is +/- 0.5% of any display. The range with different cells is from 1 cc/min to 30 lpm. Battery power will last 8 hours with continuous use. Charge for 16 hours. Can be operated from an A/C charger.

Manual Buret Bubble Meter Technique: When a sampling train requires an unusual combination of sampling media (e.g., glass fiber filter proceeding impinger), The same media/devices should be in line during calibration. Calibrate personal sampling pumps before and after each day of sampling. Allow the pump to run 5 minutes prior to voltage check and calibration. Assemble the polystyrene cassette filter holder using the appropriate filter for the sampling method.

If a cassette adapter is used, care should be taken to ensure that it does not come in contact with the back-up pad. When calibrating with a bubble meter, the use of cassette adapters can cause moderate to severe pressure drop in the sampling train, which will affect the calibration result. If adapters are used for sampling, then they should be used when calibrating. Connect the collection device, tubing, pump and calibration apparatus. A visual inspection should be made of all Tygon tubing connections. Wet the inside of a 1-liter buret with a soap solution. Turn on the pump and adjust the pump rotameter to the appropriate flow rate setting. Momentarily submerge the opening of the buret in order to capture a film of soap.

Draw two or three bubbles up the buret in order to ensure that the bubbles will complete their run. Visually capture a single bubble and time the bubble from 0 to 1000 ml for high flow pumps or 0 to 100 ml for low flow pumps. The timing accuracy must be within +1 second of the time corresponding to the desired flow rate. If the time is not within the range of accuracy, adjust the flow rate and repeat the last two steps until the correct flow rate is achieved. While the pump is still running, mark the center of the float in the pump rotameter as a reference. Repeat the procedures described above for all pumps to be used for sampling. The same cassette and filter may be used for all calibrations involving the same sampling method.

Sampling for Special Analyses

Air Samples: Respirable dust samples are analyzed for quartz and cristobalite by x-ray diffraction (XRD). XRD is the preferred analytical method due to its sensitivity, minimum requirements for sample preparation and ability to identify polymorphs (different crystalline forms) of free silica.

a. The analysis of free silica by XRD requires that the particle size distribution of the samples be matched as closely as possible to the standards. This is best accomplished by collecting a respirable sample. (1) Respirable dust samples are collected on a low ash PVC filter using a 10 mm nylon cyclone at a flow rate of 1.7 lpm. (2) A sample not collected in this manner is considered a total dust (or nonrespirable) sample. Total dust samples do not allow for an accurate analysis by XRD.

b. Quartz (or cristobalite) is identified by its major (primary) X-ray diffraction peak. Because other substances also have peaks at the same position, it is necessary to confirm quartz (or cristobalite) principally by the presence of secondary and/or tertiary peaks.

c. If they are considered to be present in the work environment, the following major chemicals which can interfere with an analysis should be noted: Aluminum phosphate, Feldspars (microcline, orthoclase, piagiodase), Graphite, Iron carbide, Lead sulfate, Micas (biotite, muscovite), Potash, Silver chloride, Talc, Zircon (Zirconium silicate).

d. A sample weight and total air volume shall accompany all filter samples. Sample weights of 0.1 to 5.0 milligrams are acceptable. Sample weights of 0.5 to 3.0 milligrams are preferred. (1) Do not submit a sample(s) unless its weight or the combined weights of ail filter's representing an individual exposure exceed 0.1 mg. (2) If heavy sample loading is noted during the sampling period, it is recommended that the filter cassette be changed to avoid collecting a sample with a weight greater than 5.0 milligrams. If a sample weight exceeds 5.0 mg, another sample of a smaller air volume, whenever possible, should be collected to obtain a sample weight of less than 5.0 mg. Laboratory results for air samples are usually reported under one of four categories: (1) Percent Quartz (or Cristobalite). Applicable for a respirable sample in which the amount of quartz (or cristobalite) in the sample was confirmed. (2)"Less than or equal to" value in units of %. Less or equal to values are used when the adjusted 8-hour exposure is found to be less than the PEL, based on the sample's primary diffraction peak. The value reported represents the maximum amount of quartz (or cristobalite) which could be present. However, the presence of quartz (or cristobalite) was not confirmed using secondary and/or tertiary peaks in the sample since the sample could not be in violation of the PEL. (3) Approximate Values in Units of Percent: The particle size distribution in a total dust sample is unknown and error in the XRD analysis may be greater than for respirable samples. Therefore, for total dust samples, an approximate result is given. (4) Nondetected: A sample reported as nondetected indicates that the quantity of quartz (or cristobalite) present in the sample is not greater than the detection limit of the instrument. The detection limit is usually 10 micrograms for quartz and 50 micrograms for cristobalite. If less than a full-shift sample was collected, one should evaluate a nondetected result to determine whether adequate sampling was performed. If the presence of quartz (or cristobalite) is suspected in this case, the Industrial Hygienist may want to sample for a longer period of time to increase the sample weights.

Bulk Samples: Bulk samples must be submitted for all silica analyses. They have two purposes: (1) For laboratory use only, to confirm the presence of quartz or cristobalite in respirable samples, or to assess the presence of other substances that may interfer in the analysis of respirable samples. (2) To determine the approximate percentage of quartz (or cristobalite) in the bulk sample. A bulk sample submitted "for laboratory use only" must be representative of the airborne free silica content of the work environment sampled: otherwise it will be of no value. The order of preference for an evaluation is:

- A high volume respirable area sample.
- A high volume area sample.
- A representative settled dust (rafter) sample.
- A bulk sample of the raw material used in the manufacturing process.

A bulk sample is the last choice and the least desirable. It should be submitted "for laboratory use only" if there is a possibility of contamination by other matter. The type of bulk sample submitted to the laboratory should be cross-reference to the appropriate air samples. A reported bulk sample analysis for quartz (or cristobalite) will be semi-quantitative in nature because: (1) The XRD analysis procedure requires a thin layer deposition for an accurate analysis. (2) The error for bulk samples analyzed by XRD is unknown because the particle size of nonrespirable bulk samples varies from sample to sample.

Samples Analyzed by Inductively Coupled Plasma (ICP): **Metals** — Where two or more of the following analytes are requested on the same filter, an ICP analysis may be conducted. However, the Industrial Hygienist should specify the metals of interest in the event samples cannot be analyzed by the ICP method. A computer print-out of the following 13 analytes may be typically reported:

Antimony, Beryllium, Cadmium, Chromium, Cobalt, Copper, Iron, Lead, Manganese, Molybdenum, Nickel, Vanadium, Zinc. **Arsenic** — Lead, cadmium, copper and iron can be analyzed on the same filter with arsenic.

Sampling for Surface Contamination

The terms "wipe sampling," "swipe sampling," and "smear sampling" are all used synonymously to describe the techniques used for assessing surface contamination. "Wipe sampling" is most often used to screen for asbestos, lead, other metals, and PCBs. The uses are:

Skin Sampling: Potential contact with skin irritants may be evaluated by wiping surfaces, which may be touched by workers. Skin wipes are not recommended for those substances which absorb rapidly through the skin. Biological monitoring for these substances or their metabolites, or biological markers, is often the only means of assessing their absorption. Wipe the inside surfaces of protective gear or other surfaces which may contact skin, instead.

Surfaces: Surfaces which may be contacted by food or other materials which are ingested or placed in the mouth (e.g., chewing tobacco, gum, cigarettes) may be wipe sampled (including hands and fingers) to show contamination. Contaminated smoking materials may allow the toxic materials, or their combustion products, to enter the body via the lungs (e.g., lead, mercury). Wiping of surfaces which smoking materials may touch (e.g., hands and fingers) may be useful in evaluating this possible route of exposure. Accumulated toxic materials may become suspended in air, and may contribute to airborne exposures (e.g., asbestos, lead or beryllium). Bulk and wipe samples may aid in determining this possibility.

Personal Protective Equipment Sampling: Effectiveness of personal protective gear (e.g., gloves, aprons, respirators, etc.) may sometimes be evaluated by wipe sampling the inner surfaces of the protective gear (and protected skin). Effectiveness of decontamination of surfaces and protective gear (e.g., respirators) may sometimes be evaluated by wipe sampling. When accompanied by dose observation of the operation in question, wipe sampling can help identify sources of contamination and poor work practices.

The following describes techniques applicable to wipe sampling for contaminants. Remember that these methods can be modified and adapted to the particular needs of the investigation and work environment.

Technique for Wipe Sampling

Direct skin wipes should not be taken when high skin absorption of a substance is expected. Under no conditions should any solvent other than distilled water be used on skin, personal protective gear which directly contacts the skin, or surfaces which contact food or tobacco products. Generally, there are two types of filters recommended for taking wipe samples: (1) Glass fiber filters (GPF) (37 mm) are usually used for materials which are analyzed by High Performance Liquid Chromatography (HPLC), and often for substances analyzed by Gas Chromatogaphy (GC). (2) Paper filters are generally used for metals, and may be used for anything not analyzed by HPLC. For convenient usage, the Whatman smear tab (or its equivalent) is commonly used. Preloading a group of vials with appropriate filters is a convenient method. (The Whatman smear tabs should be inserted with the tab end out.) Always wear clean rubber gloves when handling filters. Gloves should be disposable and should not be powdered. Follow these procedures when wipe samples are taken:

a. If multiple samples are to be taken at the worksite, prepare a rough sketch of the area(s) or room(s) which are to be wipe sampled.
b. A new set of clean impervious gloves should be used with each individual sample. This avoids contamination of the filter by the hand and the subsequent possibility for false positives, and prevents contact with the substance.
c. Withdraw the filter from the vial. If a damp wipe sample is desired, moisten the filter with distilled water or other advent. Skin, personal protective equipment or surfaces which contact food or tobacco products must either be wiped dry, or wiped with distilled water, never with organic solvents. Skin wipes should not be done for materials with high skin absorption. It is recommended that hands and fingers be the only skin surfaces wiped. Before any skin wipe is taken, explain why you want the sample and ask the employee about possible skin allergies to the chemicals in the sampling filter or media.
d. Wipe a section of the surface to be sampled using a template with an opening exactly 100 cm^2.
e. For surfaces smaller than 100 cm^2 use a template of the largest size possible. Be sure to document the size of the area wiped. For curved surfaces, the wiped area should be estimated as accurately as possible and then documented.
f. Maximum pressure should be applied when wiping. To insure that all portions of the partitioned area are wiped, start at the outside edge and

progress toward the center making concentric squares of decreasing size.

g. If the filter dries out during the wiping procedure, discard the filter, reduce area to be wiped by half, and repeat wiping procedure with a new filter.

h. Without allowing the filter to contact any other surface, fold the filter with the exposed side in, then fold it over again. Place the filter in a sample vial, cap the vial, number it, and place a corresponding number at the sample location on a sketch. Include notes with the sketch giving any further description of the sample (e.g., "Fred- Employee's respirator, Inside"; "Lunch table"; etc.).

i. At least one blank filter treated in the same fashion, but without wiping, should be submitted for each sampled area.

Due to their volatile nature, most organic solvents are not suitable for wipes. If necessary, surface contamination can be judged by other means, (e.g., by use of detector tubes, photoionization analyzers, or other similar instruments).

Some substances are not stable enough as samples to be wipe sampled reliably. Some substances should have solvent added to the vial as soon as the wipe sample is placed in the vial (e.g., Benzidine).

Do not take surface wipe samples on skin if OSHA or ACGIH shows a "skin" notation, the substance has a skin LD_{50} of 200 mg/kg or less, or an acute oral LD_{50} of 500 mg/kg or less, or the substance is an irritant, causes dermatitis, contact sensitization, or is termed corrosive.

Screening for Carcinogenic Aromatic Amines: As in the case of routine wipe sampling, wear clean, disposable impervious gloves. Wipe off exactly 100 cm² with a sheet of filter paper moistened in the center with 5 drops of methanol. After wiping the sample area, apply 3 drops of fluorescamine (a visualization reagent to the contaminated area of the filter paper. Place a drop of the visualization reagent on an area of the filter paper which has not contacted the surface. This marks a non-sample area or blank on the fitter paper adjacent to the test area.

After a reaction time of 6 minutes, irradiate the fitter paper with 366 nm ultraviolet light. Compare the color development of the contacted area with the non-sample area or blank. A positive reaction will show a discoloration as a yellow color darker than the yellow color of the fluoroescamine blank. A discoloration indicates surface contamination, possible aromatic amine carcinogen. Repeat a wipe sampling of the contaminated areas using the regular surface contamination procedure.

The following compounds are some of the suspected carcinogenic agents that can be detected by this screening procedure: Benzidine, α-Naphthylamine, β-Naphthylamine, and 4-Aminobiphenyl.

Air Sampling and Personal Safety Monitoring

On and off the job, everyone is exposed to a great variety of chemical and physical agents, most of which do no harm under ordinary circumstances, but all of which have the potential for being injurious at some level and under some conditions of exposure. How a material is used is the major determinant of the hazard potential. Any substance contacting or entering the body can be injurious at some degree of exposure and will be tolerated without effect at some lower exposure. The practice of industrial hygiene or environmental health is based on the concept that for each substance there is a safe or tolerable lower level of exposure below which significant injury, illness or discomfort will not occur. The industrial hygienist protects the health of the worker by determining this safe limit of exposure for a substance and then controlling the environmental conditions so that exposure does not exceed that limit. The toxicity, or hazard properties, of a chemical refer generally to the capacity of the substance to injure an individual. Frequently the word poison is used to mean a substance with some capability of producing adverse reaction on the health or well-being of an individual. Whether or not any ill effects occur depends on: (1) the properties of the chemical, (2) the dose (the amount of the chemical acting on the body or system), (3) the route by which the substance enters the body, and (4) the susceptibility or resistance of the exposed individual. There are four routes of entry or means by which a substance may enter or act on the body: (1) inhalation, (2) ingestion, (3) injection and (4) contact or absorption through the skin. Of these, inhalation is the most important insofar as serious and acute industrial poisoning is concerned, but contact of the skin with corrosive or irritating chemicals is the most frequently encountered. Ingestion of toxic materials occurs only through accidental or careless procedures in the industrial environment and, while it cannot be ignored, it is seldom a significant factor in exposure. However, there are multiple routes of entry to the body for some materials. When a toxic chemical acts on the body or system, the nature and extent of the injurious response depends upon the dose received, that is, the amount of the chemical actually entering the body or system. This relationship of dose and response is shown in Figure 3. The dose-response curve varies with the type of material and the response.

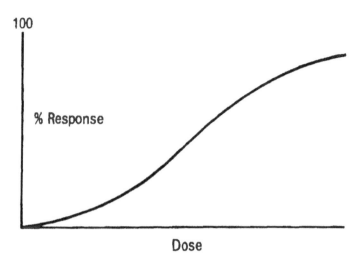

Figure 3. The dose-response curve.

Typically, there would be no response up to a certain dose, then a gradually increasing response to an increasing dose. At the upper part of the curve, the increase in response to an increase in dose would diminish, and would gradually approach 100% response in all exposed animals or individuals. However, no such curve is precise, because there is great variation in the susceptibility or resistance of individuals to a fixed dose of any material and, in addition, the biological response covers a rather wide range of effects for any given dose.

In experimental work, some accidental exposures, or in the administration of medicine, the dose may be a certain quantity of the chemical administered at one time, such as in a pill, an injection, or an accidentally swallowed poison. In industry, time is a factor in most exposures, and the dose is the result of both the concentration of the toxic agent and the duration of the exposure.

In exposures by inhalation of airborne materials, the dose is the concentration multiplied by the time (CT), and is roughly a constant for any given material and specified effect. The CT value can be used to provide a rough approximation of other combinations of concentration and time which would have about the same effect. Although this concept must be used very cautiously and cannot be applied at extreme conditions of either concentration or time, it is most important in setting limits for airborne contaminants and physical agents in respect to environmental exposures. The worker is exposed for various periods of time, day after day, to the materials in his environment, and the safe limits are set so that the combinations of concentrations and durations are below the levels which will produce injury.

Over the years, various individuals proposed different limits, and some states, as well as, and the American Conference of Governmental Industrial Hygienists (ACGIH) began to develop limits or standards. In 1945, W. A. Cook compiled a list of concentration limits for 150 substances. The ACGIH adopted this list and developed an active program which continues to this day.

In the early development of such limits, they were generally known as Maximum Allowable Concentrations or MACs, sometimes called Maximum Acceptable Concentrations, or Maximum Permissible Concentrations.

The early concept was that these were values which must not be exceeded; in other words, they were truly maximum values. As the understanding of limits and the development of the philosophy for such limits grew, it was realized that short-term exposures to somewhat higher concentrations could be permitted without undue harm if the total exposure during the day was sufficiently below the maximum limit. This lead to the development of the concept of Threshold Limit Values (TLVs) by the ACGIH.

The TLVs, as recommended and published by the ACGIH, refer to concentrations of airborne contaminants or levels of physical agents, and represent the conditions to which it is believed nearly all workers may be repeatedly exposed day after day without adverse effects. TLVs are based on the results of animal experiments, limited human experiments, some industrial experience and, when possible, a combination of all three.

The basis on which the TLVs are set may differ from substance to substance. For some, such as levels for silica dust, a guiding factor is protection against impairment of health. For others, it is the comfort level of the individual, such as freedom from irritation, nuisance, or other forms of stress; for example, the TLV for sulfur dioxide is based on irritation and not on toxicity per se. The TLV list is reviewed annually resulting in some revisions in values and some additions to the list.

It is most important that TLV data be correctly used. Misuse can occur when uninformed individuals view these levels as magic numbers, below which workers are safe and above which they become ill. It should be remembered that there is wide variation in individual susceptibility to air contaminants and physical agents. Some workers may experience some discomfort from exposures at or below the TLV, and a much smaller number may be affected more seriously by aggravation of a pre-existing condition or by development of an occupational illness. Therefore, the TLVs as published were intended to be used only as guides in the control of health hazards and not as levels which separate safe from dangerous exposures. In addition, a TLV is not intended as a relative index of hazard or toxicity, nor to be mathematically manipulated by applying physical constants to derive a relative hazard. Basically a TLV refers to a time-weighted averaged exposure for a 7 or 8-hour work day and a 40-hour work week. In other words, it is a level directed toward chronic (long-term) exposure and not toward acute (short-term) exposure. Generally, toxicity data for acute exposures is obtainable largely from animal experimentation, early medical data, and limited information from accidental exposures. In the use of TLVs, it is important to recognize that the levels are generally developed for normal individuals doing normal work. Under conditions of high heat, unusual humidity, heavy exertion, abnormal pressure, or other work factors which may place added stress on the body, the effects from exposure to an air contaminant at its TLV may be altered. Generally, most of these stresses act adversely to increase the toxic response to a substance, and proper downward adjustment of the level should be made.

The TLVs for airborne contaminants are based on the premise that although all chemical substances are toxic at some concentration for some period of time, a concentration exists for all substances from which no toxicity may be expected no matter how often the exposure is repeated. A similar premise holds for substances producing irritation, discomfort and nuisance. In using these limits, items such as excursion factors, ceiling values, "skin" notations, mixtures of substances, and inert material should be considered. These factors are discussed below.

Excursion Factors: Most TLVs refer to time-weighted average exposures for an 8-hour work day and a 40-hour work week. However, in calculating time-weighted average exposure, excursions above the limit are permitted provided they are compensated for by equivalent excursions below the limit during the same work day. The question here is: "How much of a fluctuation above the limit is permissible in developing the average?" These fluctuations above the limits are related to the magnitude of the TLV for the particular substance (refer to Table 1).

Table 1. Permissible Excursions for Time-Weighted Average Limits

TLV (ppm or mg/m^3)	Excursion factor	TLV (ppm or mg/m^3)	Excursion factor
0-1	3	10-100	1.5
1-10	2	100-1000	1.25

Note: These excursions are for a duration of only 15 min or less.

Following are two examples of the use of excursion factors:

Example 1: Carbon monoxide has a TLV of 50 ppm. Therefore, the maximum concentration permitted for a short time would be 75 ppm (50 ppm x 1.5 = 75 ppm).

Example 2: Lead has a TLV of 0.2 mg/m³. Therefore, the maximum concentration permitted for a short time would be 0.6 mg/m³ (0.2 mg/m³ x 3 = 0.6 mg/m³).

The limiting excursion factors should be considered "rule-of-thumb" guidelines for listed substances, but are not appropriate for all materials (such as those designated "C").

Ceiling "C" Values: "C" designations following the names of some substances refer to a ceiling value which should not be exceeded for that substance for any period of time. In other words, the time-weighted average exposure should fluctuate below the C value. Generally, C values are assigned to substances whose action is chiefly irritation, narcosis, or productive of serious long-term effects from a single or a few peak exposures. These are usually fast-acting substances whose TLV is more appropriately based on a ceiling value than on a time-weighted average which allows excursions above listed values.

"Skin" Notation: The designation "skin" refers to the potential contribution to the overall exposure by the cutaneous route, including mucous membranes and eyes, either by airborne, or more particularly by direct, contact with the substance. Examples of such substances are phenol (cresol and cumene), hydrogen cyanide, and mercury. The "skin" notation is intended to make known the need to prevent cutaneous absorption so that the TLV is not violated.

Mixtures: Special consideration should be given to the application of TLVs in assessing health hazards which may be associated with mixtures of two or more substances.

Generally, when two or more hazardous materials are present, their combined effect rather than either individual effect should be considered. In other words, the effects of the different hazards in a mixture should be considered additive.

An exception may be made when there is good reason to believe that the chief effects of the different harmful substances are not in fact additive, but independent, such as a combined exposure to silica dust and lead dust. In the determination of whether or not the additive effects are excessive, the following formula should be used:

$$C = C_1/T_1 + C_2/T_2 + \cdots C_n/T_n$$

where C indicates the observed concentration and T the corresponding TLV. If the sum of the fractions exceeds unity, then the mixture should be considered as being excessive.

"Inert" or Nuisance Particulates: Some materials may be classified as "inert" or nuisance particulates. Tables 2 gives some examples. Generally, these materials have a long history of little adverse effect on the lungs and do not produce significant organic disease or toxic effect when exposures are kept under reasonable control. However, excessive concentrations of "inert" or nuisance particulates in the work air may seriously reduce visibility (iron oxide), may cause unpleasant deposits in the eyes, ears and upper respiratory passages (e.g., Portland cement and limestone dust), or may cause injury to the skin or mucous membranes during rigorous cleaning or mechanical action. To control these effects, a TLV for particulates with less than 1% free silica has been set at 10 mg/m^3 or 30 ppm cf (whichever is less) of total dust. The mg/m^3 refers to a gravimetric standard and the mppcf to a particle count standard. The limits do not apply to brief exposures to higher concentrations during a normal work day, nor do they apply to substances which may cause physiologic impairment at lower concentrations, for which no TLV has been indicated. Time-weighted average exposures for an 8-hour work shift are calculated by the following formula (where E = time-weighted average exposure, C = concentration during any period of time where the concentration remains constant, and T = duration, hrs.):

$$E = \frac{(C_1 T_1) + (C_2 T_2) + ... (C_n T_n)}{Daily\ hours\ of\ work}$$

Table 2. Typical "Inert" or Nuisance Particulates

Alundum (Al_2O_3)	Glass, fiber/dust	Kaolin	Pentaerythritol
Calcium carbonate	Glycerin Mist	Limestone	Plaster of Paris
Cellulose	Graphite	Magnesite	Sawdust
Portland Cement	Gypsum	Marble	Rouge
Corundum (Al_2O_3)	Vegetable oil mists (except castor cashew nut, or similar irritant oils)	Sucrose, Tin Oxide, Titanium Dioxide	Silicon Carbide

As an example, several air samples collected during an 8-hour work shift for benzene (8-hour time-weighted average = 10 ppm) were 3 hours at 4 ppm, 1 hour at 15 ppm, 2 hours at 6 ppm and 2 hours at 9 ppm. The employee's time-weighted average exposure would be:

$$\frac{(3 \, hrs \times 4 \, ppm) + (1 \times 15) + (2 \times 6) + (2 \times 9)}{8 \, hours} =$$

$$\frac{12 + 15 + 12 + 18}{8} = \frac{57}{8} = 7 \, ppm$$

This exposure is not excessive since 7 ppm is below the 8-hour time-weighted average of 10 ppm. Mixtures of air contaminants are additive and calculated in the manner described previously; that is, the fractions (concentration divided by limit for each material) are added together and if the sum does not exceed unity, the exposure is not excessive.

The primary purpose of monitoring the air in the work environment is to determine the level of employee exposure to airborne contaminants and to protect his health. Generally, where employees may be overexposed to potential health hazards, such sampling or measurements is performed on a routine basis. In addition to the above, sampling for air contaminants may be performed for one or more of the following reasons: (1) to determine the magnitude of employee exposure at the start-up of a new process or a change in a process or material used; (2) to determine the justification of employee grievances concerning an alleged health hazard; (3) to determine the performance or effectiveness of engineering control measures; (4) for research purposes, such as to determine the chemical and/or physical characteristics of contaminants, or (5) to investigate a potential health problem on a corporate wide basis. However, the majority of the sampling with which plant personnel will be concerned will be performed because of local or federal regulations. Those health standards under the Occupational Safety and Health Act (OSHA) require monitoring, on a periodic basis, of all employees who are exposed to harmful materials.

While the concept of air sampling and the use of air monitoring devices may appear to be simple, the details of a good monitoring program may be misunderstood unless the person engaged in sampling is adequately trained and technical supervision is provided by a professional industrial hygienist. Consequently, misapplication of techniques, errors in instrument performance, and errors in interpretation of sampling data are common when monitoring is conducted by those individuals whose training is limited solely to undertaking the monitoring required by federal and state regulations without the essential professional guidance.

The major problems arise from accepting an instrument reading as reliable without determining its calibration and the reproducibility of its response; and in not obtaining representative tests. Plant personnel must have some experience with and knowledge of the resolution of these problems.

This knowledge is also important in order to determine if air tests conducted by OSHA compliance officers are valid. For example, if threshold limit value in the health standard is an 8-hour time-weighted average, the air sample should be obtained by sampling over the entire shift in the employee's breathing zone. It cannot be measured by a few short term samples, even if spaced over the full shift unless the worker is in a relatively fixed location with no variation in his work procedure or in the process. Such an event is generally the exception rather than the rule.

Unless established monitoring practices are conducted by plant personnel or the OSHA compliance officer, the results are of limited value in determining compliance with a standard, and what is more important, protecting the worker's health. The conduct of air monitoring by untrained, or poorly trained, persons relegates the program to a "numbers game" which serves no useful purpose.

A characteristic of the air contamination in most occupational environments is that there is most always a continual change in concentration with respect to time and location. This is particularly the case with respect to airborne particulates (dust, fumes and mists). Air currents within a room, process variations, changes in the work practice performed by an operator, and variation in the emission rate of contaminants are a few of the more significant factors resulting in this continual change in concentration throughout a work shift. This change in concentration will be less pronounced with gases and vapors than with particulates. However, generally speaking, marked concentration gradients can exist in most work areas which may be transient or relatively constant depending upon the contaminant sources and their number. Therefore, it is apparent that it would be most unrealistic to assume that the concentration of air contamination exists uniformly throughout a room or area.

The problem of determining the exposure of a worker to air contaminants is further complicated by the mobility of most workers who move about, in and out of, many areas of a workroom. This mobility is characteristic of many assigned jobs. Therefore, the concentration of contaminants in each work area, and the time spent in each must be considered in determining the full shift time-weighted average concentration to which each worker is exposed. Exposure is concentration averaged over a time period, which in the general case is a full 8-hour shift. However, threshold limit values in the list published in the OSHA safety and health standards are peak concentrations or "ceiling values." Such standards indicate the maximum concentration which is allowed for any time period. Although a single specific sampling strategy cannot be applicable for all air monitoring, general principles or considerations, which should be incorporated in such a strategy, can be developed.

There are a variety of air sampling instruments, as indicated previously, the normal variability of the concentration of air contaminants and the mobility of workers, the average full-shift exposure of a worker is a to different concentrations during a workday. This can be ascertained by determining each concentration and exposure period, which is an extremely time-consuming and costly procedure; however, a more effective procedure involves the use of a personal sampling device worn by the worker during his full working shift, or any portion thereof which may be under inquiry. It is important to recognize that no perfect instrument exists nor can any be worn by a worker or placed in a location without periodic observation if for no other purpose than to note variation in process operating conditions or work practices. Both people and instruments malfunction and environmental conditions can readily depart from normality. It is such observations which must be noted if sampling data is to be properly evaluated.

Standard sampling instruments and procedures have been developed by OSHA with the assistance of the National Institute of Occupational Safety and Health (NIOSH). These are developed to have OSHA industrial hygienists and compliance officers operate under a necessary standardized practice to determine compliance with standards. A principal consideration in their selection has been simplicity in operation and direct-reading response. Both of these requirements are important for any inspectorate as these individuals must engage in considerable travel and desire to ascertain compliance status as soon as possible. However, these OSHA sampling instruments and practices are not necessarily the optimum for individual plant use to provide the most accurate data, but a number of them can be so categorized as they have been selected from those long accepted by the industrial hygiene profession. Air samples to be obtained generally consist of breathing zone and fixed position samples. While both have application in hazard evaluation, the former is to be recommended for general use in determining compliance with standards. The exposure of a worker can be measured most accurately by determining the concentration of the contaminant in the air which he breathes. This does not imply that the sampling instrument must be located a few inches from his nose, as such a location would be impractical and unnecessary. The instrument should be held or located as close to an employee's nose and mouth without interfering with his freedom of movement in the normal conduct of his work. While some individuals have indicated sharp concentration gradients around a contaminant source, sample locations for exposure measurement located 1-2 feet from the nose is adequately representative. Of more significance is the need for more than one full-shift sample if representative exposure data is to be obtained. Unfortunately OSHA standards for air contaminants are based upon a single 8-hour time-weighted average except where "ceiling value" standards are involved. Such a basis of determining compliance with standards for materials such as pneumoconiosis-producing dusts do not recognize that these materials only produce long-term effects and hazard evaluation should be based on more than a single 8-hour sample. Such a single sample can be at considerable variance with the true exposure averaged over two or more days, nor is one rarely occurring excessive exposure hazardous to the employee. Nevertheless, the regulations require that samples must be based upon the sampling period prescribed in the standards. Frequently it is desirable to obtain fixed position samples. For example, a worker may spend only a fraction of an hour in a job involving exposure to a contaminant and the remaining time in uncontaminated air. Such a breathing zone sample would collect such a small quantity of the contaminant that an accurate chemical analysis cannot be made. It would be preferable to locate a sampling instrument in a fixed position within the breathing zone of a worker when in the area and to sample for the entire shift. However, the concentration derived from such a sample must be time-weighted only for the actual time spent at the job to obtain the exposure of the worker involved. There are other instances when fixed position samples are desired, such as in control cabs, pulpits, etc., where the effectiveness of engineering controls or the contribution of this location to the total exposure is desired. Another application for fixed position samples is to determine whether employees located at distances from an operation should be sampled. This can be readily determined by locating several fixed position samplers at different distances from a source of contamination. Still another type of fixed position sampling, although not directly a measure of employee exposure is the use of continuous fixed position sampling stations, such as commonly used in blast furnace divisions for carbon monoxide. These types of samplers may be the sequential multiple point sampler type or the single point sampler. They are generally arranged so that the sampling point is located in close proximity to points where accidental release of high concentration of gases having an acute effect may develop. Their principal use is to warn employees by means of an audible alarm to immediately leave the area. Such instruments also can provide a continuous recorded measurement of some gases or vapors which may or may not be exposure related.

Number of Samples: A reliable estimate of an employee's exposure requires replicate samples irrespective of their duration. This is basic whether or not one is concerned with 8-hour time-weighted concentrations, operational exposures or areal contamination. Differences involving a factor of five or more are not rare. Therefore, a minimum of three samples should be obtained, until experience dictates an upward or downward revision, based upon the variability so determined. One cannot emphasize too greatly that the objective of a sampling program is worker protection and not the collection of numbers. An occasional exposure to a concentration which exceeds the threshold limit values would result in a violation if the compliance officer is sampling on the day that such an exposure exists, even though the average of several daily samples obtained during the same week is within the standard. This demonstrates the difference between good evaluation techniques and the mere application of numbers.

In the development of a sampling schedule, one should remember that if an operation continues more than one shift, it may be prudent to collect samples during each shift, as exposure to airborne contaminants may be different for each shift. Furthermore, sampling should be performed during all seasons of the year (winter, spring, summer and fall). This is especially true for locations in areas where large temperature variations occur during the different seasons of the year. Generally, there is more natural ventilation in the warmer months with the buildings open, which tends to dilute air contaminants, than in the colder weather when natural ventilation may be limited due to closing of doors and windows. In summary, the minimum number and type of samples is dictated by OSHA standards. However, it would be highly desirable to obtain more than this minimum number of samples.

Duration of Samples: For a practical viewpoint the duration of samples will be dictated by the requirements in the OSHA standards. These are continuous 8-hour samples or short-term samples when the standard has a "ceiling value" or peak concentration limit. Scientifically speaking, the minimum volume of air to be sampled, or the duration of sampling, is based on the following considerations: (1) the threshold limit value (TLV) or regulatory standard; (2) the sensitivity of the analytical procedure; or (3) the estimated air concentration. Thus, the volume of sample needed may vary from a few liters, where the estimated concentration is high to several cubic meters where low concentrations are expected. Then, knowing the sensitivity of the analytical procedure, the TLV and the sampling rate of the particular instrument in use, one can determine the minimum time necessary for an adequate sample. However, the collected sample should represent some identifiable period of time — usually a complete cycle of an operation or so many minutes out of each hour. This will enable the worker's exposure on a time-weighted average basis to be calculated.

Preparation for Sampling: The successful application of any sampling program requires that one be knowledgeable of the processes involved, the potential hazards and be able to recognize hazardous work conditions. Therefore, the first step in evaluating the occupational environment is to become familiar with the operations in the plant. This is best obtained by a preliminary, or "walk-through," survey during which information is obtained on the job categories and the operations in each, the raw materials used, the process by-products, and the type of control measures afforded for the protection of the workers. This information should be recorded on an appropriate form following which the data is reviewed and the potential hazards. Discussions should be held with supervisory personnel and industrial engineering personnel to obtain such details. It is imperative that plant personnel responsible for environmental health tests be considerably more knowledgeable of work practices and their environmental impact than one can expect from a regulatory official. Only by such experience can plant personnel determine if samples obtained are representative and accurate.

Sampling for Gases and Vapors: Many gases and vapors can be sampled by devices which indicate the concentration of the substance during sampling or shortly thereafter, without the necessity for chemical analysis. These direct reading devices are convenient and useful when properly calibrated. Other substances cannot be sampled by this method, because no appropriate instrument is available, and indirect methods which require laboratory analysis of the sample must therefore be used. Such analyses are often delayed by days or weeks, depending upon laboratory schedules.

Direct reading samplers include simple devices such as colorimetric indicating tubes in which a color change indicates the presence of the contaminant in air passed through the tube, or instruments which are more or less specific for a particular substance. In the latter category are carbon monoxide indicators, combustible gas indicators (explosimeters) and mercury vapor meters, as well as a number of other instruments.

All instruments for sampling gases or vapors must be calibrated before use and their limitations and possible sources of error must be fully understood. Every instrument has a lower limit of sensitivity which can be too high, making the instrument useless for health hazard evaluation. For example, some explosimeters are so insensitive that they show only the presence of nearly explosive mixtures of some solvent vapors, and give no response of levels which may be harmful to health. To be useful for environmental health purposes, an instrument should give a substantial reading at or near the TLV concentration and preferably should accurately indicate the presence of air contaminants as low as

10% of this concentration. Most direct reading instruments and many colorimetric indicating tubes are not sufficiently sensitive for this kind of sampling. The manufacturer's specifications should be reviewed before a sampling device is selected. But it should be remembered that specifications may be optimistic, and that it may not be possible to detect with certainty the concentration which is listed as the lower limit of detection.

Since no device is completely specific for the substances of interest, care must be taken that interferences do not invalidate the sampling results. Many common gases and vapors react with the same chemicals, or have similar physical properties, so that the instrument may give falsely high or low readings for the substance being sampled. The manufacturer's data for colorimetric indicating tubes lists those substances which may interfere with the desired determination.

If there is reason to think that interfering substances may be present, it is advisable to sample them to determine whether their concentrations are sufficiently high to actually constitute an interference. It is very important to establish that an instrument responds properly to the substance it is designed to sample. This is generally done by calibration procedures with standard concentrations of the substance of interest.

It is also desirable to spot test the instrument's response between calibrations. For this purpose, several suppliers of compressed gas prepare cylinders containing almost any desired concentration of the gas or vapor of interest. If it is not practical to keep such cylinders on hand, other procedures may be used. For example, a carbon monoxide meter can usually be checked by exposing it to a small amount of diluted automobile exhaust; a mercury meter can be checked by holding it above an open bottle of mercury; a combustible gas indicator (explosimeter) can be checked by exposing it to a solvent mixture such as gasoline, lighter fluid, or paint thinner. Although such rough checks are not quantitative, they indicate whether the meter is responding to the substance for which it is to be used. Indicating tubes cannot be tested in this way, since they are usually designed for one-time use. If there is any doubt about the response of indicating tubes, it is advisable to sacrifice one tube from the box, to be sure that the tubes in the particular batch are, in fact, responsive to the substance being sampled. In any case, it is desirable to check one tube from each batch with a calibrating gas of known concentration.

A sampling device may use one of three basic methods for collecting gaseous air contaminants. The first involves passing air through a direct reading instrument which indicates, without further analysis, the actual concentration of the substance at the time the sample is taken. The second method involves passing a known volume of air through an absorbing solution (a liquid which takes up and retains the gas or vapor), or an adsorbing medium (a solid substance which mechanically holds a solvent or vapor on its surface), to remove the desired contaminant or contaminants from the air. The absorbing solution may be a weak alkali solution (0.01% normal sodium hydroxide) in a fritted glass bubbler and the adsorbing medium may be chemically treated silica gel or activated charcoal sealed in a glass tube. In the third method, an air sample of definite volume at known temperature and pressure is collected in a container (an evacuated flask, a bottle, or a plastic bag) which is resealed immediately to prevent sample loss. It should be noted that samples collected by the second and third methods must be sent to a laboratory for analysis. All three sampling methods should:

1. Provide an acceptable efficiency of collection for the air contaminant involved;
2. Maintain this efficiency at a specified air flow;
3. Have a high degree of reproducibility;
4. Require minimal manipulation in the field;
5. Avoid, if possible, the use of corrosive or otherwise hazardous sampling media.

The first (direct reading) method is fairly simple and results are available immediately. However, the instruments have limited sensitivity and must be recalibrated periodically. The second (absorption in a liquid or adsorption on a medium) and third (gas container) methods are generally considered more sensitive and more accurate method for trace analysis by gas chromatographs, infrared spectrophotometers, and similar instruments. However, because of their sophistication, both of these methods require careful handling to insure representative tests. Direct reading instruments enable the operator to obtain immediate indications of gas or vapor concentration by reading a meter dial or by noting the length of stain on an indicator tube. This does not mean, however, that the mere reading of a meter implies a valid test. On the contrary, the operator must be thoroughly familiar with the use and limitations of the instruments and devices.

Direct Reading Instruments: Combustible gas explosimeters are one of the most useful instruments of the direct reading type, also known as a combustible gas indicator. As the names suggest, instruments of this type were designed to detect the presence of explosive or combustible gases in the air. Safety checking is still their principal application, and many of them are suitable only for this purpose. To understand the principle on which these instruments operate, the terms lower and upper explosive limits must be defined. When certain proportions of combustible vapor are mixed with air, ignition will produce an explosion. The range of concentrations over which this will occur is called the explosive range.

It includes all concentrations in which a flash will occur or a flame will travel if the mixture is ignited. The lowest percentage at which this occurs is the lower explosive limit (LEL), and the highest percentage is the upper explosive limit (UEL). Mixtures below the LEL are too lean to ignite, and mixtures above the UEL are too rich.

Explosive limits are expressed in percent by volume of vapor in air. LELs and UELs have been determined in fire and safety, and health laboratories for all substances likely to be found in industry. Typical values for some solvents and gases are given in Table 3.

Table 3. Explosive Limits for Common Materials

Name	Flashpoint °F		Explosive Limits in air % by Volume		Autoignition Temp. °F	Density Air = 1.0
	Closed Cup	Open Cup	Lower	Upper		
Acetaldehyde	-36	—	4.0	55.0	365	1.52
Acetone	0	15	2.1	13.0	1000	2.00
Ammonia (Anhydrous)	Gas	Gas	15	28	1204	0.596
Amyl Acetate-n	76	80	1.1	7.5	714	4.49
Amyl Alcohol-n	91	120	1.2	—	572	3.04
Benzene	12	—	1.4	7.1	1044	2.77
Benzine	<0	—	1.4	5.9	550	2.50
Butyl Acetate-n	72	90	1.4	7.6	790	4.00
Butyl Alcohol-n	84	110	1.4	11.2	693	2.55
Camphor	150	200			871	5.24
Carbon Disulfide	-22	—	1.0	50	257	2.64
Carbon Tetrachloride	None	None				
Cellosolve	104	120	2.6	15.7	460	3.10
Chloroform	None	None	—	—		4.13
Coal Tar Oil	80-160	—	—			
Coal Tar Pitch	405	490	—	—		
O-Cresol	178	—	1.3 at 300° F	—	1038	3.72
Cyclohexanol	154	—	—		572	3.45
Denatured Alcohol	60	—	—		750	1.60
Ethyl Acetate	24	30	2.2	11	800	3.04
Ethylene Glycol	232	240	3.2	—	775	2.14
Ethyl n-propyl ether		—	1.9	24	—	—

Table 3. Continued

Name	Flashpoint °F Closed Cup	Open Cup	Explosive Limits in air % by Volume Lower	Upper	Autoignition Temp. °F	Density Air = 1.0
Formaldehyde, 37% in water	130	200	—	—	795	1.03
Fuel oil No. 1	114-185	—	0.6	5.6	445-560	—
Fuel oil No. 1-D	100 min	—	1.3	6	350-625	
Fuel oil No. 2	126-230	—	—	—	500-705	—
Fuel oil No. 2-D	100 min	—	1.3	6	490-545	—
Fuel oil No. 4	154-240	—	1	5	505	
Fuel oil No. 5	130-310	—	1	5	—	—
Fuel oil No. 6	150-430	—	1	5	765	—
Gasoline Automotive premium	-50+	—	1.3-1.4	6.0-7.6	770	3.0-4.0
Gasoline Automotive regular	-50 ±	—	1.3-1.4	6.0-7.6	700	3.0-4.0
Gasoline Aviation, commercial	-50 ±	—	1	6.0-7.6	800-880	3.0-4.0
Gasoline Aviation military	-50 ±	—	1	6.0-7,6	800-880	3.0-4.0
Hexane-n	-7	—	1.2	7.5	453	2.91
Hexane-iso	<-20	—	1	7	—	3.00
Hydrogen sulfide	Gas	Gas	4.3	45.5	500	1.18
Jet fuel JP-1	110-125	—	0.6	5.6	442-560	—
Jet fuel JP-4	26-36	—	0.8	6.2	468	—
Lacquer	0.86	—	—	—	—	—
Maleic Anhydride	218	240	—	—	890	3.38
Methyl Acetate	15	20	3.1	16	935	2.56
Mineral spirits	100 min	110	0.77@212°F	—	475	3.9
Naphtha	100-110	—	0.8	5	440-500	—
Naphtha VM&P	20-45	—	0.9	6.0	450-500	3.75
Naphthalene	174	190	0.9	5.9	979	4.42
Petroleum crude	20-90	—				—
Petroleum ether	<0	—	1.4	5.9	550	2.50
Phenol	175	185	—	—	1319	3.24
Phthalic Anhydride	305	330	1.7	10.5	1083	5.10
Pine oil	172	175	—	—	—	—
Propane	<-100	Gas	2.2	9.6	871	1.56
Propyl Acetate-n	58	70	1.7	8.0	842	3.52

Table 3. Continued

Name	Flashpoint °F		Explosive Limits in air % by Volume		Autoignition Temp. °F	Density Air = 1.0
	Closed Cup	Open Cup	Lower	Upper		
Stoddard solvent	100-110	—	0.8	5	440-500	—
Styrene	90	—	1.1	6.1	914	3.60
Sulfur	405	440	—	—	450	—
Toluene	40	45	1.3	7.0	997	3.14
Trichloroethylene	Weakly Flammable		10 in O_2	65 in O_2	—	4.53
Turpentine	95	—	0.8	—	488	4.84
p-Xylene	77	—	1.1	7.0	984	3.66

Several manufacturers make explosimeters or combustible gas indicators. Although they differ somewhat in design and operating features, their operation is based on the fact that a measurable amount of heat is released when a combustible gas or vapor is burned. Most meters contain a battery-operated electrical circuit known as a Wheatstone bridge, which is balanced by means of controls on the outside of the instrument.

On the simplest type of instrument (an explosimeter) only one scale is provided, usually with readings from 0 to 100% LEL. However, the detectable changes produced by combustion are too small to be measured accurately in the presence of the low concentrations of contaminants usually encountered in evaluating potential health hazards. For example, the LEL of even the most explosive gas is of the order of 1%, or 10,000 ppm, which is well in excess of the toxic limit for any gas. Therefore, explosimeters or combustible gas indicators which have only a 0-to-100% LEL explosive scale are not suitable for environmental health testing in the ppm range. More sensitive instruments, including the type used in sampling for environmental health purposes, have a dual scale, in which the second, more sensitive scale, expands the 0-to-10% LEL part of the instrument's response to full-scale reading. Refer to Figure 4.

While this permits more sensitive and accurate reading of concentrations in the 0-to-10% range, this type of instrument is not sufficiently sensitive to give precise indications of concentrations at the TLV of many toxic gases and vapors. In addition, they lack specificity, do not read directly in TLV units (ppm), and are subject to interferences. All combustible gas and vapor indicators are calibrated by the manufacturer using one specific gas or vapor such as methane, and a calibration curve is provided, in percent LEL, for the calibration gas only.

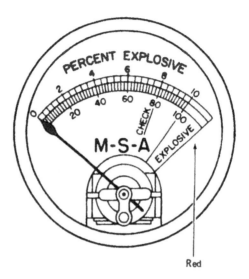

Figure 4. Scale on an explosion meter.

Obviously, accurate concentrations of other gases or vapors cannot be indicated unless the instrument has been appropriately calibrated for each of these gases. Furthermore, the manufacturer's calibration may not be sufficiently accurate and a correction may be required. For example, a meter reading of 2.0 on the 0 to 10% LEL scale of one instrument would indicate, for a solvent having an LEL of 1.4%, a concentration of 280 ppm (0.02 x 14,000 ppm). Benzene, methyl isobutyl ketone (MIBK). For example, a meter reading of 2.0 on the 0-to-10% LEL scale of one instrument would indicate, for a solvent having an LEL of 1.4%, a concentration of 280 ppm (0.02 x 14,000 ppm). Benzene, methyl isobutyl ketone (MIBK) and toluene all have LEL's of 1.4%, but actual concentrations at a meter reading of 2.0 would be 294 ppm for benzene, 350 ppm for MIBK and 231 ppm for toluene. Therefore, unless the scale reading is corrected according to a calibration chart, to indicate actual ppm values, the reading can be seriously in error.

The manufacturer's instructions for operating a combustible gas indicator should be carefully reviewed before the device is used. In general, all explosimeters require a brief initial warm-up period so that the batteries can heat the filaments. Then battery strength should be checked, and the zero scale adjusted. When zero is established and the indicating needle is stable, the instrument is ready for use.

The zero adjustment must be made by taking the instrument to a source of air which does not contain combustible gases or vapors, or by passing air into it through an activated carbon filter which will remove combustible vapors and gases except methane. Since methane is not removed by activated charcoal filters, extra caution is required if the presence of methane is suspected. In addition, the filter should be changed periodically because it tends to become saturated during prolonged use and will no longer remove many of the combustible gases and vapors. If the zero adjustments are made in fresh air, care must be taken that no combustible gas or vapor is present in an amount which would influence the instrument's response.

Most combustible gas indicators are equipped with a length of sampling tubing with a metal probe at the end. The probe is held at the sampling point (usually near the breathing zone of the worker) and, a few seconds later, the response can be read on the meter. Generally, the air is drawn through the probe and meter by means of a hand-operated rubber squeeze bulb. In some instances, however, a small electrically-operated pump in the instrument case is used for this purpose. In most work areas, the concentration of combustible gas or vapor fluctuates constantly, and it is necessary to observe the instrument carefully and to make a judgment concerning average and peak readings.

Sometimes the probe of a combustible gas indicator or explosimeter is placed in a manhole or other space not normally occupied by people, to determine if there is a potentially explosive or dangerous concentration of gas present. When the instrument is used in this way, it may show a zero response for several different reasons. Assuming that the batteries are working and the instrument is functional, the absence of meter response can mean either that there is little or no combustible gas in the space being tested, or that the concentration is so high that it is above the UEL and combustion cannot occur because of lack of oxygen. A very high concentration can be identified by carefully watching the needle as the probe is moved into and withdrawn from the space being tested. At some point during entry and withdrawal, the probe must pass through the LEL concentration and enter the flammable range. At this point, the needle will jump briefly, then settle back to zero. This "chink" or jump is a clear indication that a high concentration is present. Dual scale combustible gas indicators are rugged, relatively lightweight, and portable so they are widely used for checking work operations like solvent cleaning or painting, or locations such as coal chemical areas where organic solvents are emitted. Their limitations must be realized, however, and where there is a possibility that benzene or an unknown and possibly toxic solvent may be present, they cannot be relied upon. In such cases, sampling must be performed by other means. The user of any instrument should be thoroughly familiar with precautions to be taken in its operation; users of combustible gas indicators must also be aware of interfering gases and vapors which can create major aberrations in instrument response. One such precaution is that the 0 to 100% scale should be used first, to determine whether an explosive atmosphere exists and to prevent overloading to 0 to 10% LEL scale. Interfering gases and vapors can seriously affect instrument response, and an experienced tester recognizes the indications of their presence. The manufacturer's instructions should be thoroughly understood. For example, high concentrations of chlorinated hydrocarbons (i.e., trichloroethylene) or of an acid gas (sulfur dioxide) may cause depressed meter readings when high concentrations of combustibles are also present. Trace amounts of these interferences may not affect the readings directly, but can corrode the sensitive detector elements. High molecular alcohols in the atmosphere may burn out the filaments, making the instrument inoperative. If such limitations are understood, the tester can obtain sufficiently valid results.

For calibration of the 0 to 100% LEL scale, test kits containing known concentrations of combustible gases (usually either 2.5% methane or 2.5% natural gas) are available from the instrument manufacturers. In using these kits, it should be borne in mind that the calibration is only for that specific gas and indicates only that the meter is operational on the 0 to 100% LEL scale. To calibrate to 0 to 10% scale, it is necessary to use purchased or specifically prepared known concentrations of gases in the TLV ranges.

Carbon Monoxide Testers: Carbon monoxide is one of the most commonly encountered toxic gases and may result from many industrial processes as well as from automobile or truck exhaust. Since pure CO is colorless, odorless, and tasteless, the senses cannot be relied on to give warning of its presence. Carbon monoxide may be sampled in several ways, including the use of the colorimetric indicating tubes, but for making repeated measurements over a period of time, a direct reading meter is frequently used. Such a meter is similar to a combustible gas indicator in that its operation involves combustion of the CO and measurement of the heat released by this combustion. However, both the design of the instrument and its method of measuring heat are considerably different from those in combustible gas indicators.

Carbon monoxide is combustible, but does not burn readily at low concentrations such as the present TLV of 50 ppm. However, it will burn at low concentrations in the presence of a catalyst such as hopcalite, a granular mixture originally developed for use in gas masks. A conventional meter contains one cell of inactive chemical (reference cell) and one cell of hopcalite (detector cell), each containing a number of electric wires or thermocouples which generate a small electric current when heated. The detector cell is surrounded by two thermistors (heat sensitive resistors) which are part of a Wheatstone bridge circuit. When air containing carbon monoxide is drawn through the detector cell, it is oxidized by the hopcalite. The heat created by this oxidation causes a change in the electrical resistance of the system which is registered by an upward deflection of the meter. The reference cell compensates for all temperature changes caused by the surrounding air or by variations within the instrument, so that the net reading is a measure of the concentration of CO in the air. Since the performance of the hopcalite is hampered by water vapor, the air must be dried before entering the catalyst chamber. The drying agent built into the instrument should be checked periodically, and replaced when necessary.

Oxygen Indicators: Air normally contains about 21% oxygen, and in most working places the concentration of contaminants is so low that this oxygen level remains essentially unchanged at all times. Even if there is a relatively high concentration (such as 1,000 ppm) of a vapor in the work atmosphere, there are still 999,000 ppm of air present, and for all practical purposes, this is no different from pure air insofar as its oxygen content is concerned.

In many locations, however, such as mines, manholes, tunnels, or other confined spaces, it is possible for the oxygen content to be sufficiently low that it is hazardous to life. In such situations, it is necessary to determine the oxygen content and, in addition, to sample to determine whether combustible gases are present in dangerous concentrations.

When oxygen reaches the inside of the cell, it generates a minute electric current which is converted to a voltage and is registered on a meter as a per cent of oxygen. The instrument is basically stable, automatically compensates for temperature changes between freezing and about 100 °F, and is not affected by carbon dioxide or relative humidity. It is readily calibrated by sampling ordinary air and, according to the manufacturer, this single point calibration is adequate. Complete instructions for its use, operation and maintenance are supplied by the manufacturer.

Indicator (Colorimetric) Tubes: The use of solid chemical detectors (indicating tubes) is common practice. They are simple devices to operate which tends to cause many users to ignore their limitations which must be recognized if they are to be useful in evaluating potential hazards due to air contaminants. This apparent simplicity increases the number of people who attempt to use them, which in turn results in wide variations in both individual competence and the accuracy of the data obtained.

It is essential for all users of air sampling devices to recognize that, while collecting an air sample is a simple procedure, it is only one part of the total procedure for proper evaluation of the environment. Only minimal skills are required to operate the instrument. Far greater knowledge and skill are required to recognize its limitations, to maintain and calibrate it properly, to obtain representative tests, and to know the strategy for obtaining valid data on workers' exposures. These are problems which require experienced and well-trained personnel who either are, or are under the direction of, professional industrial hygienists.

Making reliable tests with indicating tubes requires thorough knowledge of their limitations and care in their use. Experience has shown that the following measures help to minimize some errors:

1. Test each batch of tubes with a known gas concentration.
2. Read the length of stain in a well-lighted area.
3. Read the longest length of stain if stain development is not sharp or even.
4. Observe the manufacturer's expiration date closely, and discard outdated tubes.
5. Keep detector tubes in a shirt pocket or other warm place until time to start the test.
6. Refer to the manufacturer's data for a list of interfering materials.
7. Calibrate for the proper flow rate.

Indirect Sampling Devices for Gases and Vapors: As stated earlier in this chapter, not every contaminant can be sampled by a direct reading device. In such cases (or even in some situations involving contaminants which can be measured with direct reading devices) an indirect reading sampler must be used when the sample must be sent to a laboratory for analysis of the components. For this purpose, a number of glass or plastic bubblers are available, containing water or some other liquid to capture gases or dusts when air is drawn through them. Occasionally, filter papers are used, but since ordinary filter paper cannot trap gas or vapor, the paper must be treated with chemicals which will react with the substance of interest and retain it on the paper. For many purposes, gases and vapors can also be collected in bottles, in plastic bags, or in tubes which contain activated charcoal or some other adsorbent. Each of these methods is discussed below, since the person responsible for monitoring in each plant will be required to use them at some time.

Bubblers: Bubblers are ordinarily made of glass, although some are made of clear plastic. A stated amount of absorbing liquid is placed in a sample bottle which also contains the bubbler. When air is drawn through the bubbler, the contaminant is retained in the absorbing solution. The most efficient unit for sampling gases and vapors is a fritted bubbler, which is a piece of glass with thousands of small holes which disperse the air as it bubbles through the solution.

The choice of absorbing solution, the strength (normality or pH) of the solution, the sampling rate (generally 1 to 3 liters per minute), and the size of the bottle (20 cc, 50 cc, or 125 cc), are some of the variables which must be considered in using bubblers.

It should also be emphasized that the air flow calibration of bubblers should be done before field use. The calibration should be done with the same type and amount of absorbing solution as will be used in the actual testing, since absorbing solutions vary in viscosity and can directly affect the pressure drop in the sampling system. The pressure drop is also directly affected when the pores of the fritted bubbler become clogged thus lowering the efficiency of the bubbler and materially influencing the collection rate. To avoid such clogging, a filter must be used ahead of the bubbler.

Charcoal Tubes: Reference has been made earlier to adsorption, which is the property of some solid materials, such as activated charcoal, to physically retain solvent vapors on their surfaces. In environmental health testing, the adsorbed vapors are removed, generally with a solvent, in a laboratory. The solvent is then analyzed by physical methods (gas chromatography, etc.) to determine the individual compounds whose vapors, such as benzene, were present in the sampled air. Industrial atmospheric samples can be collected in small glass tubes (4 mm ID) packed with two sections of activated charcoal, separated and retained with fiberglass plugs. To obtain an air sample, the sealed ends of the tube are broken off, and air is drawn through the charcoal at the rate of 1 liter per minute by means of a personal sampler. After sampling, the tubes are resealed and sent to a laboratory for analysis.

Four precautions must be observed when this type of collection device is used:

1. The shorter (backup, or second) section of the charcoal tube should be inserted into the sampling line so that the air is drawn through the longer section first. When analyzed, the backup section should be void of solvent vapors - in other words, there should have been no carry-over from the first section.

2. The sampling rate must be maintained at 1 liter per minute. Sampling at a higher rate may elutriate the solvent vapors from the first section, in which case they may be adsorbed in, or even elutriated from, the second section.

3. Because of the limited adsorption capacity of the tube, the sampling period should not exceed 15 to 30 minutes. The time will depend to some degree on the expertise of the observer, since high concentrations of solvent vapors could saturate the first charcoal section in a few minutes.

4.   The tubes should be carefully resealed, with the caps provided for that
      purpose, immediately after use.

Plastic Bags: Sometimes it is possible to sample simply by filling a bag, bottle,
or other container with air from the working area and sending the container to a
laboratory. Plastic bags have several advantages over bottles for this purpose.
First, of course, they require little storage space and can be kept on hand at all
times. In addition, they are available in relatively large sizes so that a much
larger volume of air can be sampled than might be practical with a bottle.
Finally, they are lightweight even when filled.

Among the materials used for these bags are Myler, Saran, and a laminated
material called Scotch Pak. These bags are, in general, strong and impermeable,
do not react with many of the common vapors which must be sampled, and do
not themselves add contaminants to the air after it is collected. Commercially
available bags come in various sizes ranging upward from 1 liter, and have built-
in tubes for sampling purposes. There are several ways to fill a sampling bag.
The most common is to attach to the bag inlet a rubber bulb containing two
valves, one to draw air in, the other to blow air out. If the bulb is squeezed
rapidly, the bag can be filled in a very short time. If a rubber bulb is not suitable
for the vapors of interest, some other means must be used to fill the bag.

Instrument Limitations: Every instrument has certain limitations which must be
recognized. A number of these have already been discussed, and it is very
important that they be kept in mind at all times. Otherwise, serious errors can
result, rendering the tests meaningless or misleading. Following is a brief
summary of limitations common to many instruments, but it must be emphasized
that any given instrument may not be susceptible to any particular error:

1.   All instruments are limited in specificity of response to the contaminant
      they are supposed to measure. Interferences may cause false readings and,
      as pointed out earlier, care must be taken to identify (and, if possible,
      compensate for) these interferences.

2.   Any instrument is suitable for only a given range of contaminant
      concentration. This means that at certain very low levels, the instrument
      will not respond while, at very high levels, the reading may be grossly
      inaccurate or even meaningless. As discussed earlier, combustible gas
      indicators can even register zero in the presence of very high
      concentrations of vapors or gases.

3.   Every instrument is likely to have a slight error, and the magnitude of this
      error, if it exists, must be known. It is meaningless, for example, to record
      an instrument reading as 51.3 ppm when in fact there is a known

instrument error of $\pm$ 5%. It is also meaningless to report the average of several readings by a number which cannot be read on the instrument itself.

4. Every instrument which requires that air be passed through it, must be operated at a specific flow rate, or within a certain range of flow. Failure to sample at the correct flow rate can produce inaccurate or meaningless results.

5. A number of limitations are imposed by the construction of the instrument and its operating parts, including such factors as fully charged or new dry cell batteries, replacement of drying agents or other chemicals and, of course, how well the moving or consumable parts such as filaments have been maintained. Most instrument failures from these causes can be prevented by routine care and maintenance, and by periodic calibrations.

Chapter 5
PERSONAL PROTECTIVE EQUIPMENT

INTRODUCTION

OSHA 1910 Subpart I App B, titled "Non-mandatory Compliance Guidelines for Hazard Assessment and Personal Protective Equipment Selection" is a standard intended to provide compliance assistance for employers and employees in implementing requirements for a hazard assessment and the selection of personal protective equipment. The abbreviation, PPE, stands for personal protective equipment. PPE devices alone should not be relied on to provide protection against hazards, but should be used in conjunction with guards, engineering controls, and sound production and manufacturing practices. To provide proper protection for employees, it is necessary to consider certain general guidelines for assessing the foot, head, eye and face, and hand hazard situations that exist in an occupational or educational operation or process, and to match the protective devices to the particular hazard. It should be the responsibility of the safety officer to exercise common sense and appropriate expertise to accomplish these tasks. In order to assess the need for PPE the following steps should be taken:

* Survey: Conduct a walkthrough survey of the areas in question. The purpose of the survey is to identify sources of hazards to workers and co-workers. Consideration should be given to the basic hazard categories: Impact, Penetration, Compression (roll-over), Chemical, Heat, Harmful dust, Light (optical) radiation.

* Sources: During the walkthrough survey the safety officer should observe:

1. sources of motion; i.e., machinery or processes where any movement of tools, machine elements or particles could exist, or movement of personnel that could result in collision with stationary objects;

2. sources of high temperatures that could result in burns, eye injury or ignition of protective equipment, etc.;

3. types of chemical exposures;

4. sources of harmful dust;

5. sources of light radiation, i.e., welding, brazing, cutting, furnaces, heat treating, high intensity lights, etc.;

6. sources of falling objects or potential for dropping objects;

7. sources of sharp objects which might pierce the feet or cut the hands;

8. sources of rolling or pinching objects which could crush the feet;

9. layout of workplace and location of co-workers; and any electrical hazards. In addition, injury/accident data should be reviewed to help identify problem areas.

Following the walkthrough survey, it is necessary to organize the data and information for use in the assessment of hazards. The objective is to prepare for an analysis of the hazards in the environment to enable proper selection of protective equipment. Having gathered and organized data on a workplace, an estimate of the potential for injuries should be made. Each of the basic hazards should be reviewed and a determination made as to the type, level of risk, and seriousness of potential injury from each of the hazards found in the area. The possibility of exposure to several hazards simultaneously should be considered. After completion of the procedures, the general procedure for selection of protective equipment is to:

•   Become familiar with the potential hazards and the type of protective equipment that is available, and what it can do; i.e., splash protection, impact protection, etc.;

•   compare the hazards associated with the environment; i.e., impact velocities, masses, projectile shape, radiation intensities, with the capabilities of the available protective equipment;

•   select the protective equipment which ensures a level of protection greater than the minimum required to protect employees from the hazards; and

•   fit the user with the protective device and give instructions on care and use of the PPE. It is very important that end users be made aware of all warning labels for and limitations of their PPE.

Careful consideration must be given to comfort and fit. PPE that fits poorly will not afford the necessary protection. Continued wearing of the device is more likely if it fits the wearer comfortably. Protective devices are generally available in a variety of sizes. Care should be taken to ensure that the right size is selected. Adjustments should be made on an individual basis for a comfortable fit that will maintain the protective device in the proper position. Particular care should be taken in fitting devices for eye protection against dust and chemical splash to ensure that the devices are sealed to the face. In addition, proper fitting of helmets is important to ensure that it will not fall off during work operations. In some cases a chin strap may be necessary to keep the helmet on an employee's head. Chin straps should break at a reasonably low force, however, so as to prevent a strangulation hazard. Where manufacturer's instructions are available, they should be followed carefully.

It is the responsibility of the safety officer to reassess the workplace hazard situation as necessary, by identifying and evaluating new equipment and processes, reviewing accident records, and reevaluating the suitability of previously selected PPE. This chapter provides helpful selection guidelines for PPE, following closely OSHA and NIOSH standards and recommendations.

EYE, FACE, AND HEAD PROTECTION

Table 1 provides a selection of chart guidelines for eye and face protection. Some occupations (not a complete list) for which eye protection should be routinely considered are: carpenters, electricians, machinists, mechanics and repairers, millwrights, plumbers and pipe fitters, sheet metal workers and tinsmiths, assemblers, sanders, grinding machine operators, lathe and milling machine operators, sawyers, welders, laborers, chemical process operators and handlers, and timber cutting and logging workers. The chart provides general guidance for the proper selection of eye and face protection to protect against hazards associated with the listed hazard "source" operations.

Table 1. Eye and Face Protection Selection Chart

Source	Hazard Assessment	Protection
IMPACT - Chipping, grinding machining, masonry work, woodworking, sawing, drilling, chiseling, powered fastening, riveting, and sanding.	Flying fragments, objects, sand, chips, particles, dirt, etc.	Spectacles with side protection, goggles, face shields. See notes (1), (3), (5), (6), (10). For severe exposure, use faceshield.
HEAT - Furnaces, pouring, casting, hot-dipping, welding.	Hot sparks	Faceshields, goggles, spectacles with side protection. See notes (1), (2), (3).
	Splashing molten metal	Faceshields worn over goggles. See notes (1), (2), (3).

	High temperature exposure	Screen face shields, reflective face shields. See notes (1), (2), (3).
CHEMICALS-Acid and chemicals handling, degreasing, plating.	Splash	Goggles, eyecup and cover types. For severe exposure, use face shield. See notes (3), (11).
	Irritating mists	Special-purpose goggles.
DUST - Woodworking, buffing, general dusty conditions.	Nuisance dust	Goggles, eyecup and cover types. See note (8).
LIGHT and/or RADIATION Welding: Electric arc	Optical radiation	Welding helmets or welding shields. See notes (9), (12).
Welding: Gas	Optical radiation	Welding goggles or welding face shield. Typical shades: gas welding, cutting, brazing. See note (9).
Cutting, Torch brazing, Torch soldering	Optical radiation	Spectacles or welding face-shield. See notes (3), (9).
Glare	Poor vision	Spectacles with shaded or special-purpose lenses. See notes (9), (10).

Notes to Eye and Face Protection Selection Chart:

1. Care should be taken to recognize the possibility of multiple and simultaneous exposure to a variety of hazards. Adequate protection against the highest level of each of the hazards should be provided. Protective devices do not provide unlimited protection.

2. Operations involving heat may also involve light radiation. As required by the standard, protection from both hazards must be provided.

3. Faceshields should only be worn over primary eye protection.

4. As required by the standard, filter lenses must meet the requirements for shade designations in 1910.133(a)(5). Tinted and shaded lenses are not filter lenses unless they are marked or identified as such.

5. As required by the standard, persons whose vision requires the use of prescription (Rx) lenses must wear either protective devices fitted with prescription (Rx) lenses or protective devices designed to be worn over regular prescription (Rx) eyewear.

6. Wearers of contact lenses must also wear appropriate eye and face protection devices in a hazardous environment. It should be recognized that dusty and/or chemical environments may represent an additional hazard to contact lens wearers.

7. Caution should be exercised in the use of metal frame protective devices in electrical hazard areas.

8. Atmospheric conditions and the restricted ventilation of the protector can cause lenses to fog. Frequent cleansing may be necessary.

9. Welding helmets or faceshields should be used only over primary eye protection (spectacles or goggles).

10. Nonsideshield spectacles are available for frontal protection only, but are not acceptable eye protection for the sources and operations listed for "impact."

11. Ventilation should be adequate, but well protected from splash entry. Eye and face protection should be designed and used so that it provides both adequate ventilation and protects the wearer from splash entry.

12. Protection from light radiation is directly related to filter lens density. See note (4). Select the darkest shade that allows task performance.

The following are selection guidelines for head protection. All head protection (helmets) is designed to provide protection from impact and penetration hazards caused by falling objects. Head protection is also available which provides protection from electric shock and burn. When selecting head protection, knowledge of potential electrical hazards is important. Class A helmets, in addition to impact and penetration resistance, provide electrical protection from low-voltage conductors (they are proof tested to 2,200 volts). Class B helmets, in addition to impact and penetration resistance, provide electrical protection from high-voltage conductors (they are proof tested to 20,000 volts). Class C helmets provide impact and penetration resistance (they are usually made of aluminum which conducts electricity), and should not be used around electrical hazards. Where falling object hazards are present, helmets must be worn. Some examples include: working below other workers who are using tools and materials which could fall; working around or under conveyor belts which are carrying parts or materials; working below machinery or processes which might cause material or objects to fall; and working on exposed energized conductors. Some examples of occupations for which head protection should be routinely considered are: carpenters, electricians, linemen, mechanics and repairers, plumbers and pipe fitters, assemblers, packers, wrappers, sawyers, welders, laborers, freight handlers, timber cutting and logging, stock handlers, and warehouse laborers.

FOOT AND HAND PROTECTION

The following are selection guidelines for foot protection. Safety shoes and boots which meet the ANSI Z41-1991 Standard provide both impact and compression protection. Where necessary, safety shoes can be obtained which provide puncture protection. In some work situations, metatarsal protection should be provided, and in other special situations electrical conductive or insulating safety shoes would be appropriate. Safety shoes or boots with impact protection would be required for carrying or handling materials such as packages, objects, parts or heavy tools, which could be dropped; and, for other activities where objects might fall onto the feet.

Safety shoes or boots with compression protection would be required for work activities involving skid trucks (manual material handling carts) around bulk rolls (such as paper rolls) and around heavy pipes, all of which could potentially roll over an employee's feet. Safety shoes or boots with puncture protection would be required where sharp objects such as nails, wire, tacks, screws, large staples, scrap metal etc., could be stepped on by employees causing a foot injury. Some occupations for which foot protection should be routinely considered are: shipping and receiving clerks, stock clerks, carpenters, electricians, machinists, mechanics and repairers, plumbers and pipe fitters, structural metal workers, assemblers, drywall installers and lathers, packers, wrappers, craters, punch and stamping press operators, sawyers, welders, laborers, freight handlers, gardeners and grounds-keepers, timber cutting and logging workers, stock handlers and warehouse laborers.

Selection guidelines for hand protection are as follows. Gloves are often relied upon to prevent cuts, abrasions, burns, and skin contact with chemicals that are capable of causing local or systemic effects following dermal exposure. OSHA is unaware of any gloves that provide protection against all potential hand hazards, and commonly available glove materials provide only limited protection against many chemicals. Therefore, it is important to select the most appropriate glove for a particular application and to determine how long it can be worn, and whether it can be reused.

It is also important to know the performance characteristics of gloves relative to the specific hazard anticipated; e.g., chemical hazards, cut hazards, flame hazards, etc. These performance characteristics should be assessed by using standard test procedures. Before purchasing gloves, the employer should request documentation from the manufacturer that the gloves meet the appropriate test standard(s) for the hazard(s) anticipated. Other factors to be considered for glove selection in general include:

- As long as the performance characteristics are acceptable, in certain circumstances, it may be more cost effective to regularly change cheaper gloves than to reuse more expensive types; and,
- The work activities of the employee should be studied to determine the degree of dexterity required, the duration, frequency, and degree of exposure of the hazard, and the physical stresses that will be applied.

With respect to selection of gloves for protection against chemical hazards:

- The toxic properties of the chemical(s) must be determined; in particular, the ability of the chemical to cause local effects on the skin and/or to pass through the skin and cause systemic effects;

- Generally, any "chemical resistant" glove can be used for dry powders;

- For mixtures and formulated products (unless specific test data are available), a glove should be selected on the basis of the chemical component with the shortest breakthrough time, since it is possible for solvents to carry active ingredients through polymeric materials; and,

- Employees must be able to remove the gloves in such a manner as to prevent skin contamination.

CHEMICAL PROTECTIVE CLOTHING

Many chemicals handled in industrial settings can cause adverse effects on unprotected skin ranging from contact dermatitis to permeation of the skin and systemic toxic effects. In addition, there are many chemicals that pose a contamination problem, where inadvertent ingestion (e.g., lead) could occur or re-entrainment in the airstream (e.g., asbestos) could lead to inhalation.

Chemical protective clothing (CPC), comprising gloves, boots, suits and other related components, can prevent direct skin contact and contamination. CPC can also prevent physical injury to the unprotected skin from thermal hazards such as from rapidly evaporating liquidified gases freezing the skin (e.g., LPG). An important reference to access is the *NIOSH Pocket Guide to Chemical Hazards, June 1997 Edition* (Publication No. 97-140), which contains recommendations on CPC for safe handling of chemicals. These recommendations are based on another published work, *Quick Selection Guide to Chemical Protective Clothing, Third Edition*, by Krister Forsberg and S.Z. Mansdorf (1997). The *Pocket Guide* provides general recommendations for skin protection according to the following designations: Prevent skin contact, Frostbite, and N.R. S. Z. Mansdorf, Ph.D., CIH, CSP (Email: mansdorf@tiac.net) has developed a special report that supplements the NIOSH publication.

In Mansdorf's publication, the "prevent skin contact" designation means that there is a dermal hazard potential. For work situations where direct contact could occur, it is recommendation that CPC providing resistance to permeation, penetration and degradation be used where the chemical has a potential to contact unprotected skin. A standard test method for permeation resistance has been devised by the American Society for Testing and Materials (ASTM) titled, *Test Method for Resistance of*

Protective Clothing Materials to Permeation by Liquids or Gases Under Conditions of Continuous Contact (method F739-91). This test determines both the breakthrough time and steady state permeation rate of chemicals through a sample of the protective barrier.

All chemicals will eventually permeate protective clothing. Breakthrough (permeation) resistance is related to temperature of the challenge material and the environment and thickness of the barrier. Therefore, higher than normal (25 °C) temperatures will result in faster breakthrough. Use of thicker materials will increase time to breakthrough of the chemical.

The ASTM method establishes the time to breakthrough under conditions of continuous liquid or gaseous contact. Hence, a breakthrough time reported as 4 hours means four hours of resistance to permeation at detection levels generally above one-tenth microgram per square centimeter per minute (0.1 $g/cm^2/min$) at standard temperature conditions. Published breakthough data from academic institutions, research organizations, trade associations, chemical manufacturers, protective clothing manufacturers and others can be used to select protective clothing in conjunction with other considerations. Not all chemicals nor barriers have been tested and the results published. For some chemicals listed in the *Pocket Guide*, no information is available.

For many of the solids, such as pesticides, the proper selection of protective clothing depends on the exact solvent system and formulation hence for many no recommendations are made. Additionally, new barriers, barriers that have not been considered, or new testing may result in recommendations different from those listed in this report. Therefore, both the chemical and the protective clothing supplier should be contacted for the most appropriate choice before any final selection is made.

The "prevent skin contact" designation will also be found for materials that are solids (e.g., dust, powder, flakes, fibers, etc.). This designation is meant to alert for the possibility of inadvertent contamination of the skin resulting in the potential for later ingestion or inhalation but no significant dermal hazard. It is recommended that these chemicals not be allowed to contact the skin to prevent cross contamination. As a dry solid, there is no permeation potential- only a penetration potential through holes, tears, loose weaves, etc. for natural fibers and polymers. Therefore, any barrier that will prevent penetration may be used for the dry chemical provided that the dry solid is not placed into solution.

The "frostbite" designation is meant to alert to the potential for freezing of the skin from direct contact with the liquidified gas through rapid evaporation. Some liquified gases may also present a direct skin hazard or a toxic hazard. For example, chlorine as a liquidified gas is corrosive, especially to wet skin, while hydrogen cyanide as a liquidified gas can permeate the skin leading to serious injury or death.

The "N.R." designation means that no recommendation can be made either because the chemical has not been shown to be a dermal hazard or inadequate information is available. Cellulose, for example, would not be expected to present a problem under normal circumstances. Nevertheless, each situation should be evaluated by a competent industrial hygienist or safety professional to determine whether protective clothing should be used.

Mansdorf provides recommendations for skin protection and CPC (Chemical Protective Clothing). Extracts of these tables are included in Table 2. In the section that follows Table 2, the OSHA designated *Levels of Protection* are described for the reader, both in terms of CPC and respiratory protection. The reader will find key references and web sites listed at the end of this chapter to supplement his reading and resources. Training sites advertised on the World Wide Web are also cited, however no attempt has been made to evaluate the programs. The reader should carefully review training curriculums and the level of experience of the trainers and training institute. Most of these programs often fall under the heading of *Hazmat* Training, following **OSHA 29 CFR** standards.

Table 2. Selection Guide for Chemical Protective Clothing Barriers as Reported by S.Z. Mansdorf (1999). *Refer to footnotes at end of table.*

CHEMICAL	SKIN PROTECTION [†]	PROTECTIVE CLOTHING BARRIERS*
Acetaldehyde	Prevent skin contact	8 hr: Butyl, Responder, Tychem; 4 hr: Teflon, PE/EVAL
Acetic acid	Prevent skin contact	8 hr: Butyl, Teflon, Viton, PE/EVAL, Responder, Tychem; 4 hr: Neoprene, Barricade
Acetic anhydride	Prevent skin contact	8 hr: Butyl, PE/EVAL, Barricade, Trellchem, Tychem; 4 hr: Teflon
Acetone	Prevent skin contact	8 hr: Butyl, PE/EVAL, Barricade, CPF3, Responder, Trellchem, Tychem
Acetone cyanohydrin	Prevent skin contact	Contact the manufacturer for recommendations
Acetonitrile	Prevent skin contact	8 hr: Butyl, Teflon, PE/EVAL, Barricade, Responder, Trellchem, Tychem
2-Acetylaminofluorene	Prevent skin contact	Contact the manufacturer for recommendations
Acetylene	Frostbite	Prevent skin freezing from direct contact

CHEMICAL	SKIN PROTECTION [†]	PROTECTIVE CLOTHING BARRIERS[*]
Acetylene tetrabromide	Prevent skin contact	Contact the manufacturer for recommendations
Acetylsalicyclic acid	Prevent skin contact	Contact the manufacturer for recommendations
Acrolein	Prevent skin contact	8 hr: Butyl, Barricade, Tychem
Acrylamide	Prevent skin contact	8 hr: Butyl, Tychem 4 hr: Nitrile, PVC, Viton, PE/EVAL
Acrylic acid	Prevent skin contact	8 hr: Butyl, Saranex, Responder, Trellchem; 4 hr: Teflon, Viton, PE/EVAL
Acrylonitrile	Prevent skin contact	8 hr: Butyl, PE/EVAL, Barricade, Responder, Tychem
Adiponitrile	Prevent skin contact	4 hr: Teflon
Aldrin	Prevent skin contact	Contact the manufacturer for recommendations
Allyl alcohol	Prevent skin contact	8 hr: Butyl, Teflon, Barricade, Responder, Tychem; 4 hr: Viton
Allyl chloride	Prevent skin contact	8 hr: Tychem; 4 hr: Teflon, PE/EVAL
Allyl glycidyl ether	Prevent skin contact	Contact the manufacturer for recommendations
Allyl propyl disulfide	Prevent skin contact	Contact the manufacturer for recommendations
Alpha-alumina	N.R.	Determine based on working conditions
Aluminum	N.R.	Determine based on working conditions
Aluminum (pyro powders and welding fumes, as Al)	N.R.	Determine based on working conditions
Aluminum (soluble salts and alkyls, as Al)	N.R.	Determine based on working conditions
4-Aminodiphenyl	Prevent skin contact	Contact the manufacturer for recommendations
2-Aminopyridine	Prevent skin contact	Contact the manufacturer for recommendations
Amitrole	Prevent skin contact	Contact the manufacturer for recommendations

CHEMICAL	SKIN PROTECTION [†]	PROTECTIVE CLOTHING BARRIERS[*]
Ammonia	Prevent skin contact	8 hr: Butyl, Teflon, Viton, Responder, Trellchem, Tychem; 4 hr: Nitrile
Ammonium chloride fume	Prevent skin contact	Any barrier that will prevent contamination from the dry chemical
Ammonium sulfamate	Prevent skin contact	Contact the manufacturer for recommendations
n-Amyl acetate	Prevent skin contact	8 hr: Barricade, Responder; 4 hr: PVA, Teflon
sec-Amyl acetate	Prevent skin contact	Contact the manufacturer for recommendations
Aniline (and homologs)	Prevent skin contact	8 hr: Butyl, PVA, PE/EVAL, Barricade, Responder, Trellchem; 4 hr: Teflon, Viton, Saranex
o-Anisidine	Prevent skin contact	Contact the manufacturer for recommendations
p-Anisidine	Prevent skin contact	Contact the manufacturer for recommendations
Antimony	Prevent skin contact	Any barrier that will prevent contamination from the dry chemical
ANTU	Prevent skin contact	Contact the manufacturer for recommendations
Arsenic (inorganic compounds, as As)	Prevent skin contact	Contact the manufacturer for recommendations
Arsenic, organic compounds (as As)	Recommendations regarding PPC vary	Contact the manufacturer for recommendations
Arsine	Prevent skin contact/ Frostbite	Contact the manufacturer for recommendations Prevent skin freezing from direct contact
Asbestos	Prevent skin contact	Any barrier that will prevent contamination from the fiber
Asphalt fumes	Prevent skin contact	Contact the manufacturer for recommendations
Atrazine	Prevent skin contact	Contact the manufacturer for recommendations
Azinphos-methyl	Prevent skin contact	8 hr: Neoprene, Nitrile
Barium chloride	Prevent skin contact	Contact the manufacturer for recommendations
Barium nitrate (as Ba)	Prevent skin contact	Contact the manufacturer for recommendations

CHEMICAL	SKIN PROTECTION [†]	PROTECTIVE CLOTHING BARRIERS[*]
Barium sulfate	Prevent skin contact	Contact the manufacturer for recommendations
Benomyl	Prevent skin contact	Contact the manufacturer for recommendations
Benzene	Prevent skin contact	8 hr: PVA, PE/EVAL, Barricade, CPF3, Responder, Tychem; 4 hr: Teflon, Viton
Benzenethiol	Prevent skin contact	Contact the manufacturer for recommendations
Benzidine	Prevent skin contact	Contact the manufacturer for recommendations
Benzoyl peroxide	Prevent skin contact	Contact the manufacturer for recommendations
Benzyl chloride	Prevent skin contact	8 hr: PE/EVAL, CPF3, Responder, Tychem; 4 hr: Teflon
Beryllium & beryllium compounds (as Be)	Prevent skin contact	Contact the manufacturer for recommendations
Bismuth telluride, doped with Selenium sulfide (as Bi_2Te_3)	Prevent skin contact	Contact the manufacturer for recommendations
Bismuth telluride, undoped	Prevent skin contact	Contact the manufacturer for recommendations
Borates, tetra, sodium salts (Anhydrous)	N.R.	Determine based on working conditions
Borates, tetra, sodium salts (Decahydrate)	N.R.	Determine based on working conditions
Borates, tetra, sodium salts (Pentahydrate)	N.R.	Determine based on working conditions
Boron oxide	Prevent skin contact	Contact the manufacturer for recommendations
Boron tribromide	Prevent skin contact	Contact the manufacturer for recommendations
Boron trifluoride	Prevent skin contact	8 hr: Teflon, Responder
Bromacil	Prevent skin contact	Contact the manufacturer for recommendations
Bromine	Prevent skin contact	4 hr: Teflon
Bromine pentafluoride	Prevent skin contact	Contact the manufacturer for recommendations

CHEMICAL	SKIN PROTECTION †	PROTECTIVE CLOTHING BARRIERS*
Bromoform	Prevent skin contact	8 hr: PVA, Viton
1,3-Butadiene	Prevent skin contact	8 hr: Viton, Saranex, Barricade, CPF3, Responder, Trellchem, Tychem; 4 hr: Teflon
n-Butane	Frostbite	Prevent skin freezing from direct contact
2-Butanone	Prevent skin contact	8 hr: Butyl, Teflon, PE/EVAL, Barricade, CPF3, Tychem; 4 hr: Responder
2-Butoxyethanol	Prevent skin contact	8 hr: Butyl, Viton, Saranex; 4 hr: PE/EVAL
2-Butoxyethanol acetate	Prevent skin contact	Contact the manufacturer for recommendations
n-Butyl acetate	Prevent skin contact	8 hr: PE/EVAL; 4 hr: PVA, Teflon
sec-Butyl acetate	Prevent skin contact	Contact the manufacturer for recommendations
tert-Butyl acetate	Prevent skin contact	Contact the manufacturer for recommendations
Butyl acrylate	Prevent skin contact	8 hr: PE/EVAL, Responder, Tychem; 4 hr: Teflon
n-Butyl alcohol	Prevent skin contact	8 hr: Butyl, Teflon, Viton, PE/EVAL, Barricade, CPF3, Responder; 4 hr: Neoprene
sec-Butyl alcohol	Prevent skin contact	8 hr: PE/EVAL; 4 hr: Butyl, Nitrile
tert-Butyl alcohol	Prevent skin contact	8 hr: Butyl, PE/EVAL, Responder
n-Butylamine	Prevent skin contact	8 hr: Responder, Tychem; 4 hr: Trellchem
tert-Butyl chromate	Prevent skin contact	Contact the manufacturer for recommendations
n-Butyl glycidyl ether	Prevent skin contact	Contact the manufacturer for recommendations
n-Butyl lactate	Prevent skin contact	Contact the manufacturer for recommendations
n-Butyl mercaptan	Prevent skin contact	Contact the manufacturer for recommendations
o-sec-Butylphenol	Prevent skin contact	Contact the manufacturer for recommendations
p-tert-Butyltoluene	Prevent skin contact	8 hr: PVA, Viton; 4 hr: Nitrile, PE/EVAL

CHEMICAL	SKIN PROTECTION [†]	PROTECTIVE CLOTHING BARRIERS[*]
n-Butyronitrile	Prevent skin contact	Contact the manufacturer for recommendations
Cadmium dust (as Cd)	Prevent skin contact	Any barrier that will prevent contamination from the dry chemical
Cadmium fume (as Cd)	Prevent skin contact	8 hr: Neoprene, Nitrile
Calcium arsenate (as As)	Prevent skin contact	Contact the manufacturer for recommendations
Calcium carbonate	N.R.	Determine based on working conditions
Calcium cyanamide	Prevent skin contact	Contact the manufacturer for recommendations
Calcium hydroxide	Prevent skin contact	8 hr: Natural, Neoprene, Nitrile
Calcium oxide	Prevent skin contact	Any barrier that will prevent contamination from the dry chemical
Calcium silicate	N.R.	Determine based on working conditions
Calcium sulfate	N.R.	Determine based on working conditions
Camphor (synthetic)	Prevent skin contact	Contact the manufacturer for recommendations
Caprolactam	Prevent skin contact	Contact the manufacturer for recommendations
Captafol	Prevent skin contact	Contact the manufacturer for recommendations
Captan	Prevent skin contact	Contact the manufacturer for recommendations
Carbaryl	Prevent skin contact	(Sevin 50 W); 4 hr: Natural, Neoprene, Nitrile, PVC
Carbofuran	Prevent skin contact	Contact the manufacturer for recommendations
Carbon black	N.R.	Determine based on working conditions
Carbon dioxide	Frostbite	Prevent possible skin freezing from direct liquid contact
Carbon disulfide	Prevent skin contact	8 hr: PVA, Viton, PE/EVAL, Barricade, Responder, Trellchem, Tychem; 4 hr: Teflon
Carbon monoxide	Frostbite	Prevent skin freezing from direct contact
Carbon tetrabromide	N.R.	Determine based on working conditions
Carbon tetrachloride	Prevent skin contact	8 hr: PVA, Viton, PE/EVAL, Barricade, Responder; 4 hr: Teflon

CHEMICAL	SKIN PROTECTION [†]	PROTECTIVE CLOTHING BARRIERS[*]
Carbonyl fluoride	Frostbite	Prevent possible skin freezing from direct liquid contact
Catechol	N.R.	Determine based on working conditions
Cellulose	N.R.	Determine based on working conditions
Cesium hydroxide	Prevent skin contact	Contact the manufacturer for recommendations
Chlordane	Prevent skin contact	8 hr: CPF3, Trellchem; 4 hr: Teflon
Chlorinated camphene	Prevent skin contact	Contact the manufacturer for recommendations
Chlorinated diphenyl oxide	Prevent skin contact	Contact the manufacturer for recommendations
Chlorine	Frostbite	Prevent possible skin freezing from direct liquid contact
Chlorine dioxide	Prevent skin contact (liquid)	Contact the manufacturer for recommendations
Chlorine trifluoride	Prevent skin contact	Contact the manufacturer for recommendations
Chloroacetaldehyde	Prevent skin contact	Contact the manufacturer for recommendations
alpha-Chloroacetophenone	Prevent skin contact	8 hr: Responder
Chloroacetyl chloride	Prevent skin contact	4 hr: Teflon
Chlorobenzene	Prevent skin contact	8 hr: Viton, Barricade, Responder, Trellchem, Tychem; 4 hr: PVA, Teflon
o-Chlorobenzylidene malononitrile	Prevent skin contact	Contact the manufacturer for recommendations
Chlorobromomethane	Prevent skin contact	4 hr: Teflon, Responder
Chlorodifluoro-methane	Frostbite	Prevent possible skin freezing from direct liquid contact
Chlorodiphenyl (42% chlorine)	Prevent skin contact	8 hr: Butyl, Neoprene, Teflon, Viton, Saranex, Barricade, Responder
		4 hr: PE/EVAL
Chlorodiphenyl (54% chlorine)	Prevent skin contact	8 hr: Butyl, Neoprene, Teflon, Viton, Saranex, Barricade, Responder; 4 hr: PE/EVAL

CHEMICAL	SKIN PROTECTION [†]	PROTECTIVE CLOTHING BARRIERS[*]
Chloroform	Prevent skin contact	8 hr: PVA, Viton, PE/EVAL, Barricade, Responder, Trellchem, Tychem; 4 hr: Teflon
bis-Chloromethyl ether	Prevent skin contact	Contact the manufacturer for recommendations
Chloromethyl methyl ether	Prevent skin contact	4 hr: Teflon
1-Chloro1-nitropropane	Prevent skin contact	Contact the manufacturer for recommendations
Chloropentafluoro-ethane	Frostbite	Prevent possible skin freezing from direct liquid contact
Chloropicrin	Prevent skin contact	Contact the manufacturer for recommendations
ß-Chloroprene	Prevent skin contact	8 hr: PVA, Viton, Responder
o-Chlorostyrene	Prevent skin contact	Contact the manufacturer for recommendations
o-Chlorotoluene	Prevent skin contact	8 hr: Barricade; 4 hr: Viton
2-Chloro-6-trichloro-methyl pyridine	Prevent skin contact	Contact the manufacturer for recommendations
Chlorpyrifos	Prevent skin contact	Contact the manufacturer for recommendations
Chromic acid and chromates	Prevent skin contact	(As Chromic Acid); 8 hr: PE, PVC, Saranex; 4 hr: Butyl, Viton
Chromium(II) compounds (as Cr)	Prevent skin contact	Contact the manufacturer for recommendations for the specific compound
Chromium(III) compounds (as Cr)	Prevent skin contact	Contact the manufacturer for recommendations for the specific compound
Chromium metal	N.R.	Determine based on working conditions
Chromyl chloride	Prevent skin contact	Contact the manufacturer for recommendations
Clopidol	N.R.	Determine based on working conditions
Coal dust	N.R.	Determine based on working conditions
Coal tar pitch volatiles	Prevent skin contact	Contact the manufacturer for recommendations
Cobalt metal dust and fume (as Co)	Prevent skin contact	Contact the manufacturer for recommendations
Cobalt carbonyl (as Co)	Prevent skin contact	Contact the manufacturer for recommendations

CHEMICAL	SKIN PROTECTION [†]	PROTECTIVE CLOTHING BARRIERS*
Cobalt hydrocarbonyl (as Co)	Prevent skin contact	Contact the manufacturer for recommendations
Coke oven emissions	Prevent skin contact	Recommendations will depend on specific nature of emissions
Copper (dusts and mists, as Cu)	Prevent skin contact	Contact the manufacturer for recommendations
Copper fume (as Cu)	N.R.	Determine based on working conditions
Cotton dust (raw)	N.R.	Determine based on working conditions
Crag® herbicide	Prevent skin contact	Contact the manufacturer for recommendations
o-Cresol	Prevent skin contact	Contact the manufacturer for recommendations
m-Cresol	Prevent skin contact	4 hr: Neoprene, Teflon
p-Cresol	Prevent skin contact	4 hr: PE/EVAL
Crotonaldehyde	Prevent skin contact	8 hr: Butyl, Responder; 4 hr: Teflon
Crufomate	Prevent skin contact	Contact the manufacturer for recommendations
Cumene	Prevent skin contact	8 hr: Tychem; 4 hr: Teflon
Cyanamide	Prevent skin contact	Contact the manufacturer for recommendations
Cyanogen	Frostbite	Prevent skin freezing from direct contact
Cyclohexane	Prevent skin contact	8 hr: Nitrile, Viton, PE/EVAL, Barricade, Responder; 4 hr: PVA, Teflon
Cyclohexanethiol	Prevent skin contact	Contact the manufacturer for recommendations
Cyclohexanol	Prevent skin contact	8 hr: Butyl, Nitrile, PVA, Teflon, Viton; 4 hr: Neoprene, PVC, PE/EVAL
Cyclohexanone	Prevent skin contact	8 hr: Butyl, PE/EVAL; 4 hr: PVA
Cyclohexene	Prevent skin contact	Contact the manufacturer for recommendations
Cyclohexylamine	Prevent skin contact	8 hr: Responder
Cyclonite	Prevent skin contact	Contact the manufacturer for recommendations

CHEMICAL	SKIN PROTECTION [†]	PROTECTIVE CLOTHING BARRIERS[*]
Cyclopentadiene	Prevent skin contact	Contact the manufacturer for recommendations
Cyclopentane	Prevent skin contact	Contact the manufacturer for recommendations
Cyhexatin	Prevent skin contact	Contact the manufacturer for recommendations
2,4-D	Prevent skin contact	8 hr: Natural, Neoprene, Nitrile, PVC
DDT	Prevent skin contact	Any barrier that will prevent contamination from the dry chemical
Decaborane	Prevent skin contact	Any barrier that will prevent contamination from the dry chemical
1-Decanethiol	Prevent skin contact	Contact the manufacturer for recommendations
Demeton	Prevent skin contact	Contact the manufacturer for recommendations
Diacetone alcohol	Prevent skin contact	8 hr: Butyl; 4 hr: Neoprene, PE/EVAL
2,4-Diaminoanisole (and its salts)	Prevent skin contact	Contact the manufacturer for recommendations
o-Dianisidine	Prevent skin contact	Contact the manufacturer for recommendations
Diazinon®	Prevent skin contact	Contact the manufacturer for recommendations
Diazomethane	Frostbite	Prevent possible skin freezing from direct liquid contact
Diborane	N.R.	Determine based on working conditions
1,2-Dibromo-3-chloro-propane	Prevent skin contact	Contact the manufacturer for recommendations
2-N-Dibutylaminoethanol	Prevent skin contact	Contact the manufacturer for recommendations
Dibutyl phosphate	Prevent skin contact	Contact the manufacturer for recommendations
Dibutyl phthalate	Prevent skin contact	8 hr: Butyl, Nitrile, PVA, Viton; 4 hr: PE/EVAL
Dichloroacetylene	Prevent skin contact	Contact the manufacturer for recommendations
o-Dichlorobenzene	Prevent skin contact	4 hr: Viton, PE/EVAL

CHEMICAL	SKIN PROTECTION [†]	PROTECTIVE CLOTHING BARRIERS[*]
p-Dichlorobenzene	Prevent skin contact	Contact the manufacturer for recommendations
3,3'-Dichlorobenzidine (and its salts)	Prevent skin contact	Contact the manufacturer for recommendations
Dichlorodifluoro-methane	Frostbite	Prevent skin freezing from direct contact
1,3-Dichloro-5,5-dimethylhydantoin	Prevent skin contact	Contact the manufacturer for recommendations
1,1-Dichloroethane	Prevent skin contact	8 hr: Tychem
1,2-Dichloroethylene	Prevent skin contact	8 hr: Teflon, Viton, PE/EVAL, Barricade, CPF3, Responder, Tychem; 4 hr: PVA
Dichloroethyl ether	Prevent skin contact	8 hr: Tychem; 4 hr: Teflon
Dichloromonofluorom ethane	Frostbite	Prevent skin freezing from direct contact
1,1-Dichloro-1-nitroethane	Prevent skin contact	Contact the manufacturer for recommendations
1,3-Dichloropropene	Prevent skin contact	8 hr: PVA, Viton, Responder; 4 hr: Teflon
2,2-Dichloropropionic acid	Prevent skin contact	Contact the manufacturer for recommendations
Dichlorotetrafluoroeth ane	Frostbite	Prevent skin freezing from direct contact
Dichlorvos	Prevent skin contact	Contact the manufacturer for recommendations
Dicrotophos	Prevent skin contact	Contact the manufacturer for recommendations
Dicyclopentadiene	Prevent skin contact	Contact the manufacturer for recommendations
Dicyclopentadienyl iron	N.R.	Determine based on working conditions
Dieldrin	Prevent skin contact	Contact the manufacturer for recommendations
Diethanolamine	Prevent skin contact	8 hr: Butyl, Neoprene, Nitrile, PVC, Viton, CPF3, Responder; 4 hr: Natural, Teflon
Diethylamine	Prevent skin contact	Contact the manufacturer for recommendations

CHEMICAL	SKIN PROTECTION [†]	PROTECTIVE CLOTHING BARRIERS[*]
2-Diethylaminoethanol	Prevent skin contact	8 hr: Butyl, Nitrile, PVA, Viton
Diethylenetriamine	Prevent skin contact	8 hr: Butyl, Neoprene, Viton, Tychem; 4 hr: PE/EVAL
Diethyl ketone	N.R.	Determine based on working conditions
Diethyl phthalate	Prevent skin contact	4 hr: PE/EVAL
Difluorodibromo-methane	Prevent skin contact	Contact the manufacturer for recommendations
Diglycidyl ether	Prevent skin contact	Contact the manufacturer for recommendations
Diisobutyl ketone	Prevent skin contact	4 hr: PVA, PE/EVAL
Diisopropylamine	Prevent skin contact	8 hr: Teflon, Viton
Dimethyl acetamide	Prevent skin contact	8 hr: Butyl, PE/EVAL, Barricade, CPF3, Responder, Tychem; 4 hr: Teflon
Dimethylamine	Prevent skin contact	8 hr: Butyl, Neoprene; 4 hr: Teflon, PE/EVAL
4-Dimethylamino-azobenzene	Prevent skin contact	Contact the manufacturer for recommendations
bis(2-(Dimethylamino)-ethyl)ether	Prevent skin contact	Contact the manufacturer for recommendations
Dimethylaminopropio-nitrile	Prevent skin contact	Contact the manufacturer for recommendations
N,N-Dimethylaniline	Prevent skin contact	8 hr: Tychem; 4 hr: PE/EVAL
Dimethyl carbamoyl chloride	Prevent skin contact	Contact the manufacturer for recommendations
Dimethyl-1,2-dibromo-2,2-dichlorethyl phosphate	Prevent skin contact	Contact the manufacturer for recommendations
Dimethylformamide	Prevent skin contact	8 hr: Butyl, Teflon, PE/EVAL, CPF3, Responder, Trellchem, Tychem
1,1-Dimethylhydrazine	Prevent skin contact	8 hr: Butyl, Barricade, Responder, Trellchem, Tychem
Dimethyl sulfate	Prevent skin contact	8 hr: Barricade, Responder, Tychem; 4 hr: Butyl, PE/EVAL

CHEMICAL	SKIN PROTECTION [†]	PROTECTIVE CLOTHING BARRIERS[*]
Dinitolmide	Prevent skin contact	Contact the manufacturer for recommendations
o-Dinitrobenzene	Prevent skin contact	Contact the manufacturer for recommendations
m-Dinitrobenzene	Prevent skin contact	Contact the manufacturer for recommendations
p-Dinitrobenzene	Prevent skin contact	Contact the manufacturer for recommendations
Dinitro-o-cresol	Prevent skin contact	Contact the manufacturer for recommendations
Dinitrotoluene	Prevent skin contact	Contact the manufacturer for recommendations
Di-sec octyl phthalate	N.R.	Determine based on working conditions
Dioxane	Prevent skin contact	8 hr: PE/EVAL, Tychem; 4 hr: Butyl, Teflon
Dioxathion	Prevent skin contact	Contact the manufacturer for recommendations
Diphenyl	Prevent skin contact	Contact the manufacturer for recommendations
Diphenylamine	Prevent skin contact	Contact the manufacturer for recommendations
Dipropylene glycol methyl ether	N.R.	Determine based on working conditions
Dipropyl ketone	Prevent skin contact	Contact the manufacturer for recommendations
Diquat (Diquat dibromide)	Prevent skin contact	4 hr: PE/EVAL
Disulfiram	Prevent skin contact	Contact the manufacturer for recommendations
Disulfoton	Prevent skin contact	Contact the manufacturer for recommendations
2,6-Di-tert-butyl-p-cresol	Prevent skin contact	Contact the manufacturer for recommendations
Diuron	Prevent skin contact	Contact the manufacturer for recommendations
Divinyl benzene	Prevent skin contact	8 hr: PVA, Viton
1-Dodecanethiol	Prevent skin contact	Contact the manufacturer for recommendations

CHEMICAL	SKIN PROTECTION †	PROTECTIVE CLOTHING BARRIERS*
Emery	N.R.	Determine based on working conditions
Endosulfan	Prevent skin contact	Contact the manufacturer for recommendations
Endrin	Prevent skin contact	Contact the manufacturer for recommendations
Enflurane	N.R.	Determine based on working conditions
Epichlorohydrin	Prevent skin contact	8 hr: Butyl, Barricade, Trellchem, Tychem; 4 hr: PE/EVAL, Responder
EPN	Prevent skin contact	Contact the manufacturer for recommendations
Ethanolamine	Prevent skin contact	8 hr: Butyl, Neoprene, Nitrile, Viton, PE/EVAL, Tychem; 4 hr: PVC
Ethion	Prevent skin contact	4 hr: Teflon
2-Ethoxyethanol	Prevent skin contact	8 hr: Butyl, Saranex, Responder; 4 hr: PE/EVAL
2-Ethoxyethyl acetate	Prevent skin contact	8 hr: Butyl, Barricade, Responder; 4 hr: PVA, PE/EVAL
Ethyl acetate	Prevent skin contact	8 hr: PE/EVAL, Barricade, CPF3, Responder, Trellchem, Tychem; 4 hr: PVA, Teflon
Ethyl acrylate	Prevent skin contact	8 hr: PVA, Teflon, Responder, Tychem; 4 hr: Butyl, PE/EVAL
Ethyl alcohol	Prevent skin contact	8 hr: Butyl, Viton, PE/EVAL; 4 hr: Neoprene, Teflon
Ethylamine	Prevent skin contact (liquid)	8 hr: Butyl; 4 hr: Teflon, Responder
Ethyl benzene	Prevent skin contact	8 hr: Viton, Barricade, Responder, Tychem; 4 hr: Teflon
Ethyl bromide	Prevent skin contact	Contact the manufacturer for recommendations
Ethyl butyl ketone	Prevent skin contact	Contact the manufacturer for recommendations
Ethyl chloride	Prevent skin contact (liquid)	4 hr: Teflon
Ethylene chlorohydrin	Prevent skin contact	8 hr: Butyl, Viton, PVA, Barricade; 4 hr: PE/EVAL
Ethylenediamine	Prevent skin contact	8 hr: Butyl, Saranex, Responder; 4 hr: Teflon

CHEMICAL	SKIN PROTECTION [†]	PROTECTIVE CLOTHING BARRIERS[*]
Ethylene dibromide	Prevent skin contact	8 hr: PVA, Teflon, Viton, PE/EVAL, Barricade, Tychem
Ethylene dichloride	Prevent skin contact	8 hr: Teflon, Viton, PE/EVAL, Barricade, CPF3, Responder, Tychem; 4 hr: PVA
Ethylene glycol	Prevent skin contact	8 hr: Butyl, Natural, Neoprene, Nitrile, PE, PVC, Teflon, Viton, Saranex, PE/EVAL, Trellchem; 4 hr: Responder
Ethylene glycol dinitrate	Prevent skin contact	Contact the manufacturer for recommendations
Ethyleneimine	Prevent skin contact	8 hr: Butyl, Tychem; 4 hr: Teflon, Responder
Ethylene oxide	Prevent skin contact (liquid)	8 hr: Barricade, Responder, Trellchem, Tychem; 4 hr: Butyl, Teflon, PE/EVAL
Ethylene thiourea	Prevent skin contact	Contact the manufacturer for recommendations
Ethyl ether	Prevent skin contact	8 hr: PVA, PE/EVAL, Barricade; 4 hr: Teflon, Responder, Trellchem
Ethyl formate	Prevent skin contact	Contact the manufacturer for recommendations
Ethylidene norbornene	Prevent skin contact	Contact the manufacturer for recommendations
Ethyl mercaptan	Prevent skin contact	4 hr: Teflon
N-Ethylmorpholine	Prevent skin contact	Contact the manufacturer for recommendations
Ethyl silicate	Prevent skin contact	Contact the manufacturer for recommendations
Fensulfothion	Prevent skin contact	Contact the manufacturer for recommendations
Fenthion	Prevent skin contact	Contact the manufacturer for recommendations
Ferbam	Prevent skin contact	Contact the manufacturer for recommendations
Fibrous glass dust	Prevent skin contact	Any barrier that will prevent contamination from the dust and fiber
Fluorine	Prevent skin contact (liquid)	8 hr: Barricade, Responder
Fluorotrichloro-methane	Prevent skin contact	Contact the manufacturer for recommendations

CHEMICAL	SKIN PROTECTION [†]	PROTECTIVE CLOTHING BARRIERS*
Fluoroxene	N.R.	Determine based on working conditions
Fonofos	Prevent skin contact	Contact the manufacturer for recommendations
Formaldehyde	Prevent skin contact	Contact the manufacturer for recommendations
Formalin (as formaldehyde)	Prevent skin contact	8 hr: Butyl, Nitrile, Viton, Saranex, Barricade, CPF3; 4 hr: Teflon, PE/EVAL, Responder
Formamide	N.R.	Determine based on working conditions
Formic acid	Prevent skin contact	8 hr: Butyl, Neoprene, Saranex, Barricade, Responder, Trellchem; 4 hr: PVC, Teflon
Furfural	Prevent skin contact	8 hr: Butyl, PE/EVAL, Barricade, CPF3, Trellchem, Tychem; 4 hr: PVA, Saranex
Furfuryl alcohol	Prevent skin contact	8 hr: PE/EVAL
Gasoline	Prevent skin contact	8 hr: Nitrile, Viton, Barricade; 4 hr: PVA, PE/EVAL, Responder
Germanium tetrahydride	N.R.	Determine based on working conditions
Glutaraldehyde	Prevent skin contact	8 hr: Butyl, Viton; 4 hr: Natural, Neoprene, Nitrile, PVC
Glycerin (mist)	Prevent skin contact	8 hr: Natural, Neoprene, Nitrile; 4 hr: PE/EVAL
Glycidol	Prevent skin contact	Contact the manufacturer for recommendations
Glycolonitrile	Prevent skin contact	Contact the manufacturer for recommendations
Grain dust (oat, wheat, barley)	N.R.	Determine based on working conditions
Graphite (natural)	N.R.	Determine based on working conditions
Graphite (synthetic)	N.R.	Determine based on working conditions
Gypsum	N.R.	Determine based on working conditions
Hafnium	Prevent skin contact	Contact the manufacturer for recommendations
Halothane	Prevent skin contact	8 hr: PVA
Heptachlor	Prevent skin contact	Contact the manufacturer for recommendations

CHEMICAL	SKIN PROTECTION †	PROTECTIVE CLOTHING BARRIERS*
n-Heptane	Prevent skin contact	8 hr: Nitrile, Viton, PE/EVAL
1-Heptanethiol	Prevent skin contact	Contact the manufacturer for recommendations
Hexachlorobutadiene	Prevent skin contact	8 hr: Responder, Tychem
Hexachlorocyclopenta diene	Prevent skin contact	8 hr: Butyl, Nitrile, PVA, Viton, Responder
Hexachloroethane	Prevent skin contact	Contact the manufacturer for recommendations
Hexachloronaph- thalene	Prevent skin contact	Contact the manufacturer for recommendations
1-Hexadecanethiol	Prevent skin contact	Contact the manufacturer for recommendations
Hexafluoroacetone	Prevent skin contact/ Frostbite	Contact the manufacturer for recommendations
Hexamethylene diisocyanate	Prevent skin contact	8 hr: Saranex, Barricade, Responder
Hexamethyl phosphoramide	Prevent skin contact	Contact the manufacturer for recommendations
n-Hexane	Prevent skin contact	8 hr: Nitrile, PVA, Teflon, Viton, PE/EVAL, CPF3, Responder, Trellchem, Tychem ; 4 hr: Barricade
Hexane isomers (excluding n-Hexane)	Prevent skin contact	Contact the manufacturer for recommendations
n-Hexanethiol	Prevent skin contact	Contact the manufacturer for recommendations
2-Hexanone	Prevent skin contact	Contact the manufacturer for recommendations
Hexone	Prevent skin contact	Contact the manufacturer for recommendations
sec-Hexyl acetate	Prevent skin contact	Contact the manufacturer for recommendations
Hexylene glycol	Prevent skin contact	Contact the manufacturer for recommendations
Hydrazine	Prevent skin contact	8 hr: Butyl, Neoprene, Nitrile, PVC, Teflon, Saranex, Barricade, Responder; 4 hr: PE/EVAL
Hydrogenated terphenyls	Prevent skin contact	Contact the manufacturer for recommendations

CHEMICAL	SKIN PROTECTION [†]	PROTECTIVE CLOTHING BARRIERS[*]
Hydrogen bromide	Prevent skin contact (solution)/ Frostbite	4 hr: Teflon
Hydrogen chloride	Prevent skin contact (solution)/ Frostbite	8 hr: Butyl, Teflon, Saranex, Barricade, Responder, Trellchem, Tychem ; 4 hr: Neoprene, PVC
	Prevent possible skin freezing from direct liquid contact	
Hydrogen cyanide	Prevent skin contact	8 hr: Teflon ; 4 hr: PE/EVAL, Responder, Tychem
Hydrogen fluoride (as F)	Prevent skin contact (liquid)	8 hr: Tychem; 4 hr: Teflon
Hydrogen peroxide (solution 30%-70%)	Prevent skin contact	8 hr: Butyl, Natural, Nitrile, PE, Viton, CPF3, Responder, Tychem; 4 hr: PVC, PE/EVAL
Hydrogen selenide	Frostbite	Prevent possible skin freezing from direct liquid contact
Hydroquinone	Prevent skin contact	4 hr: Natural, Neoprene, Nitrile, PVC, PE/EVAL
2-Hydroxypropyl acrylate	Prevent skin contact	Contact the manufacturer for recommendations
Indene	Prevent skin contact	Contact the manufacturer for recommendations
Indium	N.R.	Determine based on working conditions
Iodine	Prevent skin contact	8 hr: Saranex; 4 hr: PE
Iodoform	Prevent skin contact	Contact the manufacturer for recommendations
Iron oxide dust and fume (as Fe)	N.R.	Determine based on working conditions
Iron pentacarbonyl (as Fe)	Prevent skin contact	Contact the manufacturer for recommendations
Iron salts (soluble, as Fe)	Prevent skin contact	Contact the manufacturer for recommendations for the specific compound
Isoamyl acetate	Prevent skin contact	Contact the manufacturer for recommendations
Isoamyl alcohol (primary)	Prevent skin contact	8 hr: Butyl, Neoprene, Nitrile, Viton

CHEMICAL	SKIN PROTECTION [†]	PROTECTIVE CLOTHING BARRIERS*
Isoamyl alcohol (secondary)	Prevent skin contact	Contact the manufacturer for recommendations
Isobutane	Frostbite	Prevent possible skin freezing from direct liquid contact
Isobutyl acetate	Prevent skin contact	Contact the manufacturer for recommendations
Isobutyl alcohol	Prevent skin contact	8 hr: Butyl, Neoprene, Viton, Responder; 4 hr: Nitrile, PE/EVAL
Isobutyronitrile	Prevent skin contact	4 hr: Teflon
Isooctyl alcohol	Prevent skin contact	Contact the manufacturer for recommendations
Isophorone	Prevent skin contact	8 hr: Responder; 4 hr: PVA, PE/EVAL
Isophorone diisocyanate	Prevent skin contact	8 hr: Butyl, Nitrile, PVA, Viton, Responder
2-Isopropoxyethanol	Prevent skin contact	Contact the manufacturer for recommendations
Isopropyl acetate	Prevent skin contact	Contact the manufacturer for recommendations .
Isopropyl alcohol	Prevent skin contact	8 hr: Butyl, Nitrile, Viton, PE/EVAL, CPF3, Responder; 4 hr: Neoprene, Teflon
Isopropylamine	Prevent skin contact	8 hr: Tychem; 4 hr: Teflon
N-Isopropylaniline	Prevent skin contact	Contact the manufacturer for recommendations
Isopropyl ether	Prevent skin contact	Contact the manufacturer for recommendations
Isopropyl glycidyl ether	Prevent skin contact	Contact the manufacturer for recommendations
Kaolin	N.R.	Determine based on working conditions
Kepone	Prevent skin contact	Contact the manufacturer for recommendations
Kerosene	Prevent skin contact	8 hr: Nitrile, PE, Viton; 4 hr: Neoprene, PVA, PVC, Barricade, Responder
Ketene	N.R.	Determine based on working conditions
Lead	Prevent skin contact	Any barrier that will prevent contamination from the dust
Limestone	N.R.	Determine based on working conditions

CHEMICAL	SKIN PROTECTION [†]	PROTECTIVE CLOTHING BARRIERS[*]
Lindane	Prevent skin contact	Contact the manufacturer for recommendations
Lithium hydride	Prevent skin contact	Contact the manufacturer for recommendations
L.P.G.	Frostbite	Prevent possible skin freezing from direct liquid contact
Magnesite	N.R.	Determine based on working conditions
Magnesium oxide fume	N.R.	Determine based on working conditions
Malathion	Prevent skin contact	4 hr: Teflon, PE/EVAL
Maleic anhydride	Prevent skin contact	8 hr: Responder
Malonaldehyde	Prevent skin contact	Contact the manufacturer for recommendations
Malononitrile	Prevent skin contact	Contact the manufacturer for recommendations
Manganese compounds and fume (as Mn)	N.R.	Determine based on working conditions
Manganesecyclopen-tadienyl tricarbonyl (as Mn)	Prevent skin contact	Contact the manufacturer for recommendations
Manganese tetroxide (as Mn)	N.R.	Determine based on working conditions
Marble	N.R.	Determine based on working conditions
Mercury compounds [except (organo) alkyls] (as Hg)	Prevent skin contact	Contact the manufacturer for recommendations for the specific compound
Mercury	Prevent skin contact	Contact the manufacturer for recommendations
Mesityl oxide	Prevent skin contact	8 hr: Responder
Methacrylic acid	Prevent skin contact	8 hr: Butyl, Viton, PE/EVAL, REsponder, Tychem
Methomyl	Prevent skin contact	8 hr: Tychem
Methoxychlor	Prevent skin contact	Contact the manufacturer for recommendations

CHEMICAL	SKIN PROTECTION [†]	PROTECTIVE CLOTHING BARRIERS*
Methoxyflurane	N.R.	Determine based on working conditions
4-Methoxyphenol	Prevent skin contact	Contact the manufacturer for recommendations
Methyl acetate	Prevent skin contact	8 hr: PE/EVAL; 4 hr: Teflon
Methyl acetylene	Frostbite	Prevent possible skin freezing from direct liquid contact
Methyl acetylene-propadiene mixture	Frostbite	Prevent possible skin freezing from direct liquid contact
Methyl acrylate	Prevent skin contact	8 hr: Butyl, Tychem; 4 hr: Teflon
Methylacrylonitrile	Prevent skin contact	Contact the manufacturer for recommendations
Methylal	Prevent skin contact	Contact the manufacturer for recommendations
Methyl alcohol	Prevent skin contact	8 hr: Butyl, Teflon, Viton, Saranex, PE/EVAL, Responder, Trellchem, Tychem
Methylamine	Prevent skin contact/Frostbite Prevent possible skin freezing from direct liquid contact	8 hr: Responder, Tychem; 4 hr: Teflon
Methyl (n-amyl) ketone	Prevent skin contact	4 hr: PE/EVAL
Methyl bromide	Prevent skin contact (liquid)	8 hr: Responder, Tychem; 4 hr: Butyl, Neoprene, Teflon
Methyl Cellosolve®	Prevent skin contact	8 hr: Butyl, Tychem; 4 hr: PE/EVAL
Methyl Cellosolve® acetate	Prevent skin contact	8 hr: Butyl, Tychem; 4 hr: Saranex, PE/EVAL
Methyl chloride	Prevent skin contact/ Frostbite; Prevent possible skin freezing from direct liquid contact	8 hr: Viton, Saranex, Barricade, Responder, Trellchem, Tychem; 4 hr: Teflon
Methyl chloroform	Prevent skin contact	8 hr: PVA, Viton, PE/EVAL, Barricade, CPF3, Responder, Tychem; 4 hr: Teflon
Methyl-2-cyanoacrylate	Prevent skin contact	Contact the manufacturer for recommendations

CHEMICAL	SKIN PROTECTION [†]	PROTECTIVE CLOTHING BARRIERS*
Methylcyclohexane	Prevent skin contact	Contact the manufacturer for recommendations
Methylcyclohexanol	Prevent skin contact	Contact the manufacturer for recommendations
o-Methylcyclohex-anone	Prevent skin contact	Contact the manufacturer for recommendations
Methyl cyclopentadienyl manganese tricarbonyl (as Mn)	Prevent skin contact	Contact the manufacturer for recommendations
Methyl demeton	Prevent skin contact	Contact the manufacturer for recommendations
4,4'-Methylenebis(2-chloroaniline)	Prevent skin contact	8 hr: Saranex, Barricade; 4 hr: PE/EVAL
Methylene bis(4-cyclo-hexylisocyanate)	Prevent skin contact	Contact the manufacturer for recommendations
Methylene bisphenyl isocyanate	Prevent skin contact	8 hr: PE/EVAL, Barricade, Responder
Methylene chloride	Prevent skin contact	8 hr: PVA, PE/EVAL, Responder, Trellchem, Tychem; 4 hr: Teflon, Barricade
4,4'-Methylenedianiline	Prevent skin contact	8 hr: PE/EVAL
Methyl ethyl ketone peroxide	Prevent skin contact	Contact the manufacturer for recommendations
Methyl formate	Prevent skin contact	Contact the manufacturer for recommendations
5-Methyl-3-heptanone	Prevent skin contact	Contact the manufacturer for recommendations
Methyl hydrazine	Prevent skin contact	8 hr: Responder, Tychem
Methyl iodide	Prevent skin contact	8 hr: Viton, Responder, Tychem
Methyl isoamyl ketone	Prevent skin contact	Contact the manufacturer for recommendations
Methyl isobutyl carbinol	Prevent skin contact	Contact the manufacturer for recommendations
Methyl isocyanate	Prevent skin contact	8 hr: PVA, Barricade, Responder, Trellchem, Tychem
Methyl isopropyl ketone	Prevent skin contact	Contact the manufacturer for recommendations

CHEMICAL	SKIN PROTECTION [†]	PROTECTIVE CLOTHING BARRIERS[*]
Methyl mercaptan	Prevent skin contact (liquid)/Frostbite Prevent possible skin freezing from direct liquid contact	8 hr: Barricade, Responder, Tychem
Methyl methacrylate	Prevent skin contact	8 hr: PVA, PE/EVAL, Barricade, Trellchem; 4 hr: Teflon
Methyl parathion	Prevent skin contact	Contact the manufacturer for recommendations
Methyl silicate	Prevent skin contact	Contact the manufacturer for recommendations
alpha-Methyl styrene	Prevent skin contact	Contact the manufacturer for recommendations
Metribuzin	Prevent skin contact	Contact the manufacturer for recommendations
Mica (containing less than 1% quartz)	N.R.	Determine based on working conditions
Mineral wool fiber	Prevent skin contact	Any barrier that will prevent contamination from the dust and fiber
Molybdenum	N.R.	Determine based on working conditions
Molybdenum (soluble compounds, as Mo)	Prevent skin contact	Contact the manufacturer for recommendations for specific compounds
Monocrotophos	Prevent skin contact	Contact the manufacturer for recommendations
Monomethyl aniline	Prevent skin contact	Contact the manufacturer for recommendations
Morpholine	Prevent skin contact	8 hr: Butyl, PE/EVAL
Naphtha (coal tar)	Prevent skin contact	8 hr: Viton; 4 hr: Nitrile, PVA
Naphthalene	Prevent skin contact	8 hr: Teflon
Naphthalene diisocyanate	Prevent skin contact	Contact the manufacturer for recommendations
alpha-Naphthylamine	Prevent skin contact	Contact the manufacturer for recommendations
ß-Naphthylamine	Prevent skin contact	Contact the manufacturer for recommendations

CHEMICAL	SKIN PROTECTION [†]	PROTECTIVE CLOTHING BARRIERS[*]
Niax® Catalyst ESN	Prevent skin contact	Contact the manufacturer for recommendations
Nickel carbonyl	Prevent skin contact	8 hr: Tychem
Nickel metal and other compounds (as Ni)	Prevent skin contact	Contact the manufacturer for recommendations for specific compounds
Nicotine	Prevent skin contact	8 hr: Barricade, CPF3, Responder; 4 hr: Teflon, PE/EVAL
Nitric acid	Prevent skin contact	(<70% only); 8 hr: Butyl, Viton, Saranex, Barricade, CPF3, Trellchem, Tychem; 4 hr: Neoprene, PE, PE/EVAL, Responder
p-Nitroaniline	Prevent skin contact	Contact the manufacturer for recommendations
Nitrobenzene	Prevent skin contact	8 hr: Butyl, PVA, Teflon, Viton, PE/EVAL, Barricade, CPF3, Responder, Trellchem, Tychem
4-Nitrobiphenyl	Prevent skin contact	Contact the manufacturer for recommendations
p-Nitrochlorobenzene	Prevent skin contact	4 hr: Saranex
Nitroethane	Prevent skin contact	8 hr: Butyl, PE/EVAL, Barricade; 4 hr: Teflon
Nitrogen dioxide	Prevent skin contact	8 hr: Saranex
Nitrogen trifluoride	N.R.	Determine based on working conditions
Nitroglycerine	Prevent skin contact	4 hr: PE/EVAL
Nitromethane	Prevent skin contact	8 hr: Butyl, Teflon, PE/EVAL, Barricade, Responder, Trellchem; 4 hr: PVA
2-Nitronaphthalene	Prevent skin contact	Contact the manufacturer for recommendations
1-Nitropropane	Prevent skin contact	8 hr: Butyl, PVA
2-Nitropropane	Prevent skin contact	8 hr: Butyl, PVA, Tychem; 4 hr: Teflon, PE/EVAL
o-Nitrotoluene	Prevent skin contact	8 hr: Butyl; 4 hr: Teflon, Saranex
m-Nitrotoluene	Prevent skin contact	Contact the manufacturer for recommendations

CHEMICAL	SKIN PROTECTION [†]	PROTECTIVE CLOTHING BARRIERS[*]
p-Nitrotoluene	Prevent skin contact	Contact the manufacturer for recommendations
Nitrous oxide	Frostbite	Prevent possible skin freezing from direct liquid contact
Nonane	N.R.	Determine based on working conditions
1-Nonanethiol	Prevent skin contact	Contact the manufacturer for recommendations
Octachloronaphthalene	Prevent skin contact	Contact the manufacturer for recommendations
1-Octadecanethiol	Prevent skin contact	Contact the manufacturer for recommendations
Octane	Prevent skin contact	8 hr: Responder, Tychem; 4 hr: Nitrile, Viton
1-Octanethiol	Prevent skin contact	Contact the manufacturer for recommendations
Oil mist (mineral)	Prevent skin contact	Contact the manufacturer for recommendations
Osmium tetroxide	Prevent skin contact	Contact the manufacturer for recommendations
Oxalic acid	Prevent skin contact	8 hr: Butyl, Natural, Neoprene, Nitrile, PVC, Viton
Oxygen difluoride	N.R.	Determine based on working conditions
Ozone	N.R.	Determine based on working conditions
Paraffin wax fume	N.R.	Determine based on working conditions
Paraquat (Paraquat dichloride)	Prevent skin contact	Contact the manufacturer for recommendations
Parathion	Prevent skin contact	8 hr: Tychem; 4 hr: Teflon
Particulates not otherwise regulated	N.R.	Determine based on working conditions
Pentaborane	Prevent skin contact	Contact the manufacturer for recommendations
Pentachloroethane	Prevent skin contact	Contact the manufacturer for recommendations
Pentachloro-naphthalene	Prevent skin contact	Contact the manufacturer for recommendations
Pentachlorophenol	Prevent skin contact	8 hr: Nitrile, Viton; 4 hr: Neoprene, PE/EVAL
Pentaerythritol	N.R.	Determine based on working conditions
n-Pentane	Prevent skin contact	8 hr: Viton, PE/EVAL; 4 hr: Nitrile, PVA

CHEMICAL	SKIN PROTECTION [†]	PROTECTIVE CLOTHING BARRIERS*
1-Pentanethiol	Prevent skin contact	Contact the manufacturer for recommendations
2-Pentanone	Prevent skin contact	4 hr: Butyl
Perchloromethyl mercaptan	Prevent skin contact	Contact the manufacturer for recommendations
Perchloryl fluoride	Frostbite	Prevent possible skin freezing from direct contact
Perlite	N.R.	Determine based on working conditions
Petroleum distillates (naphtha)	Prevent skin contact	Contact the manufacturer for recommendations
Phenol	Prevent skin contact	8 hr: Viton, Saranex, Barricade, Responder, Trellchem; 4 hr: Butyl, Neoprene, Teflon, PE/EVAL
Phenothiazine	Prevent skin contact	Contact the manufacturer for recommendations
p-Phenylene diamine	Prevent skin contact	Contact the manufacturer for recommendations
Phenyl ether (vapor)	Prevent skin contact	Contact the manufacturer for recommendations
Phenyl ether-biphenyl mixture (vapor)	Prevent skin contact	Contact the manufacturer for recommendations
Phenyl glycidyl ether	Prevent skin contact	Contact the manufacturer for recommendations
Phenylhydrazine	Prevent skin contact	Contact the manufacturer for recommendations
N-Phenyl-ß-naphthylamine	Prevent skin contact	Contact the manufacturer for recommendations
Phenylphosphine	Prevent skin contact	Contact the manufacturer for recommendations
Phorate	Prevent skin contact	Contact the manufacturer for recommendations
Phosdrin	Prevent skin contact	Contact the manufacturer for recommendations
Phosgene	Prevent skin contact (liquid)	8 hr: Responder, Tychem; 4 hr: Teflon
Phosphine	Prevent skin contact/ Frostbite; Prevent possible skin freezing from direct contact	8 hr: Responder.

CHEMICAL	SKIN PROTECTION †	PROTECTIVE CLOTHING BARRIERS*
Phosphoric acid	Prevent skin contact	8 hr: Butyl, Natural, Neoprene, Nitrile, PE, PVC, Viton, Saranex, PE/EVAL, Barricade, Responder, Trellchem
Phosphorus (yellow)	Prevent skin contact* [*Note: Flame retardant equipment should be provided.]	Contact the manufacturer for recommendations
Phosphorus oxychloride	Prevent skin contact	8 hr: Responder, Trellchem; 4 hr: Teflon, PE/EVAL
Phosphorus pentachloride	Prevent skin contact	Contact the manufacturer for recommendations
Phosphorus pentasulfide	Prevent skin contact Flame retardant equipment should be provided	Contact the manufacturer for recommendations
Phosphorus trichloride	Prevent skin contact	8 hr: Barricade, Trellchem; 4 hr: Teflon
Phthalic anhydride	Prevent skin contact	4 hr: PE/EVAL
m-Phthalodinitrile	Prevent skin contact	Contact the manufacturer for recommendations
Picloram	Prevent skin contact	Contact the manufacturer for recommendations
Picric acid	Prevent skin contact	Contact the manufacturer for recommendations
Pindone	N.R.	Determine based on working conditions
Piperazine dihydrochloride	Prevent skin contact	Contact the manufacturer for recommendations
Plaster of Paris	N.R.	Determine based on working conditions
Platinum	N.R.	Determine based on working conditions
Platinum (soluble salts, as Pt)	Prevent skin contact	Contact the manufacturer for recommendations for specific compounds
Portland cement	Prevent skin contact	Any barrier that will prevent contamination from the cement
Potassium cyanide (as CN)	Prevent skin contact	(solution <30% only); 8 hr: PE
Potassium hydroxide	Prevent skin contact	(solution <70% only); 8 hr: Butyl, Natural, Nitrile, PVC, Viton; 4 hr: Teflon, PE/EVAL
Propane	Frostbite; Prevent possible skin freezing from direct contact	

CHEMICAL	SKIN PROTECTION [†]	PROTECTIVE CLOTHING BARRIERS[*]
Propane sultone	Prevent skin contact	Contact the manufacturer for recommendations
Propargyl alcohol	Prevent skin contact	Contact the manufacturer for recommendations
ß-Propiolactone	Prevent skin contact	4 hr: Teflon
Propionic acid	Prevent skin contact	4 hr: Teflon
Propionitrile	Prevent skin contact	8 hr: PVA
Propoxur	Prevent skin contact	Contact the manufacturer for recommendations
n-Propyl acetate	Prevent skin contact	8 hr: PE/EVAL
n-Propyl alcohol	Prevent skin contact	8 hr: Butyl, Nitrile, Viton; 4 hr: Neoprene, PVA
Propylene dichloride	Prevent skin contact	Contact the manufacturer for recommendations
Propylene glycol dinitrate	Prevent skin contact	Contact the manufacturer for recommendations
Propylene glycol monomethyl ether	Prevent skin contact	8 hr: Butyl; 4 hr: Neoprene, PE/EVAL
Propylene imine	Prevent skin contact	Contact the manufacturer for recommendations
Propylene oxide	Prevent skin contact	8 hr: Barricade, Tychem; 4 hr: Teflon, PE/EVAL, Responder
n-Propyl nitrate	Prevent skin contact	Contact the manufacturer for recommendations
Pyrethrum	Prevent skin contact	Contact the manufacturer for recommendations
Pyridine	Prevent skin contact	8 hr: PE/EVAL, Responder, Tychem
Quinone	Prevent skin contact	8 hr: Saranex
Resorcinol	Prevent skin contact	Contact the manufacturer for recommendations
Rhodium (metal fume and as Rh)	N.R.	Determine based on working conditions
Rhodium (soluble compounds, as Rh)	Prevent skin contact	Contact the manufacturer for recommendations for specific compound
Ronnel	Prevent skin contact	Contact the manufacturer for recommendations
Rosin core solder, pyrolysis products (as formaldehyde)	N.R.	Determine based on working conditions

CHEMICAL	SKIN PROTECTION [†]	PROTECTIVE CLOTHING BARRIERS[*]
Rotenone	Prevent skin contact	Contact the manufacturer for recommendations
Rouge	N.R.	Determine based on working conditions
Selenium hexafluoride	N.R.	Determine based on working conditions
Silica, amorphous	N.R.	Determine based on working conditions
Silica, crystalline (as respirable dust)	N.R.	Determine based on working conditions
Silicon	N.R.	Determine based on working conditions
Silicon carbide	N.R.	Determine based on working conditions
Silicon tetrahydride	N.R.	Determine based on working conditions
Silver (metal dust and soluble compounds	Prevent skin contact	Contact the manufacturer for recommendations for specific compound
Soapstone (containing less than 1% quartz)	N.R.	Determine based on working conditions
Sodium aluminum fluoride (as F)	Prevent skin contact	Contact the manufacturer for recommendations
Sodium azide	Prevent skin contact	Contact the manufacturer for recommendations
Sodium bisulfite	N.R.	Determine based on working conditions
Sodium cyanide (as CN)	Prevent skin contact	(solution > 70% only); 8 hr: Saranex, Barricade
Sodium fluoride (as F)	Prevent skin contact	8 hr: Natural, Neoprene, Nitrile, PVC, Saranex
Sodium fluoroacetate	Prevent skin contact	Contact the manufacturer for recommendations
Sodium hydroxide	Prevent skin contact	(solution > 70% only); 8 hr: Neoprene, PVC, Barricade
Sodium metabisulfite	N.R.	Determine based on working conditions
Starch	Prevent skin contact	Any barrier that will prevent skin contact
Stibine	N.R.	Determine based on working conditions
Stoddard solvent	Prevent skin contact	8 hr: Nitrile, Viton, Saranex, PE/EVAL, Barricade, Responder; 4 hr: PVA
Strychnine	Prevent skin contact	Contact the manufacturer

CHEMICAL	SKIN PROTECTION [†]	PROTECTIVE CLOTHING BARRIERS[*]
Styrene	Prevent skin contact	8 hr: Viton, PE/EVAL, Barricade, CPF3, Trellchem; 4 hr: PVA, Teflon, Responder
Subtilisins	Prevent skin contact	Contact the manufacturer for recommendations
Succinonitrile	Prevent skin contact	Contact the manufacturer for recommendations
Sucrose	N.R.	Determine based on working conditions
Sulfur dioxide	Prevent skin contact	8 hr: Saranex, Barricade, Responder; 4 hr: Teflon
Sulfur hexafluoride	Frostbite; Prevent possible skin freezing from direct contact with liquid	
Sulfuric acid (solution > 70% only)	Prevent skin contact	8 hr: Butyl, PE, Teflon, Saranex, PE/EVAL, Barricade, CPF3, Responder, Trellchem, Tychem; 4 hr: Viton
Sulfur monochloride	Prevent skin contact	Contact the manufacturer for recommendations
Sulfur pentafluoride	Prevent skin contact	Contact the manufacturer for recommendations
Sulfur tetrafluoride	Frostbite; Prevent possible skin freezing from direct liquid contact	
Sulfuryl fluoride	Frostbit; Prevent possible skin freezing from direct liquid contact	
Sulprofos	Prevent skin contact	Contact the manufacturer for recommendations
Talc (containing no asbestos and less than 1% quartz)	N.R.	Determine based on working conditions and contact with manufacturer for recommendations
Tantalum (metal and oxide dust, as Ta)	N.R.	Determine based on working conditions and contact with manufacturer for recommendations
TEDP	Prevent skin contact	Determine based on working conditions and contact the manufacturer for recommendations

CHEMICAL	SKIN PROTECTION [†]	PROTECTIVE CLOTHING BARRIERS[*]
Tellurium	N.R.	Determine based on working conditions
Tellurium hexafluoride	N.R.	Determine based on working conditions
Temephos	Prevent skin contact	Contact the manufacturer for recommendations
TEPP	Prevent skin contact	Contact the manufacturer for recommendations
o-Terphenyl	Prevent skin contact	Contact the manufacturer for recommendations
m-Terphenyl	Prevent skin contact	Contact the manufacturer for recommendations
p-Terphenyl	Prevent skin contact	Contact the manufacturer for recommendations
2,3,7,8-Tetrachlorodibenzo-p-dioxin	Prevent skin contact	Contact the manufacturer for recommendations
1,1,1,2-Tetrachloro-2,2-difluoroethane	Prevent skin contact	Contact the manufacturer for recommendations
1,1,2,2-Tetrachloro-1,2-difluoroethane	Prevent skin contact	Contact the manufacturer for recommendations
1,1,1,2-Tetrachloroethane	Prevent skin contact	8 hr: PVA, Viton
1,1,2,2-Tetrachloroethane	Prevent skin contact	8 hr: PVA, Teflon, Viton, Barricade, Tychem
Tetrachloro-ethylene	Prevent skin contact	8 hr: PVA, Teflon, Viton, PE/EVAL, Barricade, CPF3, Responder, Trellchem, Tychem
Tetrachloronaph-thalene	Prevent skin contact	Contact the manufacturer for recommendations
Tetraethyl lead (as Pb)	Prevent skin contact (>0.1%)	8 hr: CPF3, Tychem
Tetrahydrofuran	Prevent skin contact	8 hr: Teflon, PE/EVAL, Barricade, CPF3, Responder, Trellchem, Tychem
Tetramethyl lead (as Pb)	Prevent skin contact (>0.1%)	Contact the manufacturer for recommendations
Tetramethyl succinonitrile	Prevent skin contact	Contact the manufacturer for recommendations

CHEMICAL	SKIN PROTECTION [†]	PROTECTIVE CLOTHING BARRIERS[*]
Tetranitromethane	Prevent skin contact	Contact the manufacturer for recommendations
Tetrasodium pyrophosphate	Prevent skin contact	Contact the manufacturer for recommendations
Tetryl	Prevent skin contact	Contact the manufacturer for recommendations
Thallium (soluble compounds, as Tl)	Prevent skin contact	Contact the manufacturer for recommendations for the specific compound
4,4'-Thiobis(6-tert-butyl-m-cresol)	N.R.	Determine based on working conditions
Thioglycolic acid	Prevent skin contact	8 hr: Butyl, Neoprene, Viton; 4 hr: PE/EVAL
Thionyl chloride	Prevent skin contact	Contact the manufacturer for recommendations
Thiram	Prevent skin contact	Contact the manufacturer for recommendations
Tin	N.R.	Determine based on working conditions
Tin (organic compounds, as Sn)	Recommendations vary depending upon the specific compound.	Contact the manufacturer for recommendations for the specific compound
Tin(II) oxide (as Sn)	N.R.	Determine based on working conditions
Tin(IV) oxide (as Sn)	N.R.	Determine based on working conditions
Titanium dioxide	N.R.	Determine based on working conditions
o-Tolidine	Prevent skin contact	Contact the manufacturer for recommendations
Toluene	Prevent skin contact	8 hr: PVA, Teflon, Viton, PE/EVAL, Barricade, CPF3, Responder, Trellchem, Tychem
Toluenediamine	Prevent skin contact	Contact the manufacturer for recommendations
Toluene-2,4-diisocyanate	Prevent skin contact	8 hr: Butyl, Nitrile, PVA, PVC, Viton, Saranex, PE/EVAL, Barricade, CPF3, Responder; 4 hr: Teflon
o-Toluidine	Prevent skin contact	8 hr: Barricade, Tychem; 4 hr: Teflon, Saranex
m-Toluidine	Prevent skin contact	8 hr: Saranex

CHEMICAL	SKIN PROTECTION [†]	PROTECTIVE CLOTHING BARRIERS*
p-Toluidine	Prevent skin contact	Contact the manufacturer for recommendations
Tributyl phosphate	Prevent skin contact	4 hr: PE/EVAL
Trichloroacetic acid	Prevent skin contact	8 hr: Trellchem; 4 hr: Nitrile
1,2,4-Trichloro-benzene	Prevent skin contact	8 hr: Teflon, Barricade; 4 hr: Viton
1,1,2-Trichloroethane	Prevent skin contact	8 hr: Teflon, Viton, Tychem; 4 hr: PVA
Trichloroethylene	Prevent skin contact	8 hr: PVA, Viton, PE/EVAL, Barricade, Trellchem, Tychem; 4 hr: Teflon, Responder
Trichloronaph-thalene	Prevent skin contact	Contact the manufacturer for recommendations
1,2,3-Trichloropropane	Prevent skin contact	8 hr: Butyl, PVA, Viton
1,1,2-Trichloro-1,2,2-trifluoroethane	Prevent skin contact	8 hr: Nitrile, Teflon, Barricade, CPF3, Responder; 4 hr: PVA, Viton, PE/EVAL
Triethylamine	Prevent skin contact	8 hr: Nitrile, Viton, Saranex, Responder, Trellchem, Tychem
Trifluorobromo-methane	Frostbite; Prevent skin freezing from direct liquid contact	
Trimellitic anhydride	Prevent skin contact	Contact the manufacturer for recommendations
Trimethylamine	Prevent skin contact (liquid)/ Frostbite	4 hr: Teflon
1,2,3-Trimethyl-benzene	Prevent skin contact	Contact the manufacturer
1,2,4-Trimethyl-benzene	Prevent skin contact	8 hr: PVA, Viton, PE/EVAL, Barricade, CPF3, Tychem; 4 hr: Teflon, Responder
1,3,5-Trimethyl-benzene	Prevent skin contact	Contact the manufacturer for recommendations
Trimethyl-phosphite	Prevent skin contact	Contact the manufacturer for recommendations
2,4,6-Trinitrotoluene	Prevent skin contact	Contact the manufacturer for recommendations
Triorthocresyl phosphate	Prevent skin contact	Contact the manufacturer for recommendations

CHEMICAL	SKIN PROTECTION [†]	PROTECTIVE CLOTHING BARRIERS[*]
Triphenylamine	Prevent skin contact	Contact the manufacturer for recommendations
Triphenyl phosphate	N.R.	Determine based on working conditions
Tungsten	N.R.	Determine based on working conditions
Tungsten (soluble compounds, as W)	Recommendations for protective clothing vary	Contact the manufacturer for recommendations for the specific compound
Tungsten carbide	Prevent skin contact	Contact the manufacturer for recommendations
Turpentine	Prevent skin contact	8 hr: Viton, PE/EVAL, Responder; 4 hr: Nitrile, PVA, Teflon
1-Undecanethiol	Prevent skin contact	Contact the manufacturer for recommendations
Uranium (insoluble compounds, as U)	Prevent skin contact	Contact the manufacturer for recommendations
Uranium (soluble compounds, as U)	Prevent skin contact	Contact the manufacturer for recommendations for specific compound
n-Valeraldehyde	Prevent skin contact	Contact the manufacturer for recommendations
Vanadium dust	Prevent skin contact	Any barrier that will prevent contamination from the dust
Vanadium fume	N.R.	Determine based on working conditions
Vegetable oil mist	N.R.	Determine based on working conditions
Vinyl acetate	Prevent skin contact	8 hr: Teflon, PE/EVAL, Barricade; 4 hr: Responder
Vinyl bromide	Prevent skin contact (liquid)	Contact the manufacturer for recommendations
Vinyl chloride	Prevent skin contact/ Frostbite; Prevent possible skin freezing from direct liquid contact	8 hr: Tychem; 4 hr: PVA, Teflon
Vinyl cyclohexene dioxide	Prevent skin contact	Contact the manufacturer for recommendations
Vinyl fluoride	Frostbite/ Prevent possible skin freezing from direct liquid contact	

CHEMICAL	SKIN PROTECTION [†]	PROTECTIVE CLOTHING BARRIERS*
Vinylidene chloride	Prevent skin contact/ Frostbite; Prevent possible skin freezing from direct liquid contact	Contact the manufacturer for recommendations.
Vinylidene fluoride	Prevent skin contact/ Frostbite; Prevent possible skin freezing from direct liquid contact	8 hr: Butyl, Teflon, Viton
Vinyl toluene	Prevent skin contact	Contact the manufacturer for recommendations
VM & P Naphtha	Prevent skin contact	Contact the manufacturer for recommendations
Warfarin	Prevent skin contact	Contact the manufacturer for recommendations
Welding fumes	N.R.	Determine based on working conditions
o-Xylene	Prevent skin contact	Contact the manufacturer for recommendations
m-Xylene	Prevent skin contact	Contact the manufacturer for recommendations
p-Xylene	Prevent skin contact	Contact the manufacturer for recommendations
m-Xylene alpha,alpha'-diamine	Prevent skin contact	Contact the manufacturer for recommendations
Xylidine	Prevent skin contact	Contact the manufacturer for recommendations
Yitrium	N.R.	Determine based on working conditions
Zinc chloride fume	N.R.	Determine based on working conditions
Zinc oxide	N.R.	Determine based on working conditions
Zinc stearate	N.R.	Determine based on working conditions
Zirconium compounds (as Zr)	Recommendations regarding personal protective clothing vary depending upon the specific compound.	Contact the manufacturer for recommendations for the specific compound

† *Footnotes on recommendations for skin protection*

Prevent skin contact: Wear appropriate personal protective clothing to prevent skin contact. Suggested barriers for use should be confirmed with the vendor and for additional information and use limitations.

Frostbite: Wear appropriate personal protective clothing to prevent the skin from becoming frozen from contact with the evaporating liquid or from contact with vessels containing the liquid.

N.R.: No specific recommendation can be made. Actual working conditions will determine the need and type of personal protective equipment.

* *Footnotes on recommended protective clothing barriers*

Butyl = Butyl Rubber (Gloves, Suits, Boots)

Natural = Natural Rubber (Gloves)

Neoprene = Neoprene Rubber (Gloves, Suits, Boots)

Nitrile = Nitrile Rubber (Gloves, Suits, Boots)

PE = Polyethylene (Gloves, Suits, Boots)

PVA = Polyvinyl Alcohol (Gloves)

PVC = Polyvinyl Chloride (Gloves, Suits, Boots)

Teflon = Teflon™ (Gloves, Suits, Boots)

Viton = Viton® (Gloves, Suits)

Saranex = Saranex™ coated suits

PE/EVAL = 4H™ and Silver Shield™ brand gloves

Barricade = Barricade™ coated suits

CPF3 = CPF3™ suits

Responder = Responder™ suits

Trellchem = Trellchem HPS™ suits

Tychem = Tychem 10000™ suits

8 hr = *More than 8 hours of resistance to breakthrough* >0.1 g/cm²/min.;

4 hr = *At least 4 but less than 8 hours of resistance to breakthrough* >0.1 g/cm²/min.

Neoprene is a tradename and Teflon™, Barricade™ and Tychem 10000™ are trademarks of the DuPont Company. Viton® is a registered trademark of DuPont Dow Elastomers. Saranex is a tradename of the Dow Chemical Company. 4H is a trademark of the Safety 4 Company. Silver Shield is a trademark of the Siebe North Company. CPF3 and Responder are trademarks of the Kappler Company. Trellchem HPS is a trademark of the Trelleborg Company. Recommendations are NOT valid for very thin Natural Rubber, Neoprene, Nitrile, and PVC gloves (0.3 mm or less).

OSHA has amended standards for personal protective equipment (PPE) to require employers to assess the workplace to determine if there are any hazards requiring the use of PPE and certify that this assessment was performed. Protective apparel should always be worn if there is a possibility that personal clothing could become contaminated with hazardous material.

Table 3 summarizes generic PPE requirement for laboratories. Laboratory coats and gloves that have been used in the lab need to be left there to minimize the possibility of spreading chemicals and/or pathogens to other areas. Standard prescriptive eye glasses are not acceptable for eye protection. Clothing that leaves large areas of the skin exposed is inappropriate in laboratories where hazardous chemicals are in use. Shorts, short skirts, sandals, open-toed, or high-heeled shoes are dangerous attire in such laboratories. Long hair, dangling jewelry, and loose fitting clothing should be constrained.

Table 4 provides a summary of chemical resistance evaluations for different glove materials. Note that aromatic and halogenated hydrocarbons will attack all types of natural and synthetic glove materials.

Table 3. Hazard Assessment and PPE Requirements for Laboratory Operations

| Hazard | Personal Protective Equipment Required | | |
	Eye	Face	Hand/Skin/Body
Any lab use of chemicals	Safety glasses at all times		Lab coat
Use of corrosive chemicals, strong oxidizing agents, carcinogens, mutagens, etc.	Chemical splash goggles	Full face shield and goggles (for work with over 4 liters of corrosive liquids)	Resistant gloves -- Impervious lab coat, coveralls, apron, protective suit (for work with over 5 gallons corrosive liquids)
Temperature extremes			Insulated gloves for handling ovens, furnaces, cryogenic bath and other devices over 100 °C or below -1 °C
Sharp objects (broken glass, insertion of tubes or rods into stoppers)			Heavy cloth barrier or leather gloves

Should swelling occur, the user should change to fresh gloves and allow the swollen gloves to dry and return to normal. No data on the resistance to dimetyl sulfoxide of natural rubber, neoprene, nitrile rubber, or vinyl materials are available; the manufacturer of the substance recommends the use of butyl rubber gloves.

Table 4. Chemical Resistance of Common Glove Materials
(E=Excellent, G=Good, F=Fair, P=Poor)

Chemical	Natural Rubber	Neoprene	Nitrile	Vinyl
Acetaldehyde	G	G	E	G
Acetic Acid	E	E	E	E
Acetone	G	G	G	F
Acrylonitrile	P	G	-	F
Ammonium Hydroxide	G	E	E	E
Aniline	F	G	E	G
Benzaldehyde	F	F	E	G
Benzene	P	F	G	F
Benzyl Chloride	F	P	G	P
Bromine	G	G	-	G
Butane	P	E	-	P
Calcium Hypochlorite	P	G	G	G
Carbon Disulfide	P	P	G	F
Carbon Tetrachloride	P	F	G	F
Chlorine	G	G	-	G
Chlorocetone	F	E	-	P
Chloroform	P	F	G	P
Chromic Acid	P	F	F	E
Cyclohexane	F	E	-	P
Dibenzylether	F	G	-	P
Dibutyl Phthalate	F	G	-	P
Diethanolamine	F	E	-	E
Diethyl Ether	F	G	E	P
Dimethyl Sulfoxide	-	-	-	-
Ethyl Acetate	F	G	G	F

Table 4. Continued

Chemical	Natural Rubber	Neoprene	Nitrile	Vinyl
Ethylene Dichloride	P	F	G	P
Ethylene Glycol	G	G	E	E
Ethylene Trichloride	P	P	'-	P
Fluorine	G	G	-	G
Formic Acid	G	E	E	E
Formaldehyde	G	E	E	E
Glycerol	G	G	E	E
Hexane	P	E	-	P
Hydrobromic acid (40%)	G	E	-	E
Hydrofluoric acid (30%)	G	G	G	E
Hydrogen Peroxide	G	G	G	E
Iodine	G	G	-	G
Methylamine	G	G	E	E
Methyl Cellosolve	F	E	-	P
Methyl Chloride	P	E	-	P
Methyl Ethyl Ketone	F	G	G	P
Methylene Chloride	F	F	G	F
Monoethanolamine	F	E	-	E
Morpholine	F	E	-	E
Naphthalene	G	G	E	G
Nitric Acid (conc)	P	P	P	G
Perchloric Acid	F	G	F	E
Phenol	G	E	-	E
Phosphoric Acid	G	E	-	E
Potassium Hydroxide (sat)	G	G	G	E
Propylene Dichloride	P	F	-	P
Sodium Hydroxide	G	G	G	E
Sodium Hypochlorite	G	P	F	G
Sulfuric Acid (conc)	G	G	F	G
Toluene	P	F	G	F
Trichloroethylene	P	F	G	F

LEVELS OF PROTECTION

OSHA defines four levels of protection for skin and respiratory safety in the workplace. These are levels A, B, C and D, as defined below:

Level A: Level A protection should be worn when the highest available level of both respiratory, skin and eye contact protections are needed. While Level A provides the maximum degree of personal protection, it does not protect against all possible airborne or splash hazards. For example, suit materials may be rapidly permeable to certain chemicals in high air concentrations or heavy splashes.

Level B: Level B protection should be selected when the highest level of respiratory protection is needed, but cutaneous or percutaneous exposure to the small unprotected areas of the body (i.e., neck and back of head) is unlikely, or where concentrations are known within acceptable exposure standards.

Level C: Level C protection should be selected when the type(s) and concentration(s) of respirable material are known, or reasonably assumed to be not greater the protection factors associated with air-purifying respirators; and if exposure to the few unprotected areas of the body (i.e., neck and. back of the head) is unlikely to cause harm. Continuous monitoring of site and/or individuals should be established, to ensure this minimum protection level is still acceptable throughout the exposure .

Level D: Level D is the basic work uniform and should be worn for all site operations. Level D protection should only-be selected when sites are positively identified as having no toxic hazards. All protective clothing should meet applicable OSHA standards.

Table 5 provides a generic breakdown of the types of PPE worn for the different Levels of Protection.

OSHA's standard 1910.132, titled *General Requirements*, SubPart Number I (titled - *Personal Protective Equipment*) covers protective equipment, including personal protective equipment for eyes, face, head, and extremities, protective clothing, respiratory devices, and protective shields and barriers. These equipment must be provided, used, and maintained in a sanitary and reliable condition wherever it is necessary by reason of hazards of processes or environment, chemical hazards, radiological hazards, or mechanical irritants encountered in a manner capable of causing injury or impairment in the function of any part of the body through absorption, inhalation or physical contact. All personal protective equipment must be of safe design and construction for the work to be performed. Employer's must assess the workplace to determine if hazards are present, or are likely to be present (known as a hazard assessment), which necessitate the use of PPE.

If such hazards are present, or likely to be present, the employer must select, and have each affected employee use, the types of PPE that will protect the affected employee from the hazards identified in the hazard assessment; communicate selection decisions to each affected employee; and select PPE that properly fits each affected employee.

Table 5. Recommended PPE for Different Levels of Protection

PERSONAL PROTECTIVE EQUIPMENT	LEVEL OF PROTECTION			
	A	B	C	D
Hard Hat	■	■	■	
Face Shield or Safety Glasses			■	■
Boots	■	■	■	■
Inner Gloves	■	■	■	
Outer Gloves	■	■	■	
Work Coveralls				■
Chemical Resistant Coveralls			■	
Chemical Resistant Suit		■		
Fully-encapsulating Suit	■			
Air Purifying Respirator			■	
SCBA (Self-contained Breathing Apparatus)/Airline Respirator	■			
2-Way Radio	■			
Cooling System	■			

The employer needs to verify that the required workplace hazard assessment has been performed through a written certification that identifies the workplace evaluated; the person certifying that the evaluation has been performed; the date(s) of the hazard assessment; and, which identifies the document as a certification of hazard assessment. In addition, employers are required under the standard to provide training to each employee who is required by this section to use PPE. Each such employee shall be trained to know at least the following: when PPE is necessary; what PPE is necessary; how to properly don, doff, adjust, and wear PPE; the limitations of the PPE; and the proper care, maintenance, useful life and disposal of the PPE.

Standard Number 1910 Subpart I App B, titled *Non-mandatory Compliance*

Guidelines for Hazard Assessment and Personal Protective Equipment Selection, deals specifically with Personal Protective Equipment. It is intended to provide compliance assistance for employers and employees in implementing requirements for a hazard assessment and the selection of personal protective equipment. The principal areas covered are:

- Controlling hazards: PPE devices alone should not be relied on to provide protection against hazards, but should be used in conjunction with guards, engineering controls, and sound manufacturing practices.

- Assessment and selection: It is necessary to consider certain general guidelines for assessing the foot, head, eye and face, and hand hazard situations that exist in an occupational or educational operation or process, and to match the protective devices to the particular hazard. It should be the responsibility of the safety officer to exercise common sense and appropriate expertise to accomplish these tasks.

- Assessment guidelines: In order to assess the need for PPE the following steps should be taken:

1. Survey: Conduct a walk-through survey of the areas in question. The purpose of the survey is to identify sources of hazards to workers and co-workers. Consideration should be given to the basic hazard categories; namely: Impact, Penetration, Compression (roll-over), Chemical, Heat, Harmful dust, Light (optical) radiation.

2. Sources: During the walkthrough survey the safety officer should observe:

(a) sources of motion; i.e., machinery or processes where any movement of tools, machine elements or particles could exist, or movement of personnel that could result in collision with stationary objects;

(b) sources of high temperatures that could result in burns, eye injury or ignition of protective equipment, etc.;

(c) types of chemical exposures;

(d) sources of harmful dust;

(e) sources of light radiation, i.e., welding, brazing, cutting, furnaces, heat treating, high intensity lights, etc.;

(f) sources of falling objects or potential for dropping objects;

(g) sources of sharp objects which might pierce the feet or cut the hands;

(h) sources of rolling or pinching objects which could crush the feet;

(i) layout of workplace and location of co-workers; and

(j) any electrical hazards. In addition, injury/accident data should be reviewed to help identify problem areas.

3. Organize data: Following the walkthrough survey, it is necessary to organize the data and information for use in the assessment of hazards. The objective

is to prepare for an analysis of the hazards in the environment to enable proper selection of protective equipment.

4.   Analyze data: Having gathered and organized data on a workplace, an estimate of the potential for injuries should be made. Each of the basic hazards should be reviewed and a determination made as to the type, level of risk, and seriousness of potential injury from each of the hazards found in the area. The possibility of exposure to several hazards simultaneously should be considered.

The general procedure for selection of protective equipment is to:

(a)   Become familiar with the potential hazards and the type of protective equipment that is available, and what it can do; i.e., splash protection, impact protection, etc.;

(b)   compare the hazards associated with the environment; i.e., impact velocities, masses, projectile shape, radiation intensities, with the capabilities of the available protective equipment;

(c)   select the protective equipment which ensures a level of protection greater than the minimum required to protect employees from the hazards; and

(e)   fit the user with the protective device and give instructions on care and use of the PPE. It is very important that end users be made aware of all warning labels for and limitations of their PPE.

Careful consideration must be given to comfort and fit. PPE that fits poorly will not afford the necessary protection. Continued wearing of the device is more likely if it fits the wearer comfortably. Protective devices are generally available in a variety of sizes. Care should be taken to ensure that the right size is selected. Adjustments should be made on an individual basis for a comfortable fit that will maintain the protective device in the proper position.

Particular care should be taken in fitting devices for eye protection against dust and chemical splash to ensure that the devices are sealed to the face. In addition, proper fitting of helmets is important to ensure that it will not fall off during work operations. In some cases a chin strap may be necessary to keep the helmet on an employee's head. (Chin straps should break at a reasonably low force, however, so as to prevent a strangulation hazard). Where manufacturer's instructions are available, they should be followed carefully.

The safety officer must reassess the workplace hazard situation as necessary, by identifying and evaluating new equipment and processes, reviewing accident records, and reevaluating the suitability of previously selected PPE.

Respiratory Protection

In the control of those occupational diseases caused by breathing air contaminated with harmful dusts, fogs, fumes, mists, gases, smokes, sprays, or vapors, the primary objective is to prevent atmospheric contamination. This can be accomplished as far as feasible by accepted engineering control measures (for example, enclosure or confinement of the operation, general and local ventilation, and substitution of less toxic materials). When effective engineering controls are not feasible, or while they are being instituted, appropriate respirators must be used pursuant to this section. Respirators must be provided by the employer when such equipment is necessary to protect the health of the employee. The employer must provide the respirators which are applicable and suitable for the purpose intended. The following definitions are important terms used in the OSHA respiratory protection standard.

Air-purifying respirator means a respirator with an air-purifying filter, cartridge, or canister that removes specific air contaminants by passing ambient air through the air-purifying element.

Atmosphere-supplying respirator means a respirator that supplies the respirator user with breathing air from a source independent of the ambient atmosphere, and includes supplied-air respirators (SARs) and self-contained breathing apparatus (SCBA) units.

Canister or cartridge means a container with a filter, sorbent, or catalyst, or combination of these items, which removes specific contaminants from the air passed through the container.

Demand respirator means an atmosphere-supplying respirator that admits breathing air to the facepiece only when a negative pressure is created inside the facepiece by inhalation.

Emergency situation means any occurrence such as, but not limited to, equipment failure, rupture of containers, or failure of control equipment that may or does result in an uncontrolled significant release of an airborne contaminant.

Employee exposure means exposure to a concentration of an airborne contaminant that would occur if the employee were not using respiratory protection.

End-of-service-life indicator (ESLI) means a system that warns the respirator user of the approach of the end of adequate respiratory protection, for example, that the sorbent is approaching saturation or is no longer effective.

Escape-only respirator means a respirator intended to be used only for emergency exit.

Filter or air purifying element means a component used in respirators to remove solid or liquid aerosols from the inspired air.

Filtering facepiece (dust mask) means a negative pressure particulate respirator with a filter as an integral part of the facepiece or with the entire facepiece composed of the filtering medium.

Fit factor means a quantitative estimate of the fit of a particular respirator to a specific individual, and typically estimates the ratio of the concentration of a substance in ambient air to its concentration inside the respirator when worn.

Fit test means the use of a protocol to qualitatively or quantitatively evaluate the fit of a respirator on an individual. There are two types of fit tests, namely: a Qualitative fit test and Quantitative fit test.

Helmet means a rigid respiratory inlet covering that also provides head protection against impact and penetration.

High efficiency particulate air (HEPA) filter means a filter that is at least 99.97% efficient in removing monodisperse particles of 0.3 micrometers in diameter. The equivalent NIOSH 42 CFR 84 particulate filters are the N100, R100, and P100.

Hood means a respiratory inlet covering that completely covers the head and neck and may also cover portions of the shoulders and torso.

Immediately dangerous to life or health (IDLH) means an atmosphere that poses an immediate threat to life, would cause irreversible adverse health effects, or would impair an individual's ability to escape from a dangerous atmosphere.

Interior structural fire fighting means the physical activity of fire suppression, rescue or both, inside of buildings or enclosed structures which are involved in a fire situation beyond the incipient stage. Refer to 29 CFR 1910.155.

Loose-fitting facepiece means a respiratory inlet covering that is designed to form a partial seal with the face.

Negative pressure respirator (tight fitting) means a respirator in which the air pressure inside the facepiece is negative during inhalation with respect to the ambient air pressure outside the respirator.

Oxygen deficient atmosphere means an atmosphere with an oxygen content below 19.5% by volume.

Physician or other licensed health care professional (PLHCP) means an individual whose legally permitted scope of practice (i.e., license, registration, or certification) allows him or her to independently provide, or be delegated the responsibility to provide, some or all of the health care services required in a respiratory protection program.

Positive pressure respirator means a respirator in which the pressure inside the respiratory inlet covering exceeds the ambient air pressure outside the respirator.

Powered air-purifying respirator (PAPR) means an air-purifying respirator that uses a blower to force the ambient air through air-purifying elements to the inlet covering.

Pressure demand respirator means a positive pressure atmosphere-supplying respirator that admits breathing air to the facepiece when the positive pressure is reduced inside the facepiece by inhalation.

Qualitative fit test (QLFT) means a pass/fail fit test to assess the adequacy of respirator fit that relies on the individual's response to the test agent.

Quantitative fit test (QNFT) means an assessment of the adequacy of respirator fit by numerically measuring the amount of leakage into the respirator.

Respiratory inlet covering means that portion of a respirator that forms the protective barrier between the user's respiratory tract and an air-purifying device or breathing air source, or both. It may be a facepiece, helmet, hood, suit, or a mouthpiece respirator with nose clamp.

Self-contained breathing apparatus (SCBA) means an atmosphere-supplying respirator for which the breathing air source is designed to be carried by the user.

Service life means the period of time that a respirator, filter or sorbent, or other respiratory equipment provides adequate protection to the wearer.

Supplied-air respirator (SAR) or airline respirator means an atmosphere-supplying respirator for which the source of breathing air is not designed to be carried by the user.

Tight-fitting facepiece means a respiratory inlet covering that forms a complete seal with the face.

User seal check means an action conducted by the respirator user to determine if the respirator is properly seated to the face.

Under a respiratory protection program, the employer is required to develop and implement a written respiratory protection program with required worksite-specific procedures and elements for required respirator use. The program must be administered by a suitably trained program administrator. In addition, certain program elements may be required for voluntary use to prevent potential hazards associated with the use of the respirator. In any workplace where respirators are necessary to protect the health of the employee or whenever respirators are required by the employer, the employer must establish and implement a written respiratory protection program with worksite-specific procedures. The program must be updated as necessary to reflect those changes in workplace conditions that affect respirator use. The program should include the following provisions, as applicable:

- Procedures for selecting respirators for use in the workplace;
- Medical evaluations of employees required to use respirators;
- Fit testing procedures for tight-fitting respirators;
- Procedures for proper use of respirators in routine and reasonably foreseeable emergency situations;
- Procedures and schedules for cleaning, disinfecting, storing, inspecting, repairing, discarding, and otherwise maintaining respirators;
- Procedures to ensure adequate air quality, quantity, and flow of breathing air for atmosphere-supplying respirators;
- Training of employees in the respiratory hazards to which they are potentially exposed during routine and emergency situations;
- Training of employees in the proper use of respirators, including putting on and removing them, any limitations on their use, and their maintenance; and
- Procedures for regularly evaluating the effectiveness of the program.

Where respirator use is not required an employer may provide respirators at the request of employees or permit employees to use their own respirators, if the employer determines that such respirator use will not in itself create a hazard. In addition, the OSHA standards require that the employer must establish and implement those elements of a written respiratory protection program necessary to ensure that any employee using a respirator voluntarily is medically able to use that respirator, and that the respirator is cleaned, stored, and maintained so that its use does not present a health hazard to the user. The employer should designate a program administrator who is qualified by appropriate training or experience that is commensurate with the complexity of the program to administer or oversee the respiratory protection program and conduct the required evaluations of program effectiveness.

The employer must evaluate respiratory hazard(s) in the workplace, identify relevant workplace and user factors, and base respirator selection on these factors. The company must also specify appropriately protective respirators for use in IDLH atmospheres, and limit the selection and use of air-purifying respirators. Hence, the employer shall select and provide an appropriate respirator based on the respiratory hazard(s) to which the worker is exposed and workplace and user factors that affect respirator performance and reliability. The employer shall select a NIOSH-certified respirator. The respirator shall be used in compliance with the conditions of its certification. The employer shall identify and evaluate the respiratory hazard(s) in the workplace; this evaluation shall include a reasonable estimate of employee exposures to respiratory hazard(s) and an identification of the contaminant's chemical state and physical form. Where the employer cannot identify or reasonably estimate the employee exposure, the employer shall consider the atmosphere to be IDLH. One must select respirators from a sufficient number of respirator models and sizes so that the respirator is acceptable to, and correctly fits, the user. The following respirators should be provided for employee use in IDLH atmospheres: a full facepiece pressure demand SCBA certified by NIOSH for a minimum service life of thirty minutes, or a combination full facepiece pressure demand supplied-air respirator (SAR) with auxiliary self-contained air supply. Respirators provided only for escape from IDLH atmospheres shall be NIOSH-certified for escape from the atmosphere in which they will be used. All oxygen-deficient atmospheres shall be considered IDLH.

Respirators for atmospheres that are not IDLH: The employer shall provide a respirator that is adequate to protect the health of the employee and ensure compliance with all other OSHA statutory and regulatory requirements, under routine and reasonably foreseeable emergency situations. The respirator selected shall be appropriate for the chemical state and physical form of the contaminant. For protection against gases and vapors, the employer shall provide: an atmosphere-supplying respirator, or an air-purifying respirator, provided that the respirator is equipped with an end-of-service-life indicator (ESLI) certified by NIOSH for the contaminant; or if there is no ESLI appropriate for conditions in the employer's workplace, the employer implements a change schedule for canisters and cartridges that is based on objective information or data that will ensure that canisters and cartridges are changed before the end of their service life. The employer shall describe in the respirator program the information and data relied upon and the basis for the canister and cartridge change schedule and the basis for reliance on the data.

For protection against particulate matter, the employer shall provide: an atmosphere-supplying respirator; or an air-purifying respirator equipped with a filter certified by NIOSH under 30 CFR part 11 as a high efficiency particulate air (HEPA) filter, or an air-purifying respirator equipped with a filter certified for particulates by NIOSH under 42 CFR part 84; or for contaminants consisting primarily of particles with mass median aerodynamic diameters (MMAD) of at least 2 micrometers, an air-purifying respirator equipped with any filter certified for particulate matter by NIOSH.

Using a respirator may place a physiological burden on employees that varies with the type of respirator worn, the job and workplace conditions in which the respirator is used, and the medical status of the employee. Accordingly, OSHA specifies the minimum requirements for medical evaluation that employers must implement to determine the employee's ability to use a respirator. The employer must provide a medical evaluation to determine the employee's ability to use a respirator, before the employee is fit tested or required to use the respirator in the workplace. The employer may discontinue an employee's medical evaluations when the employee is no longer required to use a respirator. A physician or other licensed health care professional (PLHCP) must be used to perform medical evaluations using a medical questionnaire or an initial medical examination that obtains the same information as the medical questionnaire. The medical questionnaire and examinations shall be administered confidentially during the employee's normal working hours or at a time and place convenient to the employee. The medical questionnaire shall be administered in a manner that ensures that the employee understands its content. The employer shall provide the employee with an opportunity to discuss the questionnaire and examination results with the PLHCP. The following information must be provided to the PLHCP before the PLHCP makes a recommendation concerning an employee's ability to use a respirator:

- The type and weight of the respirator to be used by the employee;
- The duration and frequency of respirator use (including use for rescue and escape);
- The expected physical work effort;
- Additional protective clothing and equipment to be worn; and
- Temperature and humidity extremes that may be encountered.

Any supplemental information provided previously to the PLHCP regarding an employee need not be provided for a subsequent medical evaluation if the information and the PLHCP remain the same. The employer shall provide the PLHCP with a copy of the written respiratory protection program and a copy of this section. When the employer replaces a PLHCP, the employer must ensure that the new PLHCP obtains this information, either by providing the documents directly to the PLHCP or having the documents transferred from the former PLHCP to the new PLHCP. However, OSHA does not expect employers to have employees medically reevaluated solely because a new PLHCP has been selected. In determining the employee's ability to use a respirator, the employer shall obtain a written recommendation regarding the employee's ability to use the respirator from the PLHCP. The recommendation shall provide only the following information:

- Any limitations on respirator use related to the medical condition of the employee, or relating to the workplace conditions in which the respirator will be used, including whether or not the employee is medically able to use the respirator;
- The need, if any, for follow-up medical evaluations; and
- A statement that the PLHCP has provided the employee with a copy of the PLHCP's written recommendation.

If the respirator is a negative pressure respirator and the PLHCP finds a medical condition that may place the employee's health at increased risk if the respirator is used, the employer shall provide a PAPR if the PLHCP's medical evaluation finds that the employee can use such a respirator; if a subsequent medical evaluation finds that the employee is medically able to use a negative pressure respirator, then the employer is no longer required to provide a PAPR.

Fit testing: Before an employee may be required to use any respirator with a negative or positive pressure tight-fitting facepiece, the employee must be fit tested with the same make, model, style, and size of respirator that will be used. It is the employer's responsibility to ensure that employees using a tight-fitting facepiece respirator pass an appropriate qualitative fit test (QLFT) or quantitative fit test (QNFT). The employer shall ensure that an employee using a tight-fitting facepiece respirator is fit tested prior to initial use of the respirator, whenever a different respirator facepiece (size, style, model or make) is used, and at least annually thereafter. The employer shall conduct an additional fit test whenever the employee reports, or the employer, PLHCP, supervisor, or program administrator makes visual observations of, changes in the employee's physical condition that could affect respirator fit. Such conditions include, but are not limited to, facial scarring, dental changes, cosmetic surgery, or an obvious change in body weight. If after passing a QLFT or QNFT, the employee subsequently notifies the employer, program administrator, supervisor, or PLHCP that the fit of the respirator is unacceptable, the employee shall be given a reasonable opportunity to select a different respirator facepiece and to be retested. The fit test shall be administered using an OSHA-accepted QLFT or QNFT protocol. The OSHA-accepted QLFT and QNFT protocols and procedures are contained in Appendix A of the OSHA standard. QLFT may only be used to fit test negative pressure air-purifying respirators that must achieve a fit factor of 100 or less. If the fit factor, as determined through an OSHA-accepted QNFT protocol, is equal to or greater than 100 for tight-fitting half facepieces, or equal to or greater than 500 for tight-fitting full facepieces, the QNFT has been passed with that respirator. Fit testing of tight-fitting atmosphere-supplying respirators and tight-fitting powered air-purifying respirators shall be accomplished by performing quantitative or qualitative fit testing in the negative pressure mode, regardless of the mode of operation (negative or positive pressure) that is used for respiratory protection.

Qualitative fit testing of these respirators shall be accomplished by temporarily converting the respirator user's actual facepiece into a negative pressure respirator with appropriate filters, or by using an identical negative pressure air-purifying respirator facepiece with the same sealing surfaces as a surrogate for the atmosphere-supplying or powered air-purifying respirator facepiece.

Quantitative fit testing of these respirators shall be accomplished by modifying the facepiece to allow sampling inside the facepiece in the breathing zone of the user, midway between the nose and mouth. This requirement shall be accomplished by installing a permanent sampling probe onto a surrogate facepiece, or by using a sampling adapter designed to temporarily provide a means of sampling air from inside the facepiece. Any modifications to the respirator facepiece for fit testing shall be completely removed, and the facepiece restored to NIOSH-approved configuration, before that facepiece can be used in the workplace.

Use of Respirators: It is necessary to establish and implement procedures for the proper use of respirators. These requirements include prohibiting conditions that may result in facepiece seal leakage, preventing employees from removing respirators in hazardous environments, taking actions to ensure continued effective respirator operation throughout the work shift, and establishing procedures for the use of respirators in IDLH atmospheres or in interior structural fire fighting situations. The employer shall not permit respirators with tight-fitting facepieces to be worn by employees who have:
- Facial hair that comes between the sealing surface of the facepiece and the face or that interferes with valve function; or
- Any condition that interferes with the face-to-facepiece seal or valve function.

If an employee wears corrective glasses or goggles or other personal protective equipment, the employer shall ensure that such equipment is worn in a manner that does not interfere with the seal of the facepiece to the face of the user. For all tight-fitting respirators, the employer shall ensure that employees perform a user seal check each time they put on the respirator using procedures recommended by the respirator manufacturer.

Appropriate surveillance shall be maintained of work area conditions and degree of employee exposure or stress. When there is a change in work area conditions or degree of employee exposure or stress that may affect respirator effectiveness, the employer shall reevaluate the continued effectiveness of the respirator.

General Procedures for IDLH atmospheres: For all IDLH atmospheres, the employer shall ensure that:
- One employee or, when needed, more than one employee is located outside the IDLH atmosphere;
- Visual, voice, or signal line communication is maintained between the employee(s) in the IDLH atmosphere and the employee(s) located outside the IDLH atmosphere;
- The employee(s) located outside the IDLH atmosphere are trained and equipped to provide effective emergency rescue;
- The employer or designee is notified before the employee(s) located outside

the IDLH atmosphere enter the IDLH atmosphere to provide emergency rescue;

- The employer or designee authorized to do so by the employer, once notified, provides necessary assistance appropriate to the situation;
- Employee(s) located outside the IDLH atmospheres are equipped with: Pressure demand or other positive pressure SCBAs, or a pressure demand or other positive pressure supplied-air respirator with auxiliary SCBA; and either appropriate retrieval equipment for removing the employee(s) who enter(s) these hazardous atmospheres where retrieval equipment would contribute to the rescue of the employee(s) and would not increase the overall risk resulting from entry; or equivalent means for rescue where retrieval equipment is not required.

Procedures for interior structural fire fighting: In interior structural fires, the employer shall ensure that:

- At least two employees enter the IDLH atmosphere and remain in visual or voice contact with one another at all times;
- At least two employees are located outside the IDLH atmosphere; and
- All employees engaged in interior structural firefighting use SCBAs.

One of the two individuals located outside the IDLH atmosphere may be assigned to an additional role, such as incident commander in charge of the emergency or safety officer, so long as this individual is able to perform assistance or rescue activities without jeopardizing the safety or health of any firefighter working at the incident.

Maintenance and Care of Respirators: The employer shall provide each respirator user with a respirator that is clean, sanitary, and in good working order. The respirators shall be cleaned and disinfected at the following intervals:

- Respirators issued for the exclusive use of an employee shall be cleaned and disinfected as often as necessary to be maintained in a sanitary condition;
- Respirators issued to more than one employee shall be cleaned and disinfected before being worn by different individuals;
- Respirators maintained for emergency use shall be cleaned and disinfected after each use; and
- Respirators used in fit testing and training shall be cleaned and disinfected after each use.

Respirators are stored as follows: All respirators shall be stored to protect them from damage, contamination, dust, sunlight, extreme temperatures, excessive moisture, and damaging chemicals, and they shall be packed or stored to prevent deformation of the facepiece and exhalation valve. In addition, emergency respirators should be kept accessible to the work area; stored in compartments or in covers that are clearly marked as containing emergency respirators; and stored in accordance with any applicable manufacturer instructions. All respirators used in routine situations should be inspected before each use and during cleaning. They should be maintained for use in emergency situations and should be inspected at least monthly and in accordance with the manufacturer's recommendations, and shall be checked for proper function before and after each use. Emergency escape-only respirators should be inspected before being carried into the workplace for use. Respirator inspections include the following:

- A check of respirator function, tightness of connections, and the condition of the various parts including, but not limited to, the facepiece, head straps, valves, connecting tube, and cartridges, canisters or filters; and
- A check of elastomeric parts for pliability and signs of deterioration.

Self-contained breathing apparatus should be inspected monthly. Air and oxygen cylinders shall be maintained in a fully charged state and shall be recharged when the pressure falls to 90% of the manufacturer's recommended pressure level. The employer shall determine that the regulator and warning devices function properly. For respirators maintained for emergency use:

- Certify the respirator by documenting the date the inspection was performed, the name (or signature) of the person who made the inspection, the findings, required remedial action, and a serial number or other means of identifying the inspected respirator; and provide this information on a tag or label that is attached to the storage compartment for the respirator, is kept with the respirator, or is included in inspection reports stored as paper or electronic files. This information shall be maintained until replaced following a subsequent certification.

Respirators that fail an inspection or are otherwise found to be defective should be removed from service, and discarded or repaired or adjusted in accordance with the following procedures:

- Repairs or adjustments to respirators are to be made only by persons appropriately trained to perform such operations and shall use only the respirator manufacturer's NIOSH-approved parts designed for the respirator;
- Repairs should be made according to the manufacturer's recommendations and specifications for the type and extent of repairs to be performed; and
- Reducing and admission valves, regulators, and alarms should be adjusted or repaired only by the manufacturer or a technician trained by the manufacturer.

For SCBAs, ensure that compressed air, compressed oxygen, liquid air, and liquid oxygen used for respiration accords with the following specifications:

- Compressed and liquid oxygen must meet the United States Pharmacopoeia requirements for medical or breathing oxygen; and

- Compressed breathing air must meet at least the requirements for Grade D breathing air described in ANSI/Compressed Gas Association Commodity Specification for Air, G-7.1-1989, to include: oxygen content (v/v) of 19.5-23.5%; hydrocarbon (condensed) content of 5 milligrams per cubic meter of air or less; carbon monoxide (CO) content of 10 ppm or less; carbon dioxide content of 1,000 ppm or less; and lack of noticeable odor.

Compressed oxygen should not be used in atmosphere-supplying respirators that have previously used compressed air. Oxygen concentrations greater than 23.5% are used only in equipment designed for oxygen service or distribution. Ensure that cylinders used to supply breathing air to respirators meet the following requirements:

- Cylinders are tested and maintained as prescribed in the Shipping Container Specification Regulations of the Department of Transportation (49 CFR part 173 and part 178);
- Cylinders of purchased breathing air have a certificate of analysis from the supplier that the breathing air meets the requirements for Grade D breathing air; and
- The moisture content in the cylinder does not exceed a dew point of -50 °F (or -45.6 °C) at 1 atmosphere pressure.

Ensure that compressors used to supply breathing air to respirators are constructed and situated so as to:

- Prevent entry of contaminated air into the air-supply system;
- Minimize moisture content so that the dew point at 1 atmosphere pressure is 10 °F (5.56 °C) below the ambient temperature;
- Have suitable in-line air-purifying sorbent beds and filters to further ensure breathing air quality. Sorbent beds and filters shall be maintained and replaced or refurbished periodically following the manufacturer's instructions.

There should be a tag containing the most recent change date and the signature of the person authorized by the employer to perform the change. The tag must be maintained at the compressor. For compressors that are not oil-lubricated, ensure that carbon monoxide levels in the breathing air do not exceed 10 ppm. For oil-lubricated compressors, use a high-temperature or carbon monoxide alarm, or both, to monitor carbon monoxide levels. If only high-temperature alarms are used, the air supply shall be monitored at intervals sufficient to prevent carbon monoxide in the breathing air from exceeding 10 ppm. Use breathing gas containers marked in accordance with the NIOSH respirator certification standard, 42 CFR part 84. With filters, cartridges, and canisters, ensure that they are labeled and color coded with the NIOSH approval label and that the label is not removed and remains legible.

Training: It is the employer's responsibility to provide effective training to employees who are required to use respirators. The training must be comprehensive, understandable, and recur annually, and more often if necessary. This paragraph also requires the employer to provide the basic information on respirators in Appendix D of this section to employees who wear respirators when not required by this section or by the employer to do so. Employers should conduct evaluations of the workplace as necessary to ensure that the provisions of the current written program are being effectively implemented and that it continues to be effective. The employer should regularly consult employees required to use respirators to assess the employees' views on program effectiveness and to identify any problems. Any problems that are identified during this assessment shall be corrected. Factors to be assessed include, but are not limited to:

- Respirator fit (including the ability to use the respirator without interfering with effective workplace performance);
- Appropriate respirator selection for the hazards to which the employee is exposed;
- Proper respirator use under the workplace conditions the employee encounters; and
- Proper respirator maintenance.

Recordkeeping: An employer is required to establish and retain written information regarding medical evaluations, fit testing, and the respirator program. This information will facilitate employee involvement in the respirator program, assist the employer in auditing the adequacy of the program, and provide a record for compliance determinations by OSHA. Records of medical evaluations required by this section must be retained and made available in accordance with 29 CFR 1910.1020. The employer will need to establish a record of the qualitative and quantitative fit tests administered to an employee including:

- The name or identification of the employee tested;
- Type of fit test performed;
- Specific make, model, style, and size of respirator tested;
- Date of test; and
- The pass/fail results for QLFTs or the fit factor and strip chart recording or other recording of the test results for QNFTs.

Fit test records shall be retained for respirator users until the next fit test is administered. A written copy of the current respirator program shall be retained by the employer. Written materials required to be retained shall be made available upon request to affected employees and to the Assistant Secretary or designee for examination and copying.

WORKING WITH ASBESTOS AND OTHER SYNTHETIC MINERAL FIBERS

An estimated 1.3 million employees in construction and general industry face significant asbestos exposure on the job. Heaviest exposures occur in the construction industry, particularly during the removal of asbestos during renovation or demolition. Employees are also likely to be exposed during the manufacture of asbestos products (such as textiles, friction products, insulation, and other building materials) and during automotive brake and clutch repair work. Asbestos is well recognized as a health hazard and is highly regulated. It is important to note that the OSHA and EPA asbestos rules are intertwined, and hence both regulations need to be closely evaluated from a compliance standpoint in assessing legal issues for a building or site operation.

The Asbestos Advisor software (available on the OSHA Web site) is an interactive compliance assistance tool. Once installed on your PC, it can interview you about buildings and worksites, and the kinds of tasks workers perform there. It will produce guidance on how the Asbestos standard may apply to those buildings and that work. When queried "to obtain general guidance" (selection 1), the Asbestos Advisor asks if you have any employees. If you respond with "no" (selection 2), the Asbestos Advisor concludes you are "not subject to regulations" under OSHA. This is not accurate. Building owners and managers may be subject to the OSHA Asbestos Standards if employees of any employer work in the building. Also, *Medical surveillance* guidance is provided in the appendices to the OSHA Standards: Medical questionnaires. OSHA Regulation 1910.1001 App D, 1915.1001 App D, and 1926.1101 App D, Mandatory appendix. Interpretation and classification of chest roentgenograms. OSHA Regulation 1910.1001 App E, 1915.1001 App E, and 1926.1101 App E. Mandatory appendix. Medical surveillance guidelines for asbestos. OSHA Regulation 1910.1001 App H, 1915.1001 App I, and 1926.1101 App I, Nonmandatory appendix.

Exposure monitoring samples must be analyzed by Phase Contrast Microscopy (PCM) for OSHA purposes. PCM methods accurately assess fiber exposure levels, but PCM can not differentiate between asbestos and non-asbestos fibers. Transmission Electron Microscopy (TEM) methods can identify fibers, but fiber counting accuracy is unacceptably poor. Refer to the follwong documents for information:

- OSHA (Sampling) Reference Method. OSHA Regulation 1910.1001 App A, 1915.1001 App A, and 1926.1101 App A, Mandatory appendicies. Asbestos exposure sampling and analysis must meet these minimal requirements.

- Asbestos in Air. OSHA Analytical Method ID-160 (1997), 12 pages.
- Detailed procedure for asbestos sampling and analysis. OSHA Regulation 1910.1001 App B, 1915.1001 App B, and 1926.1101 App B, Non-Mandatory appendicies.
- NIOSH Manual of Analytical Methods (NMAM) includes asbestos methods 7400 and 7402. Method 7400 is a PCM procedure, equivalent to the OSHA methods. Method 7402 uses TEM to identify fibers (OSHA will accept this TEM procedure, as it uses PCM to determine exposure). These methods are available online as compressed (ZIP) WordPerfect 5.1 + files: NIOSH 7400, Asbestos and other fibers by PCM; NIOSH 7402, Asbestos fibers by TEM.

Bulk sample analysis should be done by Polarized Light Microscopy (PLM). Bulk analysis results will likely apply to both OSHA and EPA regulations. Refer to the following other resources for links to EPA rules.

- Asbestos (Bulks). Polarized Light Microscopy of Asbestos. OSHA Analytical Method ID-191, (1992).
- Polarized Light Microscopy of Asbestos. OSHA Regulation 1910.1001 App J, 1915.1001 App K, and 1926.1101 App K, non-mandatory analytical method.
- NIOSH has published two methods for the determination of asbestos in bulk materials. These methods are available online as compressed (ZIP) WordPerfect 5.1 + files: NIOSH 9000, Asbestos, Chrysotile by XRD; NIOSH 9002, Asbestos (bulk) by PLM . Refer also to the following documents: Interim Method for the Determination of Asbestos in Bulk Insulation Samples. U.S. EPA 40 CFR, Part 763, Subpart F, App. A, 408 KB PDF file, pages 48-58;   Directory of Accredited Laboratories. National Voluntary Laboratory Accreditation Program (NVLAP). This accreditation is required for analyses being performed in compliance with AHERA regulations; Compliance with OSHA's Asbestos Standard: Composite Bulk Samples. OSHA Standard Interpretation (1997). Each sample layer must be analyzed separately; "composite" analysis is not acceptable; Asbestos NESHAP Clarification Regarding Analysis of Multi-Layered Systems. EPA, Federal Register (1995, December 19), 1 page. Multi-Layer Analysis is required for all samples except "wall systems."; ANSI/ASTM E1368-96A Practice for Visual Inspection of Asbestos Abatement Projects This standard establishes accepted practices for evaluating asbestos abatement projects. This standard is available from ANSI.

The reader should also refer to the US EPA home page on the World Wide Web. Specifically, refer to EPA regulations (40 CFR), PDF files or text. The TSCA Asbestos regulations are found in 40 CFR 763; EPA COMPLI database (asbestos

NESHAP); Oklahoma State University; links to asbestos links, the Asbestos Institute, the American Lung Association; and Asbestos Bibliography, DHHS (NIOSH) Publication No. 97-162 (1997, September), 224 pages, 8.3-MB PDF files or 9 separate PDF files. This publication is a compendium of NIOSH research and recommendations on asbestos; includes full or partial text of selected NIOSH documents, a comprehensive bibliography, and a summary of asbestos information available from other agencies.

Asbestos is a widely used, mineral-based material that is resistant to heat and corrosive chemicals. Typically, asbestos appears as a whitish, fibrous material which may release fibers that range in texture from coarse to silky; however, airborne fibers that can cause health damage may be too small to see with the naked eye. Exposure to asbestos can cause asbestosis (scarring of the lungs resulting in loss of lung function that often progresses to disability and to death); mesothelioma (cancer affecting the membranes lining the lungs and abdomen); lung cancer; and cancers of the esophagus, stomach, colon, and rectum. OSHA has issued revised regulations covering asbestos exposure in general industry and construction. Both standards set a maximum exposure limit and include provisions for engineering controls and respirators, protective clothing, exposure monitoring, hygiene facilities and practices, warning signs, labeling, recordkeeping, and medical exams. Non-asbestiform tremolite, anthophyllite, and actinolite were excluded from coverage under the asbestos standard in May 1992. The following are some of the highlights of the revised rules, published in the Federal Register June 20, 1986; and on Sept. 14, 1988:

Permissible Exposure Limit: In both general industry and construction, workplace exposure must be limited to 0.2 fibers per cubic centimeter of air (0.2 f/cc), averaged over an eight-hour work shift. The excursion or short-term limit is one fiber per cubic centimeter of air (1 f/cc) averaged over a sampling period of 30 minutes.

Exposure Monitoring: In general industry, employers must do initial monitoring for workers who may be exposed above the "action level" of 0.1 f/cc. Subsequent monitoring must be conducted at reasonable intervals, in no case longer than six months for employees exposed above the action level. In construction, daily monitoring must be continued until exposure drops below the action level (0.1 f/cc). Daily monitoring is not required where employees are using supplied-air respirators operated in the positive pressure mode.

Methods of Compliance: Employers must control exposures using engineering controls, to the extent feasible. Where engineering controls are not feasible to meet the exposure limit, they must be used to reduce employee exposures to the lowest levels attainable and must be supplemented by the use of respiratory protection.

Respirators: In general industry and construction, the level of exposure determines what type of respirator is required; the standards specify the respirator to be used.

Regulated Areas: In general industry and construction, regulated areas must be established where the 8-hour TWA or 30-minute excursion values for airborne asbestos exceed the prescribed permissible exposure limits. Only authorized persons wearing appropriate respirators can enter a regulated area. In regulated areas, eating, smoking, drinking, chewing tobacco or gum, and applying cosmetics are prohibited. Warning signs must be displayed at each regulated area and must be posted at all approaches to regulated areas.

Labels: Caution labels must be placed on all raw materials, mixtures, scrap, waste, debris, and other products containing asbestos fibers.

Recordkeeping: The employer must keep an accurate record of all measurements taken to monitor employee exposure to asbestos. This record is to include: the date of measurement, operation involving exposure, sampling and analytical methods used, and evidence of their accuracy; number, duration, and results of samples taken; type of respiratory protective devices worn; name, social security number, and the results of all employee exposure measurements. Records must be kept for 30 years.

Protective Clothing: For any employee exposed to airborne concentrations of asbestos that exceed the PEL, the employer must provide and require the use of protective clothing such as coveralls or similar full-body clothing, head coverings, gloves, and foot covering. Wherever the possibility of eye irritation exists, face shields, vented goggles, or other appropriate protective equipment must be provided and worn. In construction, there are special regulated-area requirements for asbestos removal, renovation, and demolition operations. These provisions include a negative pressure area, decontamination procedures for workers, and a "competent person" with the authority to identify and control asbestos hazards. The standard includes an exemption from the negative pressure enclosure requirements for certain small scale, short duration operations provided special work practices prescribed in an appendix to the standard are followed.

Hygiene Facilities and Practices: Clean change rooms must be furnished by employers for employees who work in areas where exposure is above the TWA and/or excursion limit. Two lockers or storage facilities must be furnished and separated to prevent contamination of the employee's street clothes from protective work clothing and equipment. Showers must be furnished so that employees may shower at the end of the work shift. Employees must enter and exit the regulated area through the decontamination area. The equipment room must be supplied with impermeable, labeled bags and containers for the containment and disposal of contaminated protective clothing and equipment. Lunchroom facilities for those employees must have a positive pressure, filtered air supply and be readily accessible to employees. Employees must wash their hands and face prior to eating, drinking or smoking. The employer must ensure that employees do not enter lunchroom facilities with protective work clothing or equipment unless surface fibers have been removed from the clothing or equipment. Employees may not smoke in work areas where they are occupationally exposed to asbestos.

Medical Exams: In general industry, exposed employees must have a preplacement physical examination before being assigned to an occupation exposed to airborne concentrations of asbestos at or above the action level or the excursion level. The physical examination must include chest X-ray, medical and work history, and pulmonary function tests. Subsequent exams must be given annually and upon termination of employment, though chest X-rays are required annually only for older workers whose first asbestos exposure occurred more than 10 years ago. In construction, examinations must be made available annually for workers exposed above the action level or excursion limit for 30 or more days per year or who are required to wear negative pressure respirators; chest X-rays are at the discretion of the physician.

Synthetic Mineral Fibers

"Synthetic mineral fibers" are fibrous inorganic substances made primarily from rock, clay, slag or glass. These fibers are classified into three general groups: fiberglass (glasswool and glass filament), mineral wool (rockwool and slagwool), and refractory ceramic fibers (RCF).

There are over 225,000 workers in the U.S. exposed to synthetic mineral fibers in manufacturing and end-use applications. Synthetic mineral fibers are believed to cause respiratory cancers and other adverse respiratory effects.

The following are some key references for the reader to reveiw. Most of these can be linked to on the OSHA Home Page:

- Synthetic Mineral Fibers. OSHA Priority Planning Process (1996, February 5), 3 pages. This summary sheet includes general hazard information for synthetic mineral fibers and describes OSHA's action plan to reduce worker exposures.
- Refractory Ceramic Fibers. EPA Integrated Risk Information System (IRIS) (1993, July 1), 10 pages. Health effect information for RCF's.
- Criteria for a Recommended Standard: Occupational Exposure to Fibrous Glass. NIOSH Publication No. 77-152 (1977, April), 201 pages, contained

in 8 separate PDF files. Includes health effects, exposure, work practices, sampling, and control information for fibrous glass.

- An important publication is the *Industrial Hygiene Surveys of Occupational Exposure to Mineral Wool*, NIOSH Publication No. 80-135. This publication is not available on-line but can be ordered by calling 1-800-35-NIOSH or by e-mail from the NIOSH Publications Office.

- *Reasonably Anticipated to be a Human Carcinogen: Glasswool (Respirable Size)*. National Toxicology Program, 8th Report on Carcinogens (1998), 3 pages.

- *Reasonably Anticipated to be a Human Carcinogen: Ceramic Fibers (Respirable Size)*. National Toxicology Program, 8th Report on Carcinogens (1998), 2 pages.

- *IARC Monograph: Man-made Mineral Fibers*. International Agency for Research on Cancer (IARC) (1988), 6 pages. Includes IARC monographs for glasswool, glass filaments, rockwool, slagwool, and ceramic fibers.

Again, the reader is encouraged to explore the many references that are now available on the World Wide Web, to access information. Note that there are many sites that are not necessarily recognized by OSHA or NIOSH, and hence some degree of caution must be exercised, especially if a regulatory interpretation of the standards is needed. Table 6 provides a summary of the OSHA exposure limits for synthetic mineral fibers. Refer to footnotes on next page that are part of this table.

Table 6. Exposure Limits for Synthetic Mineral Fibers.

AGENCY/SUBSTANCE	STANDARD LEVEL
OSHA PEL - TWA	
*Mineral fibers are currently only regulated as nuisance dust	
General Industry Inert or Nuisance Dust (1910.1000, Table Z-3) Respirable fraction Total dust	5 mg/m^3 15 mg/m^3
Shipyard Fibrous Glass (1915.1000, Table Z) Respirable fraction Total dust	5 mg/m^3 15 mg/m^3

Shipyard Mineral Wool (1915.1000, Table Z) Respirable dust Total dust	5 mg/m³ 15 mg/m³

ACGIH TLV - TWA

Synthetic Vitreous Fibers (1999 Adopted TLV's) Continuous filament glass fibers**, A4 Continuous filament glass fibers*, A4 Glass wool fibers*, A3 Rock wool fibers*, A3 Slag wool fibers*, A3 Special purpose glass fibers*, A3	5 mg/m³ 1 f/cc 1 f/cc 1 f/cc 1 f/cc 1 f/cc
Synthetic Vitreous Fibers (1999 TLV - Notice of Intended Change) Continuous filament glass fibers*, A4 Refractory ceramic fibers*, A2	0.1 f/cc 0.1 f/cc
*	Fibers longer than 5 μm; diameter less than 3 μm; aspect ratio greater than 5:1 as determined by the membrane filter method at 400-450X magnification (4-mm objective) phase contrast illumination.
**	Inhalable fraction. The concentration of inhalable particulate for the application of this TLV is to be determined from the fraction passing a size-selector with characteristics defined in the "A" paragraph of Appendix D.
A2	Suspected Human Carcinogen.
A3	Confirmed Animal Carcinogen with Unknown Relevance to Humans.
A4	Not Classifiable as a Human Carcinogen.

NIOSH REL - TWA

Fibrous Glass Dust (1977 Proposal) Total dust Fibers with diameter equal or less than 3.5 μm, and length equal to or greater than 10 μm	5 mg/m³ 3 f/cc

Relevant OSHA standards are:

- 29 CFR 1910.1000, Air Contaminants (General Industry). Table Z-3 Mineral Dusts contains PELs for "Inert or Nuisance Dust" (respirable fraction and total dust); these are currently the only PELs applicable to synthetic mineral fibers for General Industry.

• 1915.1000, Air Contaminants (Shipyard Employment). Table Z lists PEL's for fibrous glass and mineral wool (total dust and respirable fraction).

Other applicable standards are the Threshold Limit Values for Chemical Substances and Physical Agents; ACGIH; Cincinnati, OH; (513) 742-3355; E-mail: acgih_pubs@pol.com.; Consensus exposure limits from organization of governmental industrial hygienists for the following Synthetic Vitreous Fibers: Continuous filament glass fibers, Continuous filament glass fibers, Glass wool fibers, Rock wool fibers, Slag wool fibers, Special purpose glass fibers, and Refractory ceramic fibers.

An employee's exposure to any substance listed in Tables Z-1, Z-2, or Z-3 of the OSHA standard shall be limited in accordance with the requirements of the following paragraphs:

Table Z-1: Substances with limits preceded by "C", or Ceiling Values. An employee's exposure to any substance in Table Z-1, the exposure limit of which is preceded by a "C", shall at no time exceed the exposure limit given for that substance. If instantaneous monitoring is not feasible, then the ceiling shall be assessed as a 15-minute time weighted average exposure which shall not be exceeded at any time during the working day.

Other substances - 8-hour Time Weighted Averages: An employee's exposure to any substance in Table Z-1, the exposure limit of which is not preceded by a C, shall not exceed the 8-hour Time Weighted Average given for that substance any 8-hour work shift of a 40-hour work week.

Table Z-2: An employee's exposure to any substance listed in Table Z-2 shall not exceed the exposure limits specified as follows: For 8-hour time weighted averages, an employee's exposure to any substance listed in Table Z-2, in any 8-hour work shift of a 40-hour work week, shall not exceed the 8-hour time weighted average limit given for that substance in Table Z-2. Refer to 1910.1000(b)(2).

Acceptable Ceiling Concentrations: An employee's exposure to a substance listed in Table Z-2 shall not exceed at any time during an 8-hour shift the acceptable ceiling concentration limit given for the substance in the table, except for a time period, and up to a concentration not exceeding the maximum duration and concentration allowed in the column under "acceptable maximum peak above the acceptable ceiling concentration for an 8-hour shift." As an example, during an 8-hour work shift, an employee may be exposed to a concentration of Substance A (with a 10 ppm TWA, 25 ppm ceiling and 50 ppm peak) above 25 ppm (but never above 50 ppm) only for a maximum period of 10 minutes. Such exposure must be compensated by exposures to concentrations less than 10 ppm so that the cumulative exposure for the entire 8-hour work shift does not exceed a weighted average of 10 ppm.

Table Z-3: An employee's exposure to any substance listed in Table Z-3, in any 8-hour work shift of a 40-hour work week, shall not exceed the 8-hour time weighted average limit given for that substance in the table.

The computation formula which is applied by OSHA to employee exposure to more than one substance for which 8-hour time weighted averages are listed in subpart Z of 29 CFR Part 1910 in order to determine whether an employee is exposed over the regulatory limit is as follows:
For the cumulative exposure for an 8-hour work shift:

$$(E = C(a)T(a) + C(b)T(b) + \ldots C(n)T(n))/8$$

Where E is the equivalent exposure for the working shift, C is the concentration during any period of time T where the concentration remains constant. T is the duration in hours of the exposure at the concentration C. The value of E shall not exceed the 8-hour time weighted average specified in Subpart Z or 29 CFR Part 1910 for the substance involved. To illustrate the formula, assume that Substance A has an 8-hour time weighted average limit of 100 ppm noted in Table Z-1. Assume that an employee is subject to the following exposure: Two hours exposure at 150 ppm; Two hours exposure at 75 ppm; Four hours exposure at 50 ppm. Substituting this information in the formula, we have

$$(2 \times 150 + 2 \times 75 + 4 \times 50)/8 = 81.25 \text{ ppm}$$

Since 81.25 ppm is less than 100 ppm, the 8-hour time weighted average limit, the exposure is acceptable.

In case of a mixture of air contaminants an employer shall compute the equivalent exposure as follows:

$$E(m) = C_1/L_1 + C_2/L_2 + \ldots + C_n/L_n$$

Where E(m) is the equivalent exposure for the mixture, C is the concentration of a particular contaminant, L is the exposure limit for that substance specified in Subpart Z of 29 CFR Part 1910. The value of E(m) shall not exceed unity (1). To illustrate the formula, consider the following exposures:

Substance	Actual Concentration of 8-hour Exposure (ppm)	8-hour TWA PEL (ppm)
B	500	1,000
C	45	200
D	40	200

Substituting in the formula, we have:

$$E(m) = 500/1,000 + 45/200 + 40/200$$

$$E(m) = 0.500 + 0.225 + 0.200$$

$$E(m) = 0.925$$

Since E(m) is less than unity (1), the exposure combination is within acceptable limits.

To achieve compliance, administrative or engineering controls must first be determined and implemented whenever feasible. When such controls are not feasible to achieve full compliance, protective equipment or any other protective measures shall be used to keep the exposure of employees to air contaminants within the limits prescribed in this section. Any equipment and/or technical measures used for this purpose must be approved for each particular use by a competent industrial hygienist or other technically qualified person. Whenever respirators are used, their use shall comply with 1910.134.

RADIOFREQUENCY/MICROWAVE RADIATION

Radiofrequency (RF) and microwave (MW) radiation are electromagnetic radiation in the frequency range 3 kilohertz (kHz) to 300 gigahertz (GHz). Usually MW radiation is considered a subset of RF radiation, although an alternative convention treats RF and MW radiation as two spectral regions. Microwaves occupy the spectral region between 300 GHz and 300 MHz, while RF or radio waves include 300 MHz to 3 kHz. RF/MW radiation are nonionizing in that there is insufficient energy (less than 10 eV) to ionize biologically important atoms. The primary health effects of RF/MW energy are considered to be thermal. The absorption of RF/MW energy varies with frequency. Microwave frequencies produce a skin effect, one can literally sense your skin starting to feel warm. RF radiation may penetrate the body and be absorbed in deep body organs without the skin effect which can warn an individual of danger. Much research has turned up other nonthermal effects. All the standards of western countries have, so far, based their exposure limits solely on preventing thermal problems. In the meantime, research continues. Use of RF/MW radiation includes: aeronautical radios, citizen's (CB) radios, cellular phones, processing and cooking of foods, heat sealers, vinyl welders, high frequency welders, induction heaters, flow solder machines, communications transmitters, radar transmitters, ion implant equipment, microwave drying equipment, sputtering equipment, glue curing, and power amplifiers used in EMC and metrology (calibration). Web links to information from the OSHA Web site include:

- Questions and Answers About the Biological Effects and Potential Hazards of Radiofrequency Radiation. Federal Communication Commission, Office of Engineering and Technology (OET) Bulletin #56 (1999, August).

- Radiofrequency (RF) and Microwave Radiation. American Industrial Hygiene Association (1994), Nonionizing Radiation Guide Series, Stock No. 187-EA-94, Fairfax, VA, 33 pages. An overview of RF and microwave properties, health effects, measurement, controls, and standards.

- Nonionizing Radiation, Fundamentals of Industrial Hygiene Chapter 11. National Safety Council, Chicago, Ill, (1988), 30 pages.

- Introduction to Radio Frequency Radiation - Training Slides; OSHA Radio Frequency Series; Bob Curtis; (1998, November). Presentation includes overview of RF Operations and Controls.

- Protection of DoD Personnel from Exposure to Radiofrequency Radiation and Military Exempt Lasers, Department of Defense Instruction 6055.11. (1995, February 21). Includes DoD exposure limits which closely follow ANSI C95.1-1991. Also addresses RF signs, electromagnetic pulses and RF protective clothing.

- Health Aspects of Exposure to Electric and Magnetic Fields from RF Sealers and Dielectric Heaters. Institute for Electrical and Electronic Engineers (IEEE, USA) (1992, May), 10 pages. Position paper declaring the need to control these operations to protect workers.

- Occupational Exposure of Police Officers to Microwave Radiation from Traffic Radar Devices. NIOSH (June, 1995), 31 pages. Includes exposure assessments, an analysis of existing record sources for possible epidemiological studies, and recommendations, including five specific recommendations to reduce or prevent exposure to microwave radiation from traffic radar devices.

- Radiofrequency (RF) Sealers and Heaters: Potential Health Hazards and Their Prevention, (NIOSH/OSHA Current 33) (1979, December 4).

- Radio Freqency Radiation Emissions and Wireless Communication Devices (CDHR). National Toxicology Program, U.S. Department of Health and Human Services. This document provides a summary of why FDA is concerned about exposures from cellular phones and has suggested that the National Toxicology Program study the issue.

- Mobile Phones and Health. House of Commons Science and Technology Select Committee (1999, September 22). This report outlines the recommendations and conclusions of the committee regarding mobile phones and health.

- Consumer Update on Mobile Phones. FDA Center for Devices and Radiological Health (1999, October 20), 3 pages. Questions and answers about the safety of mobile phones, including cellular phones and PCS phones. Includes suggestions on how to minimize any potential health risk.

- FCC RF Safety Program Website. Federal Communications Commission. Provides access to FCC rules and guidance documents related to RF safety.

- FDA Letter Regarding Cellular Phones. (1997, May 5), 16 pages. Letter to Congress from the Food and Drug Administration in response to questions regarding alleged health hazards associated with the use of cellular phones. The appendix includes brief summaries of six recent studies.

- FDA Letter Regarding Wireless Communication. (1998, January 14), 7 pages. Letter to Congress regarding the status of the Food and Drug Administration's oversight and investigation of wireless communication health effects.

- Human Exposure to RF Emissions from Cellular Radio Base Station Antennas. IEEE-USA (1992, May), 7 pages. IEEE position statement declaring these sites as safe assuming exposures are limited below the ANSI/IEEE standards.

- Cellular Phones Antennas and Human Health Internet Site. This site was established by Prof. John Moulder, U. of Wisconsin, a radiation oncologist, to assist in answering questions about hazards from cellular phone communications. It includes a nice series of FAQs.

- Radiofrequency Radiation-caused Burns. OSHA Hazard Information Bulletin (1990, September 5). Induced-current grasping hazards and burns caused by spark-discharges were found in longshoremen working on a pier in close proximity to AM radio transmitting towers. Suggested control methods are included.

- Evaluating Compliance with FCC Guidelines for Human Exposure to Radiofrequency Electromagnetic Fields. Federal Communications Commission OET Bulletin #65 (1997, August). Provides assistance in determining whether proposed or existing transmitting facilities, operations or devices comply with limits for human exposure to radiofrequency (RF) fields adopted by the FCC.

- Recommended Practice for the Measurement of Potentially Hazardous Electromagnetic Fields-RF and Microwave. ANSI/EEE C95.3 (1992). Copyright by IEEE, New York, NY 10017. For copies contact IEEE at: 1-800-678-4333.

- A Practical Guide to the Determination of Human Exposure to Radiofrequency Fields. National Council on Radiation Protection and Measurement (CRP), Report No. 119 (1993). Copyright CRP Beheads, MD 20814. For copies contact CRP Publications at : 1-800-229-2652.

- Radiofrequency Radiation Dosimetry Handbook (Fourth Edition). USAFSAM-TR-85-73 Technical Document (1986, October). Based on the earlier versions of this report, most RF standards (e.g., ACGIH, ANSI, ICNIRP) include exposure limits which change with frequency and based on Specific (frequency and species) Absorption Rates expressed in Watts/kilogram of body tissue.

- Evaluating Compliance With FCC Guidelines for Human Exposure to Radiofrequency Electromagnetic Fields. Federal Communications Commission OET Bulletin #65 (1997, August). Appendix A (p. 67) of this document provides a table and figure of RF exposure limits adopted by FCC. FCC received concurrence for these limits from other government agencies, including OSHA and NIOSH, with the reservation that induced current limits be added to the FCC standard (These are already a part of the ACGIH and ANSI standards).

- Threshold Limit Values; ACGIH; Cincinnati, OH; (513) 742-3355; E-mail: acgih_pubs@pol.com.; Consensus exposure limits from organization of governmental industrial hygienists for: Lasers, Light and Near-Infrared Radiation, Radiofrequency/Microwave Radiation, Static Magnetic Fields, Sub-Radiofrequency (30kHz and below) Magnetic Fields, Sub-Radiofrequency (30kHz and below) and Static Electric Fields, and Ultraviolet Radiation.

- International Commission on Nonionizing Radiation Protection. The current

FCC RF exposure limits are based on standards from this organization combined with the 1992 ANSI standard

- American National Standards Institute (ANSI). Publishes consensus standards on RF exposures and measurements. The Institute of Electrical and Electronics Engineers (IEEE), Standards Coordinating Committee 28 is the secretariat for ANSI for developing RF standards. It is also the parent organization for the IEEE Committee on Man and Radiation (COMAR) which publishes papers on human exposure to electromagnetic fields.

- American National Standards Institute. Safety Levels with Respect to Human Exposure to Radio Frequency Electromagnetic Fields (200 kHz-100 GHz); ANSI C95.1-1992; New York, NY. Popular U.S. consensus standard which includes different exposure limits for controlled and uncontrolled sites.

- Elements of a Comprehensive RF Protection Program: Role of RF Measurements. Robert A. Curtis, US DOL/OSHA Health Response Team (1995, April 12), 5 pages. This paper outlines the elements of a comprehensive RF Protection Program and the role of RF measurements in implementing the program.

- ElectroMagnetic Energy Evaluation and Management for Antenna Sites. Motorola (1997, October 9), 23 pages. Guidance for developing an RF safety program for RF antenna sites.

- Evaluation and Control of Personnel Exposure to Radio Frequency Fields, 3 kHz to 300 GHz . NATO Standardization Agreement (STANAG) 1.8 MB PDF File.

The specific OSHA standards that are applicable are: 1910.268, Telecommunications; 1910.97, Nonionizing Radiation. The exposure limit in this standard (10 mW/sq.cm.) is expressed in voluntary language and has been ruled unenforceable for Federal OSHA enforcement. However, some states with their own OSHA-type programs are enforcing this or other RF exposure limits. The standard does specify the design of an RF warning sign; and 1926.54(l), Construction. Limits worker exposure to 10 mW/sq.cm. for construction work (including the painting of towers).

The term "electromagnetic radiation" is restricted to that portion of the spectrum commonly defined as the radiofrequency region, which for the purpose of this specification shall include the microwave frequency region. The term "partial body irradiation" pertains to the case in which part of the body is exposed to the incident electromagnetic energy. The word "symbol" as used in OSHA's specification refers to the overall design, shape, and coloring of the rf radiation sign. The term "whole body irradiation" pertains to the case in which the entire body is exposed to the incident electromagnetic energy or in which the cross section of the body is smaller than the cross section of the incident radiation beam. For normal environmental conditions and for incident electromagnetic energy of frequencies from 10 MHz to 100 GHz, the radiation protection guide is 10 mW/cm² (milliwatt per square centimeter) as averaged over any possible 0.1-hour period. This means the following:

- Power density: 10 mW/cm² for periods of 0.1-hour or more.
- Energy density: 1 mW-hr/cm² (milliwatt hour per square centimeter) during any 0.1-hour period. This guide applies whether the radiation is continuous or intermittent. These formulated recommendations pertain to both whole body irradiation and partial body irradiation. Partial body irradiation must be included since it has been shown that some parts of the human body (e.g., eyes, testicles) may be harmed if exposed to incident radiation levels significantly in excess of the recommended levels.

The warning symbol for radiofrequency radiation hazards shall consist of a red isosceles triangle above an inverted black isosceles triangle, separated and outlined by an aluminum color border. The words "Warning - Radiofrequency Radiation Hazard" shall appear in the upper triangle. Refer to Figure 1.

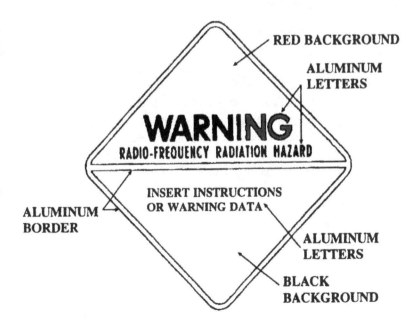

Figure 1. *Radiofrequency radiation hazard warning.*

American National Standard Safety Color Code for Marking Physical Hazards and the Identification of Certain Equipment, Z53.1-1953 which is incorporated by reference as specified in Sec. 1910.6, is used for color specification. All lettering and the border are of aluminum color. The inclusion and choice of warning information or precautionary instructions is at the discretion of the user. If such information is included it shall appear in the lower triangle of the warning symbol.

OSHA recognizes that most effective activities, including inspections, are those which encourage employers to implement their own comprehensive safety and health program. For work sites involving potentially hazardous radiofrequency radiation, OSHA compliance officers should evaluate the RF protection component of the overall program. The elements of a comprehensive RF Protection Program. include the implementation of appropriate protective policies based on the potential for excessive RF exposures. Therefore, RF exposure assessments, often requiring direct measurement, are performed to evaluate the effectiveness of RF controls; to ensure proper maintenance of RF radiating equipment; to develop work practices to minimize exposures; to obtain information to be used in training workers regarding their potential hazards and how they are controlled; to identify "RF Hazard" zones and other areas requiring signs and training: to determine the need for medical surveillance; as an alternative or enhancement of Lockout/Tagout procedures; to evaluate the effectiveness of RF personal protective equipment; and as a periodic audit of the effectiveness of the RF Protection Program. Based on literally hundreds of RF surveys conducted by the author, it is concluded that effective control of RF hazards depends primarily on the commitment to these Program elements, and not on sophisticated RF survey equipment or expertise. To minimize the risk of adverse health effects, radiofrequency (RF) fields as well as induced and contact currents must be in compliance with applicable guidelines (e.g., ICNIRP, ANSI, ACGIH). Reduction in RF exposures can be accomplished through the implementation of appropriate, administrative, work practice and engineering controls. These various controls are the elements of an RF Protection Program, and part of an employer's comprehensive safety and health program. The following outlines the principal elements of the RF Protection Program, and the role of RF measurements in implementing the program.

- Utilization of RF source equipment which meet applicable RF and other safety standards when new and during the time of use, including after any modifications. Manufacturers of RF source equipment are responsible for making equipment that complies with applicable standards, and for providing information on the hazards of operating and servicing the equipment. The information must be sufficient to alert the end-user of potential hazards and necessary controls applicable to using the equipment. Manufacturers are therefore required to make detailed RF emission measurements of their products. Appropriate RF survey results should be provided to the end-user for comparison purposes. For many low-power products, such as cellular phones, no additional measurements are required by the end-user. For other products, the users should conduct RF "screening" measurements of equipment emissions after installation, major maintenance, and any modifications which could effect RF emissions. Significant deviations from previous measurements should be resolved.

- RF hazard identification and periodic surveillance by a competent person who can effectively assess RF exposures.

- Screening measurements are normally sufficient to identify potentially hazardous RF areas which will require some control strategy, such as to determine where a fence should be located. More complex measurements are necessary if the employer intends to allow exposures to employees approaching RF standards. For example, detailed measurements are necessary if whole-body and/or time-weighted averaging of exposures is necessary to bring exposures into compliance.

- RF fields can induce currents in nearby conducting objects, such as a metal barrier or fence used to restrict access to RF hazard areas. These must be evaluated to ensure they do not constitute RF shock and burn hazards. Although detail measurements can be made, the "measurement" of startling/annoying RI spark discharge can usually be made by a quick touch.

- Controlling exposure time and the distance between the RF source and the operator are important in maintaining workers' exposures below recommended levels. When necessary due to excessive leakage, "RF hazard areas" must be identified to alert workers of areas that are not to be occupied during RF application. The location of the hazard areas must be based on exposure measurements made during maximum field generation and duty factor (i.e., ratio of RF "on" time during any 6 minute period, assuming intermittent exposure).

- Access to RF hazard areas should be controlled with standard Lockout/Tagout procedures (refer to 29 CFR 1910.147) to ensure workers are not occupying these areas during the application of RF energy. It maybe possible to use continuous monitors and/or personal monitors in lieu of, or to supplement, more traditional Lockout/Tagout procedures which lockout the RF power source.

- The RF hazard areas shall be clearly marked with appropriate signs, barricades, floor markings, etc. such that any worker who has access to the facility will be alerted not to occupy the hazardous locations. Signs shall be of standard design and shape (refer to ANSI C95.1), and of sufficient size to be recognizable and readable from a safe distance.

- Screening measurements can be used to determine where to locate signs to alert workers approaching an RF hazard area, including the appropriate warning message on the sign (e.g., **NOTICE, CAUTION, DANGER**).

- The evacuation of hazard areas prior to RF application must be strictly enforced. For example, a procedure which requires an RF sealer operator to first load the sealer, step back 2 meters to get outside the RF hazard area prior to activating the RF energy, and then walk back to unload the sealer will be difficult to enforce. The additional time required and increased operator fatigue will discourage operators from following such procedures.

- Implementation of controls to reduce RF exposures to levels in compliance with applicable guidelines (e.g., ANSI, ICNIRP), including the establishment of safe work practice procedures Reliance on averaging is normally not "recommended when establishing basic control strategies because it obligates the employer to conduct "measurement" of employee activity to ensure the averaging is applicable, such as timing an employee's access inside an area which can not be occupied for 6 minutes without exceeding the allowable time-weighted exposure. Where possible, controls should be establish under the assumption that standards are not time-weighted, i.e., assume the standards are ceiling limits which are not to be exceeded. Measurements are necessary during the development of work practices to ensure the practices are effective in preventing excessive exposures. Detailed measurements are required if exposures are approaching guideline limits as discussed above. Appropriate work practices must be followed during the repair and maintenance of RF equipment. Occasionally, cabinet panels must be removed by service personnel to allow access for maintenance. Failure to replace a panel properly may result in excessive RF leakage. RF screening measurements can be used to determine which panels can be removed during operation (assuming other hazards, such as electrical shock, are controlled), and to ensure the shielding is reinstalled properly. Detailed measurements must be made by the manufacturers' of RF personal protective equipment (PPE) to show its effectiveness and limitations. Limited measurements are necessary by the user to ensure the PPE is applicable and effective for the specific worksite conditions.

- RF safety and health training to ensure that all employees understand the RF hazards to which they may be exposed and the means by which the hazards are controlled. Measurement of worker exposures is necessary so that this information can be provided as part of employee hazard training. The scope of training, including reviews of potential biological effects, will be dependent on measured exposure levels.

- Employee involvement in the structure and operation of the program and in decisions that affect their safety and health, to make full use of their insight and to encourage their understanding and commitment to the safe work practices established. RF screening measurements should be made in the presence of employees to facilitate understanding and confidence in the program.

- Implementation of an appropriate medical surveillance program. RF measurements are necessary "to determine the need and scope of medical surveillance." For example, medical. surveillance may consists of a means to report the occurrence of RF burns, implanted medical devices (e.g., copper IUD), or the sensation of nonroutine heating as a means of identifying potential problem areas. A medical exam maybe appropriate for "accidental" exposures defined as an exposure above some measured trigger level. Although not required for compliance with existing standards, RF exposure data is necessary to enhance epidemiology studies of RF biological effects.

- Periodic (e.g., annual) reviews of the effectiveness of the program so that deficiencies can be identified and resolved. Periodic RF screening measurements are necessary to ensure conditions have not changed and that the RF Protection Program continues to be effective in preventing excessive RF exposures.

- Assignment of responsibilities, including the necessary authority and resources to implement and enforce all aspects of the RF protection program. Although this element does not directly require RF measurements, it is included for completeness of the list of RF Program elements. Without the commitment to the Program, as demonstrated by the assignment of necessary responsibility, authority and resources, the previous elements will not be effective.

A variety of RF measurements are necessary for an effective RF protection program. Usually RF screening measurements are adequate unless control strategies allow exposures approaching RF limits. Detailed RF measurements are required of manufacturers of RF products (e.g., RF transmitters, PPE, RF meters) to document their effectiveness and limitations. The effectiveness of the RF protection program depends primarily on an employer's understanding and commitment to the listed program elements, rather than on sophisticated RF survey equipment or measurement procedures.

WEB SITES FOR ADDITIONAL INFORMATION

The following is a sampling of Web sites that the reader may explore for company specific information on training and personal protective equipment. This listing is not an endorsement, and the exclusion of any company or Web site is not a negative review.

Pensacola Testing Labs., OSHA Hazwoper training, confined space training. nde, ndt, ndi training *http://www.pensacolatesting.com/ptl/hazmat1*

DOE Course Index - PX0003304: Course Number: PX0003304 Site: PANTEX Cost: Call for information Length: 24.00 hr Original Provider : PANTEX Prerequisites: Contact: Title: Advanced RCRA (29CFR 1910.120) Description: WASTE IDENTIFICATION, ADVANCED RCRA TOPICS, & LAND DISPOSAL. *http://cted.inel.gov/cted/crsindex/dp/px00033*

All American Environmental Services Inc. - Specializing in Hazardous Waste Operations, Chemical-Biological incidents response and Occupational Health and Safety Training and Consulting. Also Video Production for Training, Documentaries, Public Services Announcements, Infomercials, and Publication. *http://www.aaesi.com/staff.htm*

ONSITE Environmental Staffing - Training Courses Instructors & Equipment 40HR & 24HR HAZWOPER 29CFR 1910.120 Our HAZWOPER courses are designed for entry-level technicians as well as professionals who require a working knowledge of 29 CFR 1910.120, Hazardous Waste Operations. *http://www.onsite-inc.com/oes/employment/cour*

24 Hour HAZWOPER: "Safety Training for Life!" 24-Hour HAZWOPER: As referenced by 29CFR 1910.120 (e) (3) (ii), workers on site only occasionally for a specific limited task (such as, but not limited to, ground water monitoring, land surveying, or geophysical. *http://www.moti.cnchost.com/24Hour.htm*

Oil and Gas - Protective Clothing - Chemical/Gas Resistant: Oil and Gas - Protective Clothing - Chemical/Gas Resistant. Address, telephone, fax and internet links. Product Selector and Buyers Guide. Published by Applegate First Media *http://www.1stdirectory.com/oil/ps_2133.htm*

Lakeland Industries - Chemical Protective Clothing Menu: Lakeland manufactures quality industrial and consumer safety clothing, coveralls, apparel and gloves. Large product lines include Aluminized, Disposable, Chemical Protective Suits, Turnout Gear in Tyvek, Kevlar and Nomex materials, as well as other PPE. *http://www.lakeland.com/chemical.html*

LHR Services and Equipment, Inc. - Chemical Protection - Protective Clothing: providing chemical protection supplies and equipment for the offshore oil and gas drilling industry. *http://www.lhrservices.com/chemical_protectio*

The Safety Supply Store. Breathable protective clothing at wholesale prices http://www.frostproof.com/catalog/s27.html

IMS-PLUS G.I. Issue Chemical Protective Suits (NBC suits / MOPP suits) G.I. Issue Chemical Protective Suits (NBC Suits /MOPP Suits) GENUINE G.I. ISSUE CHEMICAL PROTECTIVE SUIT. Constructed from cotton twill outer shell with black charcoal filter inner lining. Jacket features: elastic cuffs, drawstring bottom, etc. *http://www.imsplus.com/ims47a.html*

NIOSH/Chemical Protective Clothing Page/INDEX Z Recommendations for Chemical Protective Clothing A Companion to the NIOSH Pocket Guide to Chemical Hazards TABLE Z Chemical CAS No. Recommendation for skin protection Recommended protective clothing barriers* Zinc chloride fume 7646-85-7 N.R. *http://ehs.clemson.edu/niosh/ncpc/zcpc.htm*

TRI/Environmental - Personal Protective Equipment Testing High quality analytical services - Compressed Air Analysis, Geosynthetics Testing, Protective Clothing Program and Industrial Hygiene. Plus Applied Research and Development, Materials Science and Non-Destructive Evaluation. *http://www.tri-env.com/env/protective.htm*

Chapter 6

SAFETY AND EMERGENCY PREPAREDNESS

INTRODUCTION

Chemicals can be described as the foundation of a modern, progressive society. They are an integral and ever-increasing part of our complex technological world, making it possible for us to enjoy a high standard of living. Yet, as the 1984 catastrophe in Bhopal, India dramatically and tragically demonstrated, those same chemicals are the source of danger to those in the workplace and surrounding locales who are regularly exposed to them. We have seen how their improper use and handling impact and exact unacceptable human and economic costs on families, industries, communities, and even nations. As a result, we have learned that correcting situations that could lead to disasters and catastrophes is more responsible and less expensive than hoping accidents will not occur and responding only when they inevitably do. U.S. Congress identified 14 different agencies engaged in accident prevention-related activities and determined there was a need to ". . . improve the effectiveness of accident prevention programs and reduce the burden of duplicative requirements on regulated entities." (Senate Report 1989). Not surprisingly, the chemical industry, in hearings before Congress, agreed with this finding. The regulated industry voiced its support ". . . for a coordinated Federal approach to accident prevention and suggested that an agency like the Board might most effectively carry out that responsibility" (Senate Report 1989).

After looking into the problem and alternative solutions, Congress determined there, indeed, was the need to identify and address the causes of the thousands of chemical accidents that occur annually, as well as the need to protect life, property and the environment from the costly consequences of those accidents.

As of February 1993, the U.S. Environmental Protection Agency's Resource Conservation and Recovery Information System (RCRIS) reported the existence of 278,755 facilities that generate, transport, treat, store and/or dispose of regulated hazardous waste. At these locations substances exist whose nature and quantities pose significant risk to the workers, general public and environment. As not all dangerous chemicals or wastes or facilities that handle chemicals are regulated, the actual number of locations may be much higher. In addition, according to the NTSB, "about four billion tons of regulated hazardous materials are shipped each year with more than 250,000 shipments of hazardous materials entering into the U.S. transportation system daily" (NTSB 1992).

The universe of chemical accidents within the United States cannot now be accurately tallied. No comprehensive, reliable historical records exist. Further, the EPA acknowledges that many accidents occurring today at fixed facilities and during transport are not reported to the federal government. This underreporting is documented by several studies (National Environmental Law Center et al. 1994). What is known, however, is that in 1991 the National Response Center received over 16,300 calls reporting the release or potential release of a hazardous material (US EPA 1993). Also, NTSB's statistics indicate that, in 1992, chemicals were involved in 3,500 fatal highway accidents and 6,500 railroad accidents (NTSB 1992). One study analyzed information contained in the EPA's Emergency Response Notification System (ERNS) database. ERNS (even with its significant limitations) is acknowledged to be the largest and most comprehensive United States database of chemical accident notifications, covering both transportation and fixed facility accidents. The study found that from 1988 through 1992 an average of 19 accidents occurred each day . . . 6,900 per year, with more than 34,500 accidents involving toxic chemicals occurring over the five-year period. The study's report emphasized that the findings gravely understated the severity of the United States' chemical accident picture (National Environmental Law Center et al. 1994).

Although the absolute numbers vary depending on the source of statistics and period of time examined, there is no doubt about the effects of chemical accidents on human life . . . year after year, large numbers of people are killed and injured. Added to these imprecise numbers must be those long-term consequences of exposure that are not immediately discernable and may not be reflected in studied databases . . . low-level exposure to some chemicals may result in debilitating diseases that appear only years later. During the years 1988 through 1992, six percent, or 2,070, of the 34,500 accidents that occurred resulted in immediate death, injury and/or evacuation; an average of two chemical-related injuries occurred every day during those five years (National Environmental Law Center et al. 1994). Between 1982 and 1986, the EPA's Acute Hazard Events (AHE) database, which contains information only for chemical accidents having acute hazard potential, recorded 11,048 events involving releases of extremely hazardous substances; these events resulted in 309 deaths, 11,341 injuries and, based on evacuation information for the one-half of the recorded events reporting whether such activity occurred, evacuation of 464,677 people from their homes and jobs (USEPA 1989). During the years 1987 through 1991, chemical accidents resulted in 453 deaths and 1,576 injuries at fixed facilities, while transportation accidents involving chemicals claimed 55 lives and injured 1,252 persons (US EPA 1993). Within a five-year period in the mid-1980s, the EPA's AHE database indicates there were 10,933 such accidents, of which 135 resulted in fatalities, 1,020 resulted in injuries and 500 resulted in evacuations (US EPA 1993).

With this as an introduction, this chapter focuses on emergency preparedness, including proper planning and risk management issues. Emphasis is given to the chemical industry, however, many of the concepts reviewed are general and can be applied across different industry sectors.

EMERGENCY PREPAREDNESS AND RESPONSE

The importance of an effective workplace safety and health program cannot be overemphasized. There are many benefits from such a program including increased productivity, improved employee morale, reduced absenteeism and illness, and reduced workers' compensation rates; however, incidents still occur in spite of efforts to prevent them. Therefore, proper planning for emergencies is necessary to minimize employee injury and property damage. Typical emergencies include accidental releases of toxic gases, chemical spills, fires, explosions, and bodily harm and trauma caused by workplace violence.

The effectiveness of response during emergencies depends on the amount of planning and training performed. Senior level management must show its support for plant safety programs and the importance of emergency planning. If management is not interested in employee protection and in minimizing property loss, little can be done to promote a safe workplace. It is therefore management's responsibility to see that a program is instituted and that it is frequently reviewed and updated. The input and support of all employees must be obtained to ensure an effective program. The emergency response plan should be developed locally and should be comprehensive enough to deal with all types of emergencies specific to that site. When emergency action plans are required by a particular OSHA standard, the plan must be in writing; except for firms with 10 or fewer employees, the plan may be communicated orally to employees. The plan must include, as a minimum, the following elements:

- Emergency escape procedures and emergency escape route assignments,
- Procedures to be followed by employees who remain to perform (or shut down) critical plant operations before the plant is evacuated,
- Procedures to account for all employees after emergency evacuation has been completed,
- Rescue and medical duties for those employees who are to perform them,
- The preferred means for reporting fires and other emergencies, and
- Names or regular job titles of persons or departments to be contacted for further information or explanation of duties under the plan.

The emergency action plan should address all potential emergencies that can be expected in the workplace. Therefore, it will be necessary to perform a hazard audit to determine toxic materials in the workplace, hazards, and potentially dangerous conditions. For information on chemicals, the manufacturer or supplier can be contacted to obtain Material Safety Data Sheets (MSDS). These forms describe the hazards that a chemical may present, list precautions to take when handling, storing, or using the substance, and outline emergency and first-aid procedures.

The employer must list in detail the procedures to be taken by those employees who must remain behind to care for essential plant operations until their evacuation becomes absolutely necessary. This may include monitoring plant power supplies, water supplies, and other essential services that cannot be shut down for every emergency alarm, and use of fire extinguishers.

For emergency evacuation, the use of floor plans or workplace maps that clearly show the emergency escape routes and safe or refuge areas should be included in the plan. All employees must be told what actions they are to take in emergency situations that may occur in the workplace, such as a designated meeting location after evacuation. This plan must be reviewed with employees initially when the plan is developed, whenever the employees' responsibilities under the plan change, and whenever the plan is changed. A copy should be kept where employees can refer to it at convenient times. In fact, to go a step further, the employer could provide the employees with a copy of the plan, particularly all new employees.

A chain of command should be established to minimize confusion so that employees will have no doubt about who has authority for making decisions. Responsible individuals should be selected to coordinate the work of the emergency response team. In larger organizations, there may be a plant coordinator in charge of plant-wide operations, public relations, and ensuring that outside aid is called in. Because of the importance of these functions, adequate backup must be arranged so that trained personnel are always available. The duties of the Emergency Response Team Coordinator should include the following:

- Assessing the situation and determining whether an emergency exists that

> requires activating the emergency procedures,

- Directing all efforts in the area including evacuating personnel,
- Ensuring that outside emergency services such as medical aid and local fire departments are called in when necessary, and
- Directing the shutdown of plant operations when necessary.

During a major emergency involving a fire or explosion it may be necessary to evacuate offices in addition to manufacturing areas. Also, normal services, such as electricity, water, and telephones, may be nonexistent. Under these conditions, it may be necessary to have an alternate area to which employees can report or that can act as a focal point for incoming and outgoing calls. Since time is an essential element for adequate response, the person designated as being in charge should make this the alternate headquarters so that he/she can be easily reached. Emergency communications equipment such as amateur radio systems, public address systems, or portable radio units should be present for notifying employees of the emergency and for contacting local authorities, such as law enforcement officials, private sector charitable groups, and the fire department. A method of communication also is needed to alert employees to the evacuation or to take other action as required in the plan.

Alarms must be audible or seen by all people in the plant and have an auxiliary power supply in the event electricity is affected. The alarm must be distinctive and recognizable as a signal to evacuate the work area or perform actions designated under the emergency action plan.

The employer must explain to each employee the means for reporting emergencies, such as manual pull box alarms, public address systems, or telephones. Emergency phone numbers should be posted on or near telephones, on employees' notice boards, or in other conspicuous locations. The warning plan should be in writing and management must be sure each employee knows what it means and what action is to be taken. It may be necessary to notify other key personnel such as the plant manager or physician during off-duty hours. An updated written list of key personnel should be kept listed in order of priority.

Management will need to know when all personnel have been accounted for. This can be difficult during shift changes or if contractors are on site. A responsible person in the control center must be appointed to account for personnel and to inform police or Emergency Response Team members of those persons believed missing.

Emergency Response Teams are the first line of defense in emergencies. Before assigning personnel to these teams, the employer must assure that employees are physically capable of performing the duties that may be assigned to them.

Depending on the size of the plant there may be one or several teams trained in the following areas:

- Use of various types of fire extinguishers,
- First aid, including cardiopulmonary resuscitation (CPR),
- Shutdown procedures,
- Evacuation procedures,
- Chemical spill control procedures,
- Use of self-contained breathing apparatus (SCBA),
- Search and emergency rescue procedures,
- Incipient and advanced stage fire fighting, and
- Trauma counseling.

The type and extent of the emergency will depend on the plant operations and the response will vary according to the type of process, the material handled, the number of employees, and the availability of outside resources. OSHA's Hazard Communication Standard (29 CFR part 1910.1200) is designed to ensure that the hazards of all chemicals produced or imported are evaluated and that information concerning their hazards is transmitted to employers and employees. This is done by means of comprehensive hazard communication programs including container labeling and other forms of warnings, material safety data sheets, and employee training. Emergency Response Teams should be trained in the types of possible emergencies and the emergency actions to be performed. They are to be informed about special hazards such as storage and use of flammable materials, toxic chemicals, radioactive sources, and water-reactive substances to which they may be exposed during fire and other emergencies. It is important to determine when not to intervene. For example, team members must be able to determine if the fire is too large for them to handle or whether search and emergency rescue procedures should be performed. If there is the possibility of members of the Emergency Response Team receiving fatal or incapacitating injuries, they should wait for professional fire fighters or emergency response groups.

Training is important to the effectiveness of an emergency plan. Before implementing an emergency action plan, a sufficient number of persons must be trained to assist in the safe and orderly evacuation of employees. Training for each type of disaster response is necessary so that employees know what actions are required. In addition to the specialized training for Emergency Response Team members, all employees should be trained in the following:

- Evacuation plans,
- Alarm systems,
- Reporting procedures for personnel,

- Shutdown procedures, and
- Types of potential emergencies.

These training programs must be provided as follows:

- Initially when the plan is developed, .
- For all new employees,
- When new equipment, materials, or processes are introduced,
- When procedures have been updated or revised,
- When exercises show that employee performance must be improved, and
- At least annually.

The emergency control procedures should be written in concise terms and be made available to all personnel. A drill should be held for all personnel, at random intervals at least annually, and an evaluation of performance made immediately by management and employees. When possible, drills should include groups supplying outside services such as fire and police departments. In buildings with several places of employment, the emergency plans should be coordinated with other companies and employees in the building. Finally, the emergency plan should be reviewed periodically and updated to maintain adequate response personnel and program efficiency.

Effective personal protection is essential for any person who may be exposed to potentially hazardous substances. In emergency situations employees may be exposed to a wide variety of hazardous circumstances, including:

- Chemical splashes or contact with toxic materials,
- Falling objects and flying particles,
- Unknown atmospheres that may contain toxic gases, vapors or mists, or inadequate oxygen to sustain life,
- Fires and electrical hazards, and
- Violence in the workplace.

It is extremely important that employees be adequately protected in these situations. Some of the safety equipment that may be used includes:

- Safety glasses, goggles, or face shields for eye protection,
- Hard hats and safety shoes for head and foot protection,
- Proper respirators for breathing protection,
- Whole body coverings chemical suits, gloves, hoods, and boots for body protection from chemicals, and
- Body protection for abnormal environmental conditions such as extreme temperatures.

The equipment selected must meet the criteria contained in the OSHA standards or described by a nationally recognized standards producing organization. The choice of proper equipment is not a simple matter and consultation should be made with health and safety professionals before making any purchases. Manufacturers and distributors of health and safety products may be able to answer questions if they have enough information about the potential hazards involved. Professional consultation will most likely be needed in providing adequate respiratory protection. Respiratory protection is necessary for toxic atmospheres of dust, mists, gases, or vapors and for oxygen-deficient atmospheres. There are four basic categories of respirators:

1.   Air-purifying devices (filters, gas masks, and chemical cartridges), which remove contaminants from the air but cannot be used in oxygen-deficient atmospheres.

2.   Air-supplied respirators (hose masks, air line respirators), which should not be used in atmospheres that are immediately dangerous to life or health.

3.   Positive-pressure self-contained breathing apparatus (SCBA), which are required for unknown atmospheres, oxygen-deficient atmospheres, or atmospheres immediately dangerous to life or health.

4.   Escape masks.

Before assigning or using respiratory equipment the following conditions must be met:

• A medical evaluation should be made to determine if the employees are physically able to use the respirator.

• Written procedures must be prepared covering safe use and proper care of the equipment, and employees must be trained in these procedures and in the use and maintenance of respirators.

• A fit test must be made to determine a proper match between the facepiece of the respirator and the face of the wearer. This testing must be repeated periodically. Training must provide the employee an opportunity to handle the respirator, have it fitted properly, test its facepiece-to-face seal, wear it in normal air for a familiarity period, and wear it in a test atmosphere.

• A regular maintenance program must be instituted including cleaning, inspecting, and testing of all respiratory equipment. Respirators used for emergency response must be inspected after each use and at least monthly to assure that they are in satisfactory working condition. A written record of inspection must be maintained.

• Distribution areas for equipment used in emergencies must be readily accessible to employees.

• A positive-pressure self-contained breathing apparatus (SCBA) offers the best protection to employees involved in controlling emergency situations. It must have a minimum service life rating of at least 30 minutes. Conditions that require a positive-pressure SCBA include the following: (1) Leaking cylinders or containers, smoke from chemical fires, or chemical spills that indicate high potential for exposure to toxic substances; (2) Atmospheres with unknown contaminants or unknown contaminant concentrations, confined spaces that may contain toxic substances, or oxygen-deficient atmospheres.

Emergency situations may involve entering confined spaces to rescue employees who are overcome by toxic compounds or who lack oxygen. These permit-required confined spaces include tanks, vaults, pits, sewers, pipelines, and vessels. Entry into permit-required confined spaces can expose the employee to a variety of hazards, including toxic gases, explosive atmospheres, oxygen deficiency, electrical hazards, and hazards created by mixers and impellers that have not been deactivated and locked out. Personnel must never enter a permit-required confined space unless the atmosphere has been tested for adequate oxygen, combustibility, and toxic substances. Conditions in a permit-required confined space must be considered immediately dangerous to life and health unless shown otherwise. If a permit-required confined space must be entered in an emergency, the following precautions must be adhered to:

- All lines containing inert, toxic, flammable, or corrosive materials must be disconnected or blocked off before entry.
- All impellers, agitators, or other moving equipment inside the vessel must be locked out.
- Appropriate personal protective equipment must be worn by employees before entering the vessel. Mandatory use of harnesses must be stressed.
- Rescue procedures must be specifically designed for each entry. A trained stand-by person must be present. This person should be assigned a fully charged, positive-pressure, self-contained breathing apparatus with a full facepiece. The stand-by person must maintain unobstructed lifelines and communications to all workers within the permit-required confined space and be prepared to summon rescue personnel if necessary. The stand-by person should not enter the confined space until adequate assistance is present. While awaiting rescue personnel, the stand-by person may make a rescue attempt utilizing lifelines from outside the permit-required confined space.

A more complete description of procedures to follow while working in confined spaces may be found in the OSHA standard for permit-required confined spaces, 29 CFR 1910.145 and the National Institute for Occupational Safety and Health

(NIOSH) Publication Number 80-106, *Criteria for a Recommended Standard...Working in Confined Spaces*.

In a major emergency, time is critical factor in minimizing injuries. Most small businesses do not have a formal medical program, but they are required to have the following medical and first-aid services:

- In the absence of an infirmary, clinic, or hospital in close proximity to the workplace that can be used for treatment of all injured employees, the employer must ensure that a person or persons are adequately trained to render first aid. The first aid is to begin within 3 to 4 minutes of the incident if the injury is of a serious nature.

- Where the eyes or body of any employee may be exposed to injurious corrosive materials, eye washes or suitable equipment for quick drenching or flushing must be provided in the work area for immediate emergency use. Employees must be trained to use the equipment.

- The employer must ensure the ready availability of medical personnel for advice and consultation on matters of employees' health. This does not mean that health care must be provided, but rather that, if health problems develop in the workplace, medical help will be available to resolve them.

To fulfill the above requirements, the following actions should be considered:

- Survey the medical facilities near the place of business and make arrangements to handle routine and emergency cases. A written emergency medical procedure should then be prepared for handling accidents with minimum confusion.

- If the business is located far from medical facilities, at least one and preferably more employees on each shift must be adequately trained to render first aid. The American Red Cross, some insurance carriers, local safety councils, fire departments, and others may be contacted for this training.

- First-aid supplies should be provided for emergency use. This equipment should be ordered through consultation with a physician.

- Emergency phone numbers should be posted in conspicuous places near or on telephones.

- Sufficient ambulance service should be available to handle any emergency. This requires advance contact with ambulance services to ensure they become familiar with plant location, access routes, and hospital locations.

During an emergency, it is often necessary to secure the area to prevent unauthorized access and to protect vital records and equipment. An off-limits area must be established by cordoning off the area with ropes and signs. It may be necessary to notify local law enforcement personnel or to employ private security personnel to secure the area and prevent the entry of unauthorized personnel. Certain records also may need to be protected, such as essential accounting files, legal documents, and lists of employees' relatives to be notified in case of emergency. These records may be stored in duplicate outside the plant or in protected secure locations within the plant.

The following is a list of some of the OSHA requirements pertaining to emergency response. These references refer to appropriate sections of the Occupational Safety and Health Standards (*Title 29, Code of Federal Regulations, Part 1910*, which are the OSHA General Industry Standards).

Subpart E - Means of Egress
910.37 Means of egress
1910.38 Employee emergency plans and fire prevention plans
Appendix to Subpart E: Means of egress

Subpart H - Hazardous Materials
1910.119 Process safety management of highly hazardous chemicals
1910.120 Hazardous waste operations and emergency response.

Subpart I - Personal Protective Equipment
1910.132 General requirements for personnel protection
1910.133 Eye and face protection
1910.134 Respiratory protection
1910.135 Occupational head protection
1910.136 Occupational foot protection
1910.138 Hand protection

Subpart J - General Environmental Controls
1910.146 Permits for required confined spaces
1910.147 Control of hazardous energy sources

Subpart K - Medical and First Aid
1910.151 Medical services and first aid

Subpart L - Fire Protection
1910.155-156 Fire protection and fire brigades
1910.157- 163 Fire suppression equipment
1910.164 Fire detection systems
1910.165 Employee alarm systems
Appendix A-E of Subpart L

Subpart R - Special Industries, Electrical Power
Generation, Transmission, and Distribution

Subpart Z - Toxic and Hazardous Substances
1910.1030 Bloodborne pathogens
1910.1200 Hazard communication

Much of the planning and program development for responding to occupational emergencies will require professional assistance. Many public and private agencies provide information and services free or at minimal cost (e.g., Federal, State, and local health and labor departments, insurance carriers, and local universities). After having exhausted these sources, consider using a private consultant selected by matching his/her specialty with your specific needs. If there is a carrier for workers' compensation insurance, that company probably has safety and health specialists on staff who are familiar with minimum standards and technical information currently available and may be quite helpful in advising about accident and illness prevention and control. Trade associations often have technical materials, programs, and industry data available for specific needs. The Department of Labor through the Occupational Safety and Health Administration (OSHA) provides information in interpreting the law and on meeting the applicable standards. This information is available free of charge or obligation. The OSHA Area Office or State Plan Office nearest to the plant may be contacted for this information.

The Department of Health and Human Services through the National Institute for Occupational Safety and Health (NIOSH) provides printed material relating to employee safety and health in the workplace. Staff from this agency will perform industrial hygiene surveys of plants upon request of employers or employees. Machine or product manufacturers can be helpful in providing additional information on precautions to take in using their products. Any special problems should be referred to them first. Professional societies in the safety, industrial hygiene, and medical fields issue publications in the form of journals, pamphlets, and books that may be quite useful (e.g., American Society of Safety Engineers or the Occupational Health Institute). They can also recommend individuals from their societies to serve as consultants.

Effective management of worker safety and health protection is a decisive factor in reducing the extent and severity of work-related injuries and their related costs. To assist employers and employees in developing effective safety and health programs, OSHA published recommended *Safety and Health Management Guidelines* [*Federal Register* 54(18): 3908-3916, January 26, 1988]. These voluntary guidelines apply to all places of employment covered by OSHA. The guidelines identify four general elements that are critical to the development of a successful safety and health management program:

- Management commitment and employee involvement;

- Worksite analysis;
- Hazard prevention and control; and
- Safety and health training.

The guidelines recommend specific actions, under each of these general elements, to achieve an effective safety and health program. A copy of the guidelines can be obtained from the OSHA Publications Office, U.S. Department of Labor, 200 Constitution Avenue, N.W., Room N3101, Washington DC 20210. The *Occupational Safety and Health Act of 1970*, under Section 18(b), encourages States to develop and operate their own State job safety and health plans under the approval and monitoring of OSHA. Twenty-five states and territories operate such plans. They are required to set standards that are at least as effective as the federal, conduct inspections to enforce those standards (including inspections in response to workplace complaints), cover State and local government employees, and operate occupational safety and health training and education programs. In addition, all States provide on-site consultation to help employers to identify and correct workplace hazards. Such consultation may be provided either under the plan or through a special agreement under section 7(c)(1) of the Act. Federal OSHA does not conduct enforcement activities in the state plan States, except in very limited circumstances. A listing of those States that operate approved State plans can be obtained from your local OSHA Area Office. A comprehensive customer service poster listing OSHA services and how to contact agency Regional, Area, and District offices is available from OSHA's Publications Office, 200 Constitution Avenue, N.W. Washington D.C. 20210, Rm N3101. Telephone (202) 219-4667. Free on-site safety and health consultation services are available to employers in all states who want help in establishing and maintaining a safe and healthful workplace. This service is largely funded by OSHA. Primarily developed for smaller employers with more hazardous operations, the consultation service is delivered by state governments employing professional safety consultants and health consultants. Comprehensive assistance includes an appraisal of all mechanical systems, physical work practices, and environmental hazards of the workplace and all aspects of the employer's present job safety and health program. This program is completely separate from OSHA's inspection efforts. No penalties are proposed or citations issued for any safety or health problems identified by the consultant. The service is confidential. OSHA's area offices offer a variety of informational services, such as publications, audiovisual aids, technical advice, and speakers for special events. OSHA's Training Institute in Des Plaines, IL, provides basic and advanced courses in safety and health for federal and state compliance officers, state consultants, federal agency personnel, and private sector employers, employees, and their representatives. OSHA also provides funds to nonprofit organizations, through grants, to conduct workplace training and education in subjects where OSHA believes there is a lack of workplace training. Grants are awarded annually. Grant recipients are expected to contribute 20 percent of the total grant cost. The following is a list of references where the reader can obtain detailed information:

AIHA Hygienic Guide Series. American Industrial Hygiene Association, 2700 Prosperity Ave., Fairfax, VA 22031. Separate data sheets on specific substances giving hygienic standards, properties, industrial hygiene practices, specific procedures, and references.

ANSI Standards, Z37 Series, Acceptable Concentrations of Toxic Dusts and Gases. American National Standards Institute, 11 West 42nd Street, New York, NY 10036. These guides represent a consensus of interested parties concerning minimum safety requirements for the storage, transportation, and handling of toxic substances; they are intended to aid manufacturers, consumers, and the public.

ASTM Standards with Related Material. American Society for Testing and Materials, 1916 Race Street, Philadelphia, PA 19103.

The following is a list of standards and specification groups where additional information can be obtained:

American National Standards Institute, 11 West 42nd Street, New York, NY 10036, coordinates and administers the federated voluntary standardization system in the United States.

American Society for Testing and Materials, 1916 Race Street, Philadelphia, PA 10103. World's largest source of voluntary consensus standards for materials, products, systems, and services.

The following is a list of fire protection organizations:

Factory Insurance Association, 85 Woodland Street, Hartford, CT 06105. Composed of capital stock insurance companies to provide engineering, inspections, and loss adjustment service to industry.

Factory Mutual System, 1151 Boston-Providence Turnpike, Norwood, MA 02062. An industrial fire protection, engineering, and inspection bureau established and maintained by mutual fire insurance companies.

National Fire Protection Association, 470 Batterymarch Park, Quincy, MA 02269. The clearinghouse for information on fire protection and fire prevention also writes NFPA standards. Nonprofit technical and educational organization.

Underwriter Laboratories, Inc., 207 East Ohio Street, Chicago, IL 60611. Not-for-profit organization whose laboratories publish annual lists of manufacturers whose products proved acceptable under appropriate standards.

The following is a list of key references for more in-depth reading:

Chemical Industries Association, Chemical Industry Safety and Health Council. *Recommended Procedures for Handling Major Emergencies*. Alembic House. 93 Albert Embarkment. London, SEIO 7TU, July 1976.

Krikorian, Michael. "Advanced Planning is the Key to Controlling Emergencies and Disasters in the Workplace." *Prof Safety*: 39-42, December 1977.

Lee, W.R. Sources of Consultation and Reference Aids. Section XI, in H.M. Key, A.F. Henschel, J.Butler, R.N. Ligo, I.R. Tabershaw, and L. Ede (Eds): *Occupational Diseases: A Guide to Their Recognition*. NIOSH Publication No. 77-181. Cincinnati, 1977. Pp. 523-556. Also available as Lee, W.R. Consultation and Reference Sources for Occupational Health. *J Occu Med* 17(7): 446-456, July 1975.

National Safety Council. *Accident Prevention Manual for Industrial Operations Administration and Programs*. 8th ed. Chicago, 1981. Pp. 439-471.

U.S. Department of Labor. Occupational Safety and Health Administration. *OSHA Handbook for Small Businesses*. OSHA 2209. Washington, DC, 1996.

Occupational Safety and Health Administration. *Principal Emergency Response and Preparedness Requirements in OSHA Standards and Guidance for Safety and Health Programs*. OSHA 3122. Washington, DC, 1990.

U.S. Department of Health and Human Services. National Institute for Occupational Safety and Health, *Safety and Health Alert: Request for Assistance In Preventing Homicide In the Workplace*. U.S. Department of Health and Human Services, Cincinnati, Ohio, September 1993, Number 93-109.

Public Health Service. National Institute for Occupational Safety and Health. *A Guide to Industrial Respiratory Protection*. NIOSH Publication No. 76-189. Cincinnati, 1976.

Criteria for a Recommended Standard...Working in Confined Spaces. NIOSH Publication. No.80-106. Cincinnati, 1980.

Respiratory Protection...An Employer's Manual. NIOSH Publication No. 78-198A. Cincinnati, October 1978.

Self-Evaluation of Occupational Safety and Health Programs. NIOSH Publication No. 78-187. Cincinnati, 1978.

We will now focus attention on site security issues as these can represent an essential element in emergency planning, as well as protecting the assets of an operation.

Site Security Issues

In the past, the major concern for site security evolved around issues whereby the public might be exposed directly to hazardous materials. Facilities that handle chemicals are actively engaged in managing risks to ensure the safety of their workers and the community. Most of their efforts focus on ensuring that the facility is designed and operated safely on a daily basis, using well-designed equipment, preventive maintenance, up-to-date operating procedures, and well-trained staff. Because of today's increased concern about terrorism and sabotage, companies are also paying increased attention to the physical security of facility sites, chemical storage areas, and chemical processes. All companies, regardless of the size of their operations, have some measure of site security in place to minimize crime and to protect company assets. This is especially true for facilities that handle extremely hazardous materials. Under section 112(r) of the Clean Air Act (CAA), EPA developed Risk Management Program (RMP) regulations that require facilities to examine their chemical accident risk and develop a formalized plan to address it. The increased concern for the physical security of facilities that handle extremely hazardous substances is also reflected in recent government actions. Highlighting site security, the Chemical Safety Information, Site Security and Fuels Regulatory Relief Act contains a major provision that requires the Department of Justice to prepare reports to be submitted to Congress describing the effectiveness of RMP regulations in reducing the risk of criminally caused releases, the vulnerability of facilities to criminal and terrorist activity, and the security of transportation of listed toxic and flammable substances. Threats may come in different forms and from different sources. Threats from outside the facility could affect people and the facility itself, and may involve trespassing, unauthorized entry, theft, burglary, vandalism, bomb threats, or terrorism. Threats from inside the facility may arise from inadequate designs, management systems, staffing or training, or other internal problems. These may include theft, substance abuse, sabotage, disgruntled employee or contractor actions, and workplace violence, among others. Threats are not restricted to people and property, but could also involve sensitive facility information. Both facility outsiders and employees or contractors could pose threats to data storage and data transmission of, for example, confidential information, privacy data, and contract information. They could also pose a threat to computer-controlled equipment. These threats may include breaches in data access and storage, uncontrolled dissemination of information, destruction of information or threats to automated information systems. Most security measures are intended to prevent intruders from gaining access to the site or to limit damage.

Most facilities have some measures that are intended to prevent intruders from entering the grounds or buildings. These measures may include fences, walls, locked doors, or alarm systems. The location of the facilities and the types of structures will determine how much and what type of protection a facility needs. In addition to basic measures, some facilities also provide physical protection of site utilities at the fence perimeter. Security lighting (good lighting around buildings, storage tanks, and storage areas) can also make it very difficult for someone to enter the facility undetected. Some facilities find the need to augment these measures with intrusion detection systems - video surveillance, security guards at fixed posts, rounds/mobile patrols, alarm stations, and detectors for explosives and metal.

To protect against unauthorized people coming in through normal entrances, security clearances, badges, procedures for daily activities and abnormal conditions, as well as vehicular and pedestrian traffic control, can provide efficient access for employees while ensuring that any visitors are checked and cleared before entering. Many facilities have procedures to recover keys from employees who leave and to immediately remove the employee's security codes from systems. At times it may be wise to consider additional measures, such as changing locks, when a disgruntled employee leaves.

In addition to protecting a facility from intruders, it is important to limit the damage that an intruder (whether physically at the site or "hacking" into the company's computers) or an employee could do. Most of the steps to limit damage are probably things you already do as part of good process safety management, because they also limit the loss of chemicals if management systems or equipment fails or an operator makes a mistake. These steps can be related to either the design of the facility and its processes or to procedures implemented.

A well-designed facility, by its layout, limits the possibility that equipment will be damaged and, by its process design, limits the quantity of chemical that could be released. Facility and process design (including chemicals used) determine the need for safety equipment, site security, buffer zones, and mitigation planning. Eliminating or attenuating to the extent practicable any hazardous characteristic during facility or process design is generally preferable to simply adding on safety equipment or security measures. The option of locating processes with hazardous chemicals in the center of a facility can thwart intruders and vandals who remain outside the facility fenceline. Transportation vehicles, which are usually placarded to identify the contents, may be particularly vulnerable to attack if left near the fenceline or unprotected. However, for some facilities and processes, the option of locating the entire process at the center of the site may not be feasible. Some facilities may need to consider external versus internal threats, such as the threat to workers if an accidental release occurs, or the access to the process in case of an emergency response. Where feasible, providing layers of security will protect equipment from damage. These layers could include, for example, blast resistant buildings or structures. Enclosing critical valves and pumps (behind fences or in buildings) can make it less likely that an intruder will be able to reach them, a vehicle will be able to collide with them, or that releases are compounded because of damage to neighboring equipment. Chlorine tanker valves are an example of equipment design with several layers of security: (1) a heavy steel dome with lid; (2) a heavy cable sealing system that requires cable cutters to remove; (3) a heavy duty valve that can withstand abuse without leaking; and (4) a seal plug in each valve. As many as three different tools would be needed to breach the container's integrity. If equipment is located where cars, trucks, forklifts, or construction equipment could collide with it or drop something on it, the equipment should be constructed from materials that could stand some abuse. In general, you should give consideration to collision protection to any equipment containing hazardous chemicals with, for example, collision barriers. Layers of security may also be applied to communications/computer security. Some companies have developed alternate capabilities and systems to protect receipt and transmission of confidential information. Backup power systems and/or conditioning systems can be important, particularly if processes are computer controlled. Access to computer systems used to control processes may need to be controlled so that unauthorized users cannot break in; appropriate computer authentication and authorization mechanisms on all computer systems and remote access may prove useful; entrance into control rooms may need to be monitored and limited to authorized personnel. For emergency communications, some companies use radios and cell phones as a backup to the regular phone system. Well-designed equipment will usually limit the loss of materials if part of a process fails. Excess flow check valves, for example, will stop flow from an opened valve if the design flow rate is exceeded. These valves are commonly installed on chlorine tankcars and some anhydrous ammonia trailers, as well as on many chemical processes. Like excess flow valves, fail-safe systems can ensure that if a release occurs, the valves in the system will close, shutting off the flow. Breakaway couplings, for example, shut off flow in transfer systems, such as loading hoses, to limit the amount released to the quantity in the hose. If hazardous liquids are staged or stored on-site, it's prudent to consider containment systems (e.g., buildings, dikes, and trenches) that can slow the rate at which the chemical evaporates and provide time to respond. Double-walled vessels can also protect against attempts to rupture a tank. The installation of chemical monitors that automatically notify personnel of off-hour releases could be important if your facility is not staffed during certain periods (e.g., overnight). Such monitors, however, are not available for all chemicals. The appropriateness of monitors, and any other equipment design solutions, will depend on site-specific conditions.

A company's policies and procedures can also limit the damage caused by a release. As with design issues, the procedural steps you routinely take to operate safely also help protect your facility from attacks. Maintaining good labor relations may protect your facility from actions by either employees or contractors. Open negotiations, workplace policies emphasizing that violence and substance abuse are not tolerated, and adequate training and resources to support these policies are important considerations. The goal is to develop a workforce and management capacity to identify and solve problems by working together. Following are several examples of specific areas where procedures and policies can prevent or limit the damage of a release. As a matter of good practice, as well as site security, you may consider disconnecting storage tanks and delivery vehicles from connecting piping, transfer hoses, or distribution systems when not in use. Leaving the tanks linked to the process or pipeline increases the chance of a release because the hoses or pipes are often more vulnerable than the tanks. In addition to accurately monitoring your inventory, another practice you may want to adopt is limiting the inventory of hazardous materials to the minimum you need for your process. This policy limits the quantity of a hazardous material that could be released. You could also consider actions such as substituting less hazardous substances when possible to make processes inherently safer. Written procedures are also an important tool in protecting your facility. As part of your regular operating procedures, you probably have emergency shutdown procedures. These procedures, and workers trained in their use, can limit the quantity released. The procedures are particularly important if you have processes that operate under extreme conditions (high or low pressures, temperature) where rapid shutdown can create further hazards if done improperly.

In reviewing a contingency plan, consider, if necessary, revisions to address vandalism, bomb threats, burglary - including evaluating the desirability of your facility as a target - working with local law enforcement, and providing extra security drills and audits. Many companies find that working with local law enforcement is an effective means of evaluating security risks. As a matter of good practice, for both process and response equipment, it is important to have a program that ensures that all equipment is subject to inspection and to corrective and preventive maintenance. In this way, you can be sure that the safety systems you install will operate as designed.

Steps taken to operate safely will often serve to address security concerns as well. Considering inherent safety in the design and operation of any facility will have the benefit of helping to prevent and/or minimize the consequences of any release. Before taking steps to improve site security, evaluate the current system and determine whether it is adequate. Factors to consider include: (1) The chemicals stored at your site; some chemicals may be particularly attractive targets because of the potential for greater consequences if released. (2) The location of the site;

sites in densely populated areas may need more security than those at a distance from populations. (3) The accessibility of the site; are the existing security systems (e.g., fences, security lighting, security patrols) adequate to limit access to the site? (4) The age and type of buildings; older buildings may be more vulnerable because they have more windows; some newer building are designed for easy access. (5) Hours of operation; a facility that operates 24-hours day may need less security than a facility that is unoccupied at night. Decisions about improving site security should be made after evaluating how vulnerable your site is to threats and what additional measures, if any, are appropriate to reduce your vulnerability. Each facility should make its own decision based on its circumstances.

If a facility produces, processes, handles, or stores extremely hazardous substances, under the Clean Air section 1 12(r)(1), it has a general duty "to identify hazards which may result from such releases, using appropriate hazard assessment techniques, to design and maintain a safe facility taking such steps as are necessary to prevent releases, and to minimize the consequences of accidental releases which do occur." Several organizations (e.g., ASTM, ANSI) have standards for site security or include site security issues in their codes. The National Fire Protection Association (NFPA) has a standard NFPA- 601, *Standard for Site Security Services for Fire Loss Prevention*. The American Petroleum Institute addresses security issues in RP 554, *Process Instrumentation and Control*. Likewise, the Chemical Manufacturers Association addresses this issue through the *Responsible Care Employee Health and Safety Code Site Security Management Practice*. Protocols developed under the Responsible Distribution Process cover security concerns. You can contact the following websites for additional security information:

www.energysecuritycouncil.org The Energy Security Council is a national industry association to assist law enforcement agencies and energy companies in combating all types of criminal activity.

www.nfpa.org The National Fire Protection Association provides standards, research, training, and education to reduce the burden of fire and other hazards.

www.nsc.org The National Safety Council provides general safety information on chemical and environmental issues.

www.asisonline.org and www.securitymanagement.com The American Society for Industrial Security develops educational programs and materials that address security concerns. Its Security Management Magazine site provides an online version of its magazine.

www.siaonline.org The Security Industry Association provides general security information.

www.atsdr.cdc.gov The Agency for Toxic Substances and Disease Registry site provides a 10-step procedure to analyze, mitigate, and prevent public health hazards resulting from terrorism involving industrial chemicals.

www.aiche.org/ccps The Center for Chemical Process Safety (CCPS) is an industry-driven, nonprofit professional organization affiliated with the American Institute of Chemical Engineers (AIChE). It is committed to developing engineering and management practices to prevent or mitigate the consequences of catastrophic events involving the release of chemicals that could harm employees, neighbors and the environment.

www.cdc.gov/niosh The National Institute for Occupational Safety and Health provides multiple resources on workplace violence prevention.

The Complete Manual of Corporate and Industrial Security, by Russell L. Bintliff (Prentice Hall, 1992) provides detailed discussions of the advantages and disadvantages

EPA's *Risk Management Programs for Chemical Accident Release Prevention* (40 CFR 68) requires regulated facilities to develop and implement appropriate risk management programs to minimize the frequency and severity of chemical plant accidents. In keeping with recent regulatory trends, EPA is requiring a performance-based approach towards compliance with the risk management program rule. In recent developments, the amendments to the RMP rule, proposed on April 17, 1998 were signed by Administrator Browner on December 29th and published in the Federal Register on January 6, 1998. The RMP that is required to be developed by facilities must include a description of the hazard assessment, prevention program, and the emergency response program. Facilities are required to submit the RMP to governmental agencies, the state emergency response commission, the local emergency planning committees, and as needed the RMP would be communicated to the public. The final rule defines the worst-case release as the release of the largest quantity of a regulated substance from a vessel or process line failure, including administrative controls and passive mitigation that limit the total quantity involved or release rate. For gases, the worst-case release scenario assumes the quantity is released in 10 minutes. For liquids, the scenario assumes an instantaneous spill and that the release rate to the air is the volatilization rate from a pool 1 cm deep unless passive mitigation systems contain the substance in a smaller area. For flammables, the scenario assumes an instantaneous release and a vapor cloud explosion using a 10 percent yield factor. For alternative scenarios (note: EPA is using the term *alternative scenario* as compared to the term *more-likely scenario* used earlier in the proposed rule), facilities may take credit for both passive and active mitigation systems. Appendix A of the final rule lists endpoints for toxic substances to be used in worst-case and alternative scenario assessment. The toxic endpoints are based on ERPG-2 or level of concern data compiled by EPA. The flammable endpoints represent vapor cloud explosion distances based on overpressure of 1 psi or radiant heat distances based on exposure to 5 kW/m² for 40 seconds. The promulgation of the final rule follows some recent actions by EPA to amend the final list rule promulgated earlier by EPA on January 31, 1994. These recent actions include the following proposed amendments to the list rule: (1) Exemption of crude oil prior to initial processing and gasoline; (2) The risk management program rule does not apply to sources in the outer continental shelf, Entire weight of the mixture containing a flammable substance shall be treated as a regulated substance unless experimental measurements can demonstrate that the mixture does not meet NFPA 4 criteria, and Exemption of explosives (as defined under DOT Class 1, Division 1. 1 - 49 CFR 172.101) from coverage by the rule.

The EPA has proposed a stay of effectiveness of the risk management program rule requirements for those facilities impacted by the proposed amendments to the list rule.

ACCIDENT INVESTIGATION PRINCIPLES

An accident is any unplanned event that results in personal injury or in property damage. When the personal injury requires little or no treatment, it is minor. If it results in a fatality or in a permanent total, permanent partial, or temporary total (lost-time) disability, it is serious. Similarly, property damage may be minor or serious. Investigate all accidents regardless of the extent of injury or damage. Thousands of industrial accidents occur every day.

The failure of people, equipment, supplies, or surroundings to behave or react as expected causes most of the accidents. Accident investigations determine how and why these failures occur. By using the information gained through an investigation, a similar or perhaps more disastrous accident may be prevented. Conduct accident investigations with accident prevention in mind. The objectives of investigations are not to place blame.

Accidents are part of a broad group of events that adversely affect the completion of a task. These events are incidents. For simplicity, the procedures discussed below refer only to accidents. They are, however, also applicable to incidents. Accidents are usually complex. An accident may have 10 or more events that can be causes.

A detailed analysis of an accident will normally reveal three cause levels: basic, indirect, and direct (refer to Figure 1).

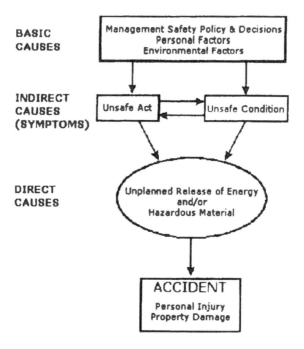

Figure 1. *The three cause levels of any accident.*

At the lowest level, an accident results only when a person or object receives an amount of energy or hazardous material that cannot be absorbed safely. This energy or hazardous material is the direct cause of the accident. The direct cause is usually the result of one or more unsafe acts or unsafe conditions, or both.

Unsafe acts and conditions are the indirect causes or symptoms. In turn, indirect causes are usually traceable to poor management policies and decisions, or to personal or environmental factors. These are the basic causes. In spite of their complexity, most accidents are preventable by eliminating one or more causes.

Accident investigations determine not only what happened, but also how and why. The information gained from these investigations can prevent recurrence of similar or perhaps more disastrous accidents. Accident investigators are interested in each event as well as in the sequence of events that led to an accident. The accident type is also important to the investigator.

The recurrence of accidents of a particular type or those with common causes shows areas needing special accident prevention emphasis.

The actual procedures used in a particular investigation depend on the nature and results of the accident. The agency having jurisdiction over the location determines the administrative procedures. In general, responsible officials will appoint an individual to be in charge of the investigation. The investigator uses most of the following steps:

1. Define the scope of the investigation.
2. Select the investigators. Assign specific tasks to each (preferably in writing).
3. Present a preliminary briefing to the investigating team, including:
   a. Description of the accident, with damage estimates.
   b. Normal operating procedures.
   c. Maps (local and general).
   d. Location of the accident site.
   e. List of witnesses.
   f. Events that preceded the accident.
4. Visit the accident site to get updated information.
5. Inspect the accident site.
   a. Secure the area. Do not disturb the scene unless a hazard exists.
   b. Prepare the necessary sketches and photographs. Label each carefully and keep accurate records.
6. Interview each victim and witness. Also interview those who were present before the accident and those who arrived at the site shortly after the accident. Keep accurate records of each interview. Use a tape recorder if desired and if approved.
7. Determine what was not normal before the accident; where the abnormality occurred; when it was first noted; and how it occurred.
8. Analyze the data obtained in step 7. Repeat any of the prior steps, if necessary.
9. Determine why the accident occurred; a likely sequence of events and probable causes (direct, indirect, basic); and alternatives.
10. Check each sequence against the data from step 7.
11. Determine the most likely sequence of events and the most probable causes.
12. Conduct a post-investigation briefing.
13. Prepare a summary report, including the recommended actions to prevent a recurrence. Distribute the report according to applicable instructions.

An investigation is not complete until all data are analyzed and a final report is completed. In practice, the investigative work, data analysis, and report preparation proceed simultaneously over much of the time spent on the investigation.

Gather evidence from many sources during an investigation. Get information from witnesses and reports as well as by observation. Interview witnesses as soon as possible after an accident. Inspect the accident site before any changes occur. Take photographs and make sketches of the accident scene. Record all pertinent data on maps. Get copies of all reports. Documents containing normal operating procedures, flow diagrams, maintenance charts, or reports of difficulties or abnormalities are particularly useful. Keep complete and accurate notes in a bound notebook. Record pre-accident conditions, the accident sequence, and post-accident conditions. In addition, document the location of victims, witnesses, machinery, energy sources, and hazardous materials.

In some investigations, a particular physical or chemical law, principle, or property may explain a sequence of events. Include laws in the notes taken during the investigation or in the later analysis of data. In addition, gather data during the investigation that may lend itself to analysis by these laws, principles, or properties. An appendix in the final report can include an extended discussion.

In general, experienced personnel should conduct interviews. If possible, the team assigned to this task should include an individual with a legal background. In conducting interviews, the team should:

1. Appoint a speaker for the group.
2. Get preliminary statements as soon as possible from all witnesses.
3. Locate the position of each witness on a master chart (including the direction of view).
4. Arrange for a convenient time and place to talk to each witness.
5. Explain the purpose of the investigation (accident prevention) and put each witness at ease.
6. Listen, let each witness speak freely, and be courteous and considerate.
7. Take notes without distracting the witness. Use a tape recorder only with consent of the witness.
8. Use sketches and diagrams to help the witness.
9. Emphasize areas of direct observation. Label hearsay accordingly.
10. Be sincere and do not argue with the witness.
11. Record the exact words used by the witness to describe each observation. Do not "put words into a witness' mouth."
12. Word each question carefully and be sure the witness understands.
13. Identify the qualifications of each witness (name, address, occupation, years of experience, etc.)
14. Supply each witness with a copy of his or her statements. Signed statements are desirable.

After interviewing all witnesses, the team should analyze each witness' statement. They may wish to re-interview one or more witnesses to confirm or clarify key points. While there may be inconsistencies in witnesses' statements, investigators should assemble the available testimony into a logical order. Analyze this information along with data from the accident site. Not all people react in the same manner to a particular stimulus. For example, a witness within close proximity to the accident may have an entirely different story from one who saw it at a distance. Some witnesses may also change their stories after they have discussed it with others. The reason for the change may be additional clues. A witness who has had a traumatic experience may not be able to recall the details of the accident. A witness who has a vested interest in the results of the investigation may offer biased testimony. Finally, eyesight, hearing, reaction time, and the general condition of each witness may affect his or her powers of observation. A witness may omit entire sequences because of a failure to observe them or because their importance was not realized.

Accidents represent problems that must be solved through investigations. Several formal procedures solve problems of any degree of complexity. This section discusses two of the most common procedures: Change Analysis and Job Safety Analysis.

Change Analysis: As its name implies, this technique emphasizes change. To solve a problem, an investigator must look for deviations from the norm. Consider all problems to result from some unanticipated change. Make an analysis of the change to determine its causes. Use the following steps in this method:

1. Define the problem (What happened?).
2. Establish the norm (What should have happened?).
3. Identify, locate, and describe the change (What, where, when, to what extent).
4. Specify what was and what was not affected.
5. Identify the distinctive features of the change.
6. List the possible causes.
7. Select the most likely causes.

Job Safety Analysis: Job safety analysis (JSA) is part of many existing accident prevention programs. In general, JSA breaks a job into basic steps, and identifies the hazards associated with each step. The JSA also prescribes controls for each hazard. A JSA is a chart listing these steps, hazards, and controls. Review the JSA during the investigation if a JSA has been conducted for the job involved in an accident. Perform a JSA if one is not available. Perform a JSA as a part of the investigation to determine the events and conditions that led to the accident. As noted above, an accident investigation is not complete until a report is prepared and submitted to proper authorities.

Special report forms are available in many cases. Other instances may require a more extended report. Such reports are often very elaborate and may include a cover page, a title page, an abstract, a table of contents, a commentary or narrative portion, a discussion of probable causes, and a section on conclusions and recommendations.

The following outline is useful in developing the information to be included in the formal report:

1. Background Information (a. Where and when the accident occurred; b. Who and what were involved; c. Operating personnel and other witnesses).
2. Account of the Accident (What happened?) (a. Sequence of events; b. Extent of damage; c. Accident type; d. Agency or source (of energy or hazardous material)).
3. Discussion (Analysis of the Accident - **HOW; WHY**): a. Direct causes (energy sources; hazardous materials); b. Indirect causes (unsafe acts and conditions); c. Basic causes (management policies; personal or environmental factors).
4. Recommendations (to prevent a recurrence) for immediate and long-range action to remedy: a. Basic causes; b. Indirect causes; c. Direct causes (such as reduced quantities or protective equipment or structures).

Thousands of accidents occur daily throughout the United States. These result from a failure of people, equipment, supplies, or surroundings to behave as expected. A successful accident investigation determines not only what happened, but also finds how and why the accident occurred.

Investigations are an effort to prevent a similar or perhaps more disastrous sequence of events. Most accident investigations follow formal procedures. This discussion covered two of the most common procedures: Change Analysis and Job Safety Analysis.

An investigation is not complete however, until completion of a final report. Responsible officials can then use the resulting information and recommendations to prevent future accidents.

ASSESSING HAZARDS ON THE JOB

Job-related injuries occur every day in the workplace. Often these injuries occur because employees are not trained in the proper job procedure. One way to prevent workplace injuries is to establish proper job procedures and train all employees in safer and more efficient work methods Establishing proper job procedures is one of the benefits of conducting a job hazard analysis - that is, carefully studying and recording each step of a job, identifying existing or potential job hazards (both safety and health), and determining the best way to perform the job to reduce or eliminate these hazards. Improved job methods can reduce costs resulting from employee absenteeism and workers' compensation, and can often lead to increased productivity.

It is important to note that the job procedures in this booklet are for illustration only and do not necessarily include all steps, hazards, or protections for similar jobs in industry. In addition, standards issued by the Occupational Safety and Health Administration (OSHA) should be referred to as part of the overall job hazard analysis. There are OSHA standards that apply to most job operations, and compliance with these standards is mandatory.

A job hazard analysis can be performed for all jobs in the workplace, whether the job is "special" (nonroutine) or routine. Even one step jobs, such as those in which only a button is pressed, can and perhaps should be analyzed by evaluating surrounding work conditions. To determine which jobs should be analyzed first, review job injury and illness reports. Obviously, a job hazard analysis should be conducted first for jobs with the highest rates of accidents and disabling injuries. Also, jobs where "close calls" or "near misses" have occurred should be given priority. Analyses of new jobs and jobs where changes have been made in processes and procedures should follow. Eventually, a job hazard analysis should be conducted and made available to employees for all jobs in the workplace. Once a job has been selected for analysis, discuss the procedure with the employee performing the job and explain its purpose. Point out that you are studying the job itself not checking on the employee's job performance. Involve the employee in all phases of the analysis - from reviewing the job steps to discussing potential hazards and recommended solutions. Before actually beginning the job hazard analysis, take a look at the general conditions under which the job is performed and develop a checklist. The following are some sample questions to ask:

Are there materials on the floor that could trip a worker?

Is lighting adequate?

Are there any live electrical hazards at the job site?

Are there any explosive hazards associated with the Job or likely to develop?

Are tools, including hand tools, machines, and equipment in need of repair?

Is there excessive noise in the work area hindering worker communication and increasing risk of hearing loss?

Is fire protection equipment readily accessible and have employees been trained to use it?

Are emergency exits clearly marked?

Are trucks or motorized vehicles properly equipped with brakes, overhead guards, backup signals, horns, steering gear and identification, as necessary?

Are all employees operating vehicles and equipment properly trained and authorized?

Are employees wearing proper personal protective equipment (PPE) for the jobs they are performing?

Have any employees complained of headaches, breathing problems, dizziness or strong odors?

Is ventilation adequate?

Does the job involve entry into a confined space?

Have tests been made for oxygen deficiency and toxic fumes?

Naturally, this list is by no means complete, because each worksite has its own requirements and environmental conditions. It is recommended to take photographs of the workplace, if appropriate, for use in making a more detailed analysis of the work environment. Nearly every job can be broken down into steps. In the first part of the job hazard analysis, list each step of the job in order of occurrence as you watch the employee performing the job. Be sure to record enough information to describe each job action, but do not make the breakdown too detailed. Later, go over the job steps with the employee. After recording the job steps, next examine each step to determine the hazards that exist or that might occur. Ask these kinds of questions:

Is the worker wearing clothing or jewelry that could get caught in the machinery?

Are there fixed objects that may cause injury, such as sharp, machine edges?

Is the worker required to make movements that could cause hand or foot injuries, repetitive motion injuries, or strain from lifting?

Can the worker be struck by an object, lean against or strike a machine part or object?

Do suspended loads or potential energy (such as compressed springs, hydraulics or jacks) pose hazards?

Can the worker fall from one level to another?

Can the worker be injured from lifting objects, or from carrying heavy objects?

Do environmental hazards, such as dust, chemicals, radiation, welding rays, heat or excessive noise result from the performance of the job?

Is the worker at any time in an off-balance position? Is the worker positioned to the machine in a way that is potentially dangerous?

Can the worker get caught in or between machine parts? Can the worker be injured by reaching over moving machinery parts or materials?

Repeat the job observation as often as necessary until all hazards nave been identified. The next step is to look into what would cause these hazards. You need to think about what events could lead to an injury or illness for each hazard you identified. Typical questions are:

Is the worker wearing protective clothing and equipment, including safety belts or harnesses that are appropriate fir the job? Does it fit properly?

Has the worker been trained to use appropriate PPE?

Are work positions, machinery, pits or holes, and hazardous operations adequately guarded?

Are lockout procedures used for machinery deactivation during maintenance procedures?

Is the flow of work improperly organized (e.g., is the worker required to make movements that are too rapid)?

How are dusts and chemicals dispersed in the air? What are the sources of noise, radiation and heat?

What causes a worker to contact sharp surfaces?

Why would a worker be tempted to reach into moving machine parts?

Recommendations should be based on the reliability of the solution. In general, the most reliable protection is to eliminate the source or cause of the hazard. Hazards might be eliminated by redesigning equipment, changing tools, installing ventilation, or adding machine guards. If the hazard cannot be eliminated, the danger should be reduced as much as possible. Improving the procedure or using personal protective equipment are some of the primary ways to reduce the danger. These changes should be accompanied by training programs that are aimed at covering the procedures and equipment in detail. Note that some OSHA standards require formal training for employees.

After you have listed each hazard or potential hazard and have reviewed them with the employee performing the job, determine whether the job could be performed in another way to eliminate the hazards, such as combining steps or changing the sequence, whether safety equipment and precautions are needed to reduce the hazards, or whether training is needed to recognize hazards. If safer and better job steps can be used, list each new step, such as describing a new method for disposing of material. List exactly what the worker needs to know to perform the job using a new method. Do not make general statements about the procedure, such as "Be careful." Be as specific as you can in your recommendations. If hazards are still present, try to reduce the necessity for performing the job or the frequency of performing it. Review the recommendations with all employees performing the job and ask for their suggestions. Their ideas about the hazards and proposed recommendations may be valuable. Be sure that they understand what they are required to do and the reasons for the changes in the job procedure.

A job hazard analysis can do much toward reducing accidents and injuries in the workplace, but it is only effective if it is reviewed and updated periodically. Even if no changes have been made in a job, hazards that were missed in an earlier analysis could be detected. If an accident or injury occurs on a specific job, the job hazard analysis should be reviewed immediately to determine whether changes are needed in the job procedure. In addition, if an accident results from an employee's failure to follow job procedures, this should be discussed with all employees performing the job. Any time a job hazard analysis is revised, training in the new job methods or protective measures should be provided to all employees affected by the changes. A Job hazard analysis also can be used to train new employ on job steps and job hazards. To show how a job hazard analysis form is prepared, a sample worksheet for grinding castings is given below. Both safety and health hazards are noted, as well as recommendations for safer methods and protection. Employees have the right to file a complaint with their employers, their unions, OSHA, or another government agency about workplace safety and health hazards. Section 11 (c) of the Occupational Safety and Health (OSH) Act makes it illegal for employees to be discriminated against for exercising this right and for participating in other job safety and health-related employee activities. These projected activities include: Submitting complaints individually or with other directly to management concerning job safety conditions; Filing formal complaints with government agencies such as OSHA or state safety and health agencies, fire departments, etc. (An employee's name can be withheld from the complaint, if so requested.); Testifying before any panel, agency or court of law concerning job hazards; Participating in walk-around inspections.

Employees also cannot be punished for refusing a work assignment if they have a reasonable belief that it would put them in real danger of death or serious physical injury, provided that, if possible, they have requested the employer to remove the danger and the employer has refused; and provided that the danger cannot be eliminated quickly enough through normal OSHA enforcement procedures. If an employee is punished or discriminated against in any way for exercising his or her rights under the OSH Act, the employee should report it to OSHA within 30 days. OSHA will investigate, and if the employee has been illegally punished OSHA will seek appropriate relief for the employee. If necessary, OSHA will go to court to protect the rights of the employee.

In all states, free onsite consultation services are available to help employers identify job safety and health hazards and to recommend solutions. The information gathered by the consultant, including the employers' identity, is kept confidential and is not made available to OSHA enforcement personnel - with two exceptions. If a consultant observes an imminent danger or serious violations of OSHA standards and the employer fails to correct it within the time period recommended by the consultant, OSHA or the responsible state agency will be notified. Employers may, on an anonymous basis, contact the nearest OSHA office to find out how to take advantage of the free consultation service, or may write for a copy of *Consultation Services for the Employer* (OSHA 3047) from OSHA Publications Office, U.S. Department of Labor, 200 Constitution Avenue, N.W., Washington, DC 20210.

The following publications provide more information on job hazard analysis and related issues and on complying with OSHA standards. A single free copy of the OSHA publications can be obtained by sending a self addressed mailing label to the OSHA Publications Office, U.S. Department of Labor, 200 Constitution Avenue, N.W., Room N3 10 1, Washington, DC 20210.

The Occupational Safety and Health Act of 1970, PL - 91-956 (OSHA 2001).

All About OSHA (OSHA 2056)

Consultation Services for the Employer (OSHA 3047)

Control of Hazardous Energy (OSHA 3120)

Hand and Power Tools (OSHA 3080)

How to Prepare for Workplace Emergencies (OSHA 3088)

ASSESSING CONFINED SPACE OPERATIONS

Many operations involved job functions in confined spaces. Examples of such locations are boilers, a cupola, degreaser, furnace, pipeline, pit, pumping station, reaction or process vessel, septic tank, sewage digester, sewer, silo, storage tank, ship's hold, utility vault, vat, or similar type enclosure. These locations present unique hazards for workers and therefore require special attention when evaluating the risks and management of the operations required in them. A confined space is a space which has any one of the following characteristics:

- limited openings for entry and exit
- unfavorable natural ventilation
- it is not designed for continuous worker occupancy
- it has limited openings for entry and exit

Confined space openings are limited primarily by size or location. Openings are usually small in size, perhaps as small as 18 inches in diameter, and are difficult to move through easily. Small openings may make it very difficult to get needed equipment in or out of the spaces, especially protective equipment such as respirators needed for entry into spaces with hazardous atmospheres, or life-saving equipment when rescue is needed. However, in some cases openings may be very large, for example open-topped spaces such as pits, degreasers, excavations, and ships' holds. Access to open-topped spaces may require the use of ladders, hoists, or other devices, and escape from such areas may be very difficult in emergency situations.

Because air may not move in and out of confined spaces freely due to the design, the atmosphere inside a confined space can be very different from the atmosphere outside. Deadly gases may be trapped inside, particularly if the space is used to store or process chemicals or organic substances which may decompose. There may not be enough oxygen inside the confined space to support life, or the air could be so oxygen-rich that it is likely to increase the chance of fire or explosion if a source of ignition is present. Most confined spaces are simply not designed for workers to enter and work in them on a routine basis. They are designed to store a product, enclose materials and processes, or transport products or substances. Therefore, occasional worker entry for inspection, maintenance, repair, cleanup, or similar tasks is often difficult and dangerous due to chemical or physical hazards within the space. A confined space found in the workplace may have a combination of these three characteristics, which can complicate working in and around these spaces as well as rescue operations during emergencies. Among the list of hazards associated with confined space operations is a hazardous atmosphere. The atmosphere in a confined space may be extremely hazardous because of the lack of natural air movement. This characteristic of confined spaces can result in oxygen -deficient atmospheres, flammable atmospheres, and/or toxic atmospheres. An oxygen-deficient atmosphere has less than 19.5% available oxygen (O_2). Any atmosphere with less than 19.5% oxygen should not be entered without an approved self-contained breathing apparatus (SCBA). The oxygen level in a confined space can decrease because of work being done, such as welding, cutting, or brazing; or, it can be decreased by certain chemical reactions (rusting) or through bacterial action (fermentation). The oxygen level is also decreased if oxygen is displaced by another gas, such as carbon dioxide or nitrogen. Total displacement of oxygen by another gas, such as carbon dioxide, will result in unconsciousness, followed by death. In the case of a flammable atmosphere, there are two contributing factors: the oxygen in air; and a flammable gas, vapor, or dust in the proper mixture. Different gases have different flammable ranges. If a source of ignition (e.g., a sparking or electrical tool) is introduced into a space containing a flammable atmosphere, an explosion will result. An oxygen-enriched atmosphere (above 21%) will cause flammable materials, such as clothing and hair, to burn violently when ignited. Therefore, never use pure oxygen to ventilate a confined space. One should always ventilate with normal air. Toxic atmospheres in confined spaces result from a variety of situations. Examples include product storage, work being performed in the confined space, and toxicants produced by work in the area of confined spaces can enter and accumulate.

The product stored in the space: The product can be absorbed into the walls and give off toxic gases when removed or when cleaning out the residue of a stored product, toxic gases can be given off. Example: Removal of sludge from a tank - decomposed material can give off deadly hydrogen sulfide gas.

The work being performed in a confined space: Examples of such include welding, cutting, brazing, painting, scraping, sanding, degreasing, etc. Toxic atmospheres are generated in various processes. For example, cleaning solvents are used in many industries for cleaning/degreasing. The vapors from these solvents are toxic in a confined space.

It is important to understand that some gases or vapors are heavier than air and will settle to the bottom of a confined space. Also, some gases are lighter than air and will be found around the top of the confined space. Therefore, it is necessary to test all areas (top, middle, bottom) of a confined space with properly calibrated testing instruments to determine what gases are present.

If testing reveals oxygen-deficiency, or the presence of toxic gases or vapors, the space must be ventilated and retested before workers enter. If ventilation is not possible and entry is necessary (for emergency rescue, for example), workers must have appropriate respiratory protection.

Never trust your senses to determine if the air in a confined space is safe to breathe. Many hazardous gases cannot be smelled, or can be masked by other odors. Figure 2 illustrates the proper approach to assessing the atmosphere in a confined space.

Ventilation by a blower or fan may be necessary to remove harmful gases and vapors from a confined space. There are several methods for ventilating a confined space. The method and equipment chosen are dependent upon the size of the confined space openings, the gases to be exhausted (e.g., are they flammable?), and the source of makeup air. Under certain conditions where flammable gases or vapors have displaced the oxygen level, but are too rich to burn, forced air ventilation may dilute them until they are within the explosive range. Also, if inert gases (e.g. carbon dioxide, nitrogen, argon) are used in the confined space, the space should be well ventilated and retested before a worker may enter.

A common method of ventilation requires a large hose, one end attached to a fan and the other lowered into a manhole or opening. For example, a manhole would have the ventilating hose run to the bottom to blow out all harmful gases and vapors. The air intake should be placed in an area that will draw in fresh air only. Ventilation should be continuous where possible, because in many confined spaces the hazardous atmosphere will form again when the flow of air is stopped.

Isolation: Isolation of a confined space is a process where the space is removed from service by locking out (electrical sources, preferably at disconnect switches remote from the equipment), blanking and bleeding pneumatic and hydraulic lines, disconnecting belt and chain drives, and mechanical linkages on shaft-driven equipment where possible, and securing mechanical moving parts within confined spaces with latches, chains, chocks, blocks, or other devices. Refer to Figures 3 and 4 for common examples.

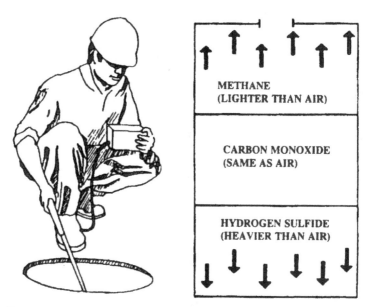

Figure 2. *Measure form outside the confined space, top to bottom.*

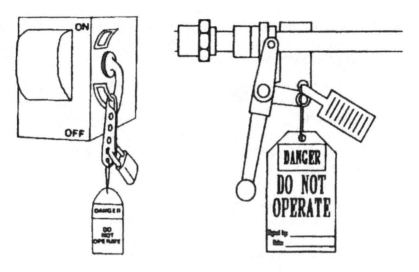

Figure 3. *Lockout examples.*

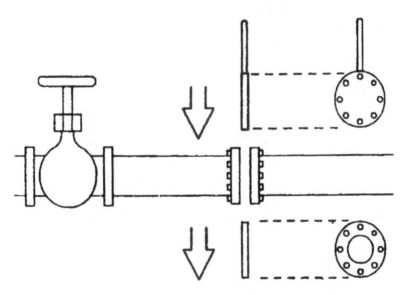

Figure 4. *Blanking hydraulic and pneumatic lines.*

Respiratory Protection: Respirators were already discussed in Chapter 5. Respirators are devices that can allow workers to safely breathe without inhaling toxic gases or particles. Two basic types are air-purifying, which filter dangerous substances from the air, and air-supplying, which deliver a supply of safe breathing air from a tank or an uncontaminated area nearby. Only air-supplying respirators should be used in confined spaces where there is not enough oxygen. Selecting the proper respirator for the job, the hazard, and the person is very important, as is thorough training in the use and limitations of respirators.

Questions regarding the proper selection and use of respirators should be addressed to a certified industrial hygienist, or to the NIOSH Division of Safety Research, 944 Chestnut Ridge Rd., Morgantown, West Virginia 26505. Respirators are discussed in greater detail in the *Handbook of Industrial Toxicology and Hazardous Materials*, Marcel Dekker, 1999. Common examples of both types of respirators are illustrated in Figure 5.

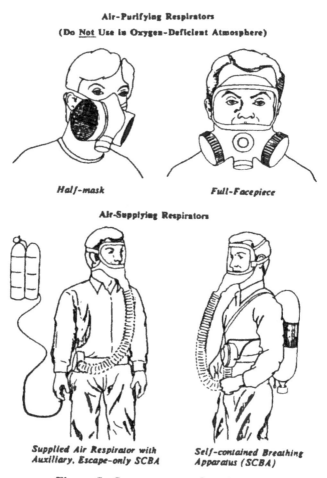

Air-Purifying Respirators

(Do Not Use in Oxygen-Deficient Atmosphere)

Half-mask   *Full-Facepiece*

Air-Supplying Respirators

Supplied Air Respirator with   *Self-contained Breathing*
Auxiliary, Escape-only SCBA   *Apparatus (SCBA)*

Figure 5. *Common types of respirators.*

Standby and Rescue: A standby person should be assigned to remain on the outside of the confined space and be in constant contact (visual or speech) with the workers inside. The standby person should not have any other duties but to serve as standby and know who should be notified in case of emergency. Standby personnel should not enter a confined space until help arrives, and then only with proper protective equipment, life lines, and respirators. Over 50% of the workers who die in confined spaces are attempting to rescue other workers. Rescuers must be trained in and follow established emergency procedures and use appropriate equipment and techniques (lifelines, respiratory protection, standby persons, etc.). Steps for safe rescue should be included in all confined space entry procedures. Rescue should be well planned and drills should be frequently conducted on emergency procedures. Unplanned rescue, such as when someone instinctively rushes in to help a downed co-worker, can easily result in a double fatality, or even multiple fatalities if there are more than one would-be rescuers. Table 1 provides a confined space entry check list that can be used to assess safe entry conditions.

Table 1. Confined Space Safe Entry Checklist.

Check One		ASSESSMENT CRITERIA
Yes	No	
		Is entry into the confined space necessary for personnel?
Air Quality Testing Questions		
		Are instruments to be used to test the atmosphere appropriate and are they properly calibrated?
		Was the atmosphere in the confined space tested?
		Is the oxygen level at least 19.5 %, but not more than 21 %?
		Were any toxic, flammable, or oxygen-displacing gases/vapors detected? If yes, what are they and their concentrations?_____
		Was hydrogen sulfide detected during air sampling?
		Was carbon monoxide detected during air sampling?
		Was carbon dioxide detected during air sampling?
		Was methane detected during air sampling?

Table 1. Continued

Air Quality Monitoring Questions		
		Will it be necessary to monitor the atmosphere in the confined space while work is being implemented?
		Will monitoring be done on a continuous basis?
		Will monitoring be performed intermittently? If so, provide details on the sampling interval and required measurement time per sample:_____
		Will sampling be performed at more than one location within the confined space? If so, specify locations:_____
Pre-entry Conditions		
		Has the confined space been cleaned prior to issuing a permit to enter?
		Has the confined space been purged of any solvents or hazardous vapors prior to issuing a permit?
		Has the confined space been steamed?
		If the space has been steamed, is there time to cool before entry?
Ventilation Questions		
		Has the space been ventilated before entry?
		Will ventilation be continued during entry?
		Is the air intake for the ventilation system located in an area that is free of combustible dusts, vapors and toxic materials?
		If the atmosphere was found to be unacceptable, and then ventilated, was it retested before entry?
Isolation Questions		
		Has the confined space been isolated from other systems?
		Has electrical equipment been locked out?
		Have disconnects been used?
		Has all mechanical equipment been blocked, chocked, and disengaged?

Table 1. Continued

		Have lines under pressure been blanked and bled?
PPE Questions		
		Is special CPC required for the operation? If so, specify:
		Is special equipment required (e.g., rescue equipment, communications equipment, etc.)? Specify the equipment:
		Are special tools required (e.g., sparkproof)? If so, specify:_____
		Is respiratory protection required and if so, specify types and numbers?_____
		Can a worker fit through the opening of the confined space suited up in full protection gear, including respirator?
		Have the workers been trained in the proper use of the respirator?
		Have the workers been properly trained on the use of all safety and rescue gear?
		Have the workers been trained on the operations to be performed?
Standby and Rescue		
		Will there be a standby person on the outside in constant visual or auditory communications with the person inside the confined space?
		Will the standby person be able to see and/or hear the person on the inside?
		Have the standby personnel been trained in rescue procedures?
		Will safety lines and harness be required to remove a person?
		Are there written rescue procedures available and have the workers been trained/drilled on them?
		Do the workers know who and how to notify a responsible party in the event of an emergency?
		Has rescue gear been checked, tested and serviced recently? Make notation for last check.

Table 1. Continued

Permit: The permit is an authorization in writing that states that the space has been tested by a qualified person; that the space is safe to enter; what precautions, equipment, etc. are required; and what work is to be performed.		
		Has the confined space entry permit been issued? Specify the duration (time/date):_____
		Does the permit include a list of emergency telephone numbers and names?

USING THE INTERNET

The Internet is a continuously expanding source of safety information that can provide instantaneous access to emergency response information, and resources that can assist in properly assessing the hazards of an operation of facility. One example is RTECS (Registry of Toxic Chemical Substances. By accessing www.doc.gov/niosh/97-119html, we can obtain the Comprehensive Guide to RTECS. The comprehensive toxicological data that RTECS contains is available in pdf format and can be downloaded for the reviewer in a manner of minutes. At the same site, the user can access the following NIOSH publications to assist in establishing safety and monitoring programs:

Compendium of NIOSH Mining Research 2000 (7,223K; 83 pages): 4/3/00

Curriculum Guide for Public-Safety and Emergency- Response Workers (4,425K; 178 pages): 3/5/99

A Guide for Evaluating the Performance of Chemical Protective Clothing (1,561K; 96 pages): 7/28/98

Guide for the Management, Analysis, and Interpretation of Occupational Mortality Data (2,464K; 89 pages): 2/23/99

Guide to Safety in Confined Spaces (325K; 22 pages): 2/8/99 Guide to the Work-Relatedness of Disease (4,784K; 272 pages): 4/10/98

A Guide To Working Safely With Silica, Joint MSHA/NIOSH Publication (213 KB; 21 pages): 1/31/97

A Model for Research on Training Effectiveness (TIER); (198K; 28 pages): 11/4/99

Analyzing Workplace Exposure Using Direct Reading Instruments and Video Exposure Monitoring Techniques (2,500K; 90 pages): 4/23/99

National Strategy for Occupational Musculoskeletal Injuries: Implementation Issues and Research Needs (1,462K, 27 pages): 4/23/99

An Evaluation of Glove Bag Containment in Asbestos Removal (3,587K; 136 pages): 2/8/99

Applications Manual for the Revised NIOSH Lifting Equation (1994) (3,915K; 164 pages): 6/18/98

A Practical Guide to Effective Hearing Conservation Programs in the Workplace (1,971K; 70 pages): 4/10/98

"Are You a Working Teen?": 6/2/97

Asbestos Bibliography (Revised)-September, 1997 (8,252K; 224 pages): 10/1/98

Second-Generation Remote Optical Methanomete (517K; 13 pages): 1/21/99

Assessing Occupational Safety and Health Training-A Literature Review; (621K; 164 pages): 9/16/98

Atlas of Respiratory Disease Mortality, United States: 1982- 1993 (August, 1998) (10,930K; 84 pages): 5/3/99

Back Belts: Do They Prevent Injury? (2 pages): 6/13/97

Building Air Quality: Action Plan: 9/30/98

Building Air Quality: A Guide for Building Owners and Facility Managers: 1/13/97

Control of Health and Safety Hazards in Commercial Dry Cleaners: Chemical Exposures, Fire Hazards, and Ergonomic Risk Factors: 9/24/98

Control Technology for Ethylene Oxide Sterilization in Hospitals (4,133K; 179pages): 1/8/99

Criteria for a Recommended Standard-Occupational Noise Exposure: 8/1/98

Cumulative Trauma Disorders in the Workplace, Bibliography (7,512K; 218 pages): 10/14/98

Detection of Downed Trolley Lines Using Arc Signature Analysis (483K; 14 pages): 1/21/99

Development of a Mine Hoist and Ore Pass Research Facility (595K; 16 pages): 1/21/99

Directory of Occupational Safety and Health Contacts for State and Territorial Health Departments: 9/8/98

Distinguishing Motor Starts From Short Circuits Through Phase-Angle Measurements (247K; 18 pages): 8/14/98

Elements of Ergonomics Programs: A Primer Based on Workplace Evaluations of Musculoskeletal Disorders: 11/17/97; (6,879K; 146 pages): 4/6/00

Documentation for Immediately Dangerous to Life or Health Concentrations (IDLHs): 6/5/97

Ergonomic Intervention for the Soft Drink Beverage Delivery Industry (3,129K; 94 pages): 10/19/98

Fatal Injuries to Workers in the U.S., 1980-1989: A Decade of Surveillance, National Profile (1,501K; 53 pages): 4/23/99

Fatal Injuries to Workers in the U.S., 1980-1989: A Decade of Surveillance, National and State Profiles (8,787K; 364 pages): 4/15/99

Guidelines for Protecting the Health and Safety of Health Care Workers: 4/24/98;(17,401K; 549 pages): 11/6/98

Guidelines: Minimum and Comprehensive State-Based Activities in Occupational Safety and Health (292K; 10 pages): 2/5/98

Health Hazard Evaluations: Noise and Hearing Loss, 1986-1997: (75K; 12 pages): 12/14/98

Histoplasmosis: Protecting Workers at Risk: September 1997

Homicide in U.S. Workplaces: A Strategy for Prevention and Research (435K; 13 pages): 1/21/99

Identifying High-Risk Small Business Industries-The Basis for Preventing Occupational Injury, Illness, and Fatality (May 1999; Pub. No. 99-107).(761K; 155 pages): 5/27/99

Industrial Noise Control Manual (December 1978, Pub. No. 79-117) (9,55OK; 356 pages): 3/3/99

Injuries Among Farm Workers in the United States, 1993 (6,781K; 368 pages): 8/7/98

Laboratory Evaluations and Performance Reports for the Proficiency Analytical Testing (PAT) and Environmental Lead' Proficiency Analytical Testing (ELPAT)Programs (940K; 34 pages): 1/25/99

Latex Allergy: A Prevention Guide-2nd printing: 5/4/98;)14PDF.pdf (158K; 2 pages): 1/8/99

Manual for Measuring Occupational Electric and Magnetic Field Exposures (904K, 150 pages): 4/28/99

Mortality by Occupation, Industry, and Cause of Death: 24 Reporting States: 10/1/97

Musculoskeletal Disorders (MSDs) and Workplace Factors-A Review of Epidemiologic Evidence for Work-Related Musculoskeletal Disorders of the Neck Upper Extremity, and Low Back; (2,874K; 590 pages): 7/1/97

National Mortality Profile of Active Duty Personnel in the U.S. Armed Forces 1980-1983: 4/4/97

National Occupational Exposure Survey Analysis of Management Interview Responses (17,133K; 754 pages): 7/2/98

National Occupational Exposure Survey Field Guidelines (4,604K; 240 pages): 7/13/98

National Occupational Exposure Survey and Methodology (1,935K; 88 pages): 7/2/98

National Occupational Research Agenda (NORA)-Update May 1999 (DHHS/NIOSH Publication No. 99-124) (489K; 44 pages): 5/12/99

New Directions at NIOSH (740K; 31 pages): 8/14/98

NIOSH Certified Equipment as of September 30, 1993 (4,889K; 409 pages): 2/23/98

Guide to Industrial Respiratory Protection (11,185K; 305 pages): 4/10/98

NIOSH Guide to the Selection and Use of Particulate Respirators Certified Under 42 CFR 84 : 3/4/96

NIOSH Fact Sheet: Exploding Flashlights: Are They a Serious Threat to Worker's Safety: 3/17/98

NIOSH Manual of Analytical Methods (NMAM) : 11/20/96

NIOSH Pocket Guide: 8/14/98

NIOSH Publications Catalog, FY 1986 - FY 1997: 9/24/98

NIOSH Publications on Video Display Terminals, Third Edition (6,331K; 141 pages): 10/25/99

NIOSH Recommendations For Occupational Safety and Health- Compendium of Policy Documents and Statements (6,414K; 217 pages): 9/8/98

NIOSH Report of Activities for Fiscal Year 1997; (208K; 27 pages): 4/15/99

NIOSH Research Projects: Agriculture, April 1977 (85K; 19 pages): 9/29/97

NIOSH Research Projects: Musculoskeletal Disorders January 1977 (75K; 15 pages): 9/29/97

NIOSH Respirator Decision Logic (2,64OK; 61 pages): 5/87

NIOSH Respiratory Protection Program In Health Care Facilities, Administrator's Guide (September 1999, Pub. No. 99-143): 11/3/99

Occupational Diseases: A Guide to Their Recognition (16,89OK; 617 pages): 3/17/98

Occupational Exposure Sampling Strategy Manual (5,348K- 150 pages): 4/10/98

Occupational Health Guidelines to Chemical Hazards (chemical pages range from 123K to 360K in file size, and contain from three to eight pages each): 1/81; Supplement I (files range in size from 267K to 348K); Supplement II (files range in size from 256K to 334K); Supplement III (files range in size from 262K to 427K); and Supplement IV (files range in size from 174K to 638K)

Occupational Injury Deaths of 16 and 17 Year Olds in the United States: Trends and Comparisons to Older Workers: 4/17/98

Occupational Mortality in Washington State, 1950-1989 (731K; 175 pages): 6/29/98

Occupational Noise and Hearing Conservation Selected Issues : 4/8/96

Occupational Respiratory Diseases, September, 1986 (40,432K; 855 pages): 2/27/98

Hazardous Waste Site Activities (5,116K; 142 pages): 12/22/98

Papers and Proceedings of the Surgeon Generals Conference on Agricultural Safety and Health (27,194K; 682 pages): 11/23/98

Participatory Ergonomic Interventions in Meat-packing Plants (5,368K; 226 pages): 7/28/98

Plain language About Shiftwork (572K; 47 pages): 3/26/98

Preventing Injuries and Deaths From Skid Steer Loaders: 2/98

Preventing Occupational Hearing Loss - A Practical Guide (823K; 106 pages): 7/28/98

Preventing Worker Injuries and Deaths from Traffic-Related Motor Vehicle Crashes: 7/31/98

Proceedings of a NIOSH Workshop: A Strategy for Industrial Power Hand Tool Ergonomic Research-Design, Selection, Installation, and Use in Automotive Manufacturing (4.8MB): 8/95

Proceedings of the Second International Workshop on Coal Pillar Mechanics and Design-Information Circular 9448 (3,647K; 190 pages): 7/77

Proceedings of the Scientific Workshop on the Health Effects of Electric and Magnetic Fields on Workers (6,082K; 229 pages): 2/23/98

Proceedings of the VIIth International Pneumoconiosis Conference Part I: (31,943K; 850 pages): 9/11/98

Proceedings of the VIIth International Pneumoconiosis Conference Part II: (28,713K; 767 pages): 3/24/99

Promoting Safe Work for Young Workers; (60 pages): 1/25/00

Protect Your Family-Reduce Contamination At Home: 9/11/97 (16 pages): 9/29/97

Protect Yourself Against Tuberculosis: A Respiratory Protection Guide for Health Care Workers: 3/4/97; (462K; 32 pages): 1/21/99

Protecting Workers Exposed to Lead-Based Paint Hazards: A Report to Congress: 1/97;(444K; 86 pages): 2/8/99

Recommendations for Chemical Protective Clothing: A Companion to the *NIOSH Pocket Guide to Chemical Hazards* 8/14/98

Registry of Toxic Effects of Chemical Substances: RTECS and Comprehensive

Guide to the RTECS (227K; 80 pages): 7/28/98

Report to congress on Workers' Home Contamination Study Conducted Under The Workers' Family Protection Act (29 U.S.C. 671a) (10,209K; 308 pages): 1/8/99

Results From the National Occupational Health Survey of Mining (NOHSM) (5,182K; 224 pages): 12/22/98

Safe Grain and Silo Handling: NIOSH Pub. No. 95-109 : 12/8/95

Safe Maintenance Guidelines for Robotic Workstations (1,259K; 56 pages): 7/2/98

Selected Topics in Surface Electromyociraphy for Use in the Occupational Setting: Expert Perspective (6,748K; 189 pages): 6/16/98

Selecting, Evaluating, and Using Sharps Disposal Containers: NIOSH Pub. No. 97-111: 1/98; (947K; 29 pages): 1/21/99

Silica ... It's Not Just Dust: 7/31/98

Strategic Plan For NIOSH as Envisioned by the Government Performance and Results Act-1997-2002: 6/1/98

Stress at Work; (350K; 32 pages): 1/5/99

Stress Management in Work Settings (3.71MB): 5/87

Study of the Prevalence of Chronic, Non-Specific Lung Disease and Related Health Problems in the Grain Handling Industry (6,556K; 360 pages): 2/3/99

The Effects of Workplace Hazards on Female Reproductive Health (293K; 23 pages): 3/24/99

The Industrial Environment: Its Evaluation and Control (36,192K; 729 pages): 5/6/98

The NIOSH Compendium of Head and Protection Devices (4,802K; 83 pages): 2/5/98

Traumatic Occupational Injury Research Needs and Priorities: 8/14/98

Updated Guidance on How to Design Safe Lifting Jobs : 7/31/96

What You Need to Know about Occupational Exposure to Metalworking Fluids (238K; 44 pages): 8/14/98

Work Related Lung Disease Surveillance Report (October 1991, Pub. No. 91-113) (1,898K; 85 pages): 3/30/99

Work-Related Lung Disease Surveillance Report Supplement 1992 (September 1992, Pub. No. 91-113s) (45 pages): 3/30/99

Worker Deaths by Electrocution: A Summary of Surveillance Findings and Investigative Case Reports; (137K; 51 pages): 9/29/98

Worker Deaths in Confined Spaces: A Summary of NIOSH Surveillance and Investigative Findings (10,212K; 282 pages): 1/8/99

Material Safety Data Sheets and Other Health Risk Information

MSDSs are now readily available on the Internet. There are a wide variety of data bases from which one can research the hazardous properties of the chemicals being handled at a site. The reader should recall that a MSDS is designed to provide both workers and emergency personnel with the proper procedures for handling or working with a substance. An MSDS will include such information as physical properties (e.g., melting point, boiling point, flash point temperature, specific gravity, solubility, and other), toxicity, health effects, first aid, chemical reactivity, storage, disposal, recommended protective equipment, emergency procedures for spills and fires.

Most of the sources for MSDS information on the Internet is free. A partial list of sites, along with the size of each data base, and some useful comments are provided in Table 2.

Table 2. Survey of Internet Sites with Material Safety Data Sheets (Reader can link to all sites in this table from *http://www.ilpi.com/msds/*)

Internet Site	No. of MSDS	Comments
General MSDS and Health Risk Databases		
MSDS Online	300,000	User required to register or sign up as a gateway domain. Searches for chemicals can be made by chemical name, CAS number to get HTML and/or PDF files on OSHA and/or ANSI format. Site contains other safety information, including free MSDS software.
Cornell University	250,000	Keyword searchable. Not all MSDSs are complete. Site also contains Toxic Substances Control Act Inventory.
Vermont SIRI hazard.com (main site), siri.com (alternate site)	180,000	Keyword searchable MSDS database. Text file format. Site contains many additional safety links and information.
ECDIN (Environmental Chemicals Data and Information Network)	120,000	Italian site - Searchable database by ECDIN #, CAS #, chemical name, formula, other. Database is not strictly MSDS. Database includes case studies and toxicological references. HTML format.

Table 2. Continued

Cambridge Soft Web site	75,000	Substructure-searchable small-molecule database with two- and three-dimensional structures. Many links to other sites.
Oxford University	4,225	Sections listed A-Z in HTML format. Optional search on full text of database. Some entries appear incomplete.
ChemExper	4,000	Belgium site - Search for chemical then click "MSDS". Site also lists suppliers of the chemicals.
University of Akron	1,991	Keyword search. HTML format. Data provided are only partial MSDSs.
Scorecard Environmental Defense Fund	6,800	Site contains human health risk information, risk assessment values, health risk rankings, environmental release data, regulatory information. Search by chemical name or CAS #. HTML format.
North American Emergency Response Guidebook	3,720	Does not provide MSDS but rather guides for emergency services personnel at transportation incidents involving hazardous materials. PDF format.
The National Toxicological Program (NIOSH)	3,000	Searchable health and safety database.
CDC/NIOSH/WHO	869	HTML format. U.S. and International versions (available in 8 different languages). Not strictly MSDS
International Agency for Research on Cancer, IARC	800	Not MSDS, but reports on cancer risks to humans. HTML format.
NIOSH Pocket Guide to Chemical Hazards	677	Tabular summaries of MSDS-type information for chemicals, along with NIOSH exposure limits and OSHA permissible exposure limits (PEL). HTML format.
Envirofacts Chemical Reference (U.S.EPA)	320	Links to EPA chemical fact sheets.

Table 2. Continued

Health Canada	169	MSDS for biohazards.
National Safety Council, EHC Division	85	Provides chemical descriptions, properties, health effects and economics information.
U.S. Environmental Protection Agency	31	Fact sheets in text format. Some in PDF format.
Databases Prepared by Chemical Manufacturers and Suppliers		
Sigma, Aldrich Fluka, Supelco, RdH-Lab	90,000	User registers for free and can obtain MSDS's by region or by key-word search. HTML format. Available in several languages.
Acros Chemicals, Fisher Scientific, Curtin Matheson Scientific	61,000	Keyword search one or all categories at once. HTML format.
Alfa Aesar	14,000	User can browse through an alphabetical listing of chemicals or search a catalogue by product name, catalogue number, CAS #. HTML files in ANSI format.
EM Sccine (U.S.) And BDH Inc. (Canada)	7,000	Search a catalogue for products. HTML format.
J. T. Baker, Inc.	2,100	Search catalogue by product name, chemical name or formula, CAS #. HTML format.
Eastman Kodak	1,500	Search by product name or catalogue number. HTML format. MSDS available in several languages.
Pesticides, Herbicides, Insecticides, Rodenticides, Fertilizers, etc.		
Automatic Rain Company	1,345	Pesticide and adjuvant MSDS. Search an alphabetical list of chemicals. Text file format.
C&P Press; Greenbrook.net	1,300	Pesticide and adjuvant MSDS, searchable by manufacturer and/or trade name. The site has regular updates by manufacturers. PDF format.
Crop Data management Systems	1,187	MSDS and pesticide labels from over 70 manufacturers. Search by name to get MSDS in PDF format. Includes links to manufacturers.

Fire Safety

A good Internet site for fire safety information is the Fire Safety Institute (http://middleburry.net/firesafe/). The Fire Safety Institute is a not-for-profit information, research, and educational corporation that focuses on innovative approaches to fire safety science and engineering. The Institute was founded in 1981. Its purpose is to encourage an integrated approach to the reduction of life and property loss from fire through rational fire safety decision making. The Institute pursues this goal by application of 1) information science to collect and organize current and developing fire safety concepts, 2) researching methods of decision analysis to develop better ways to utilize fire safety technology, and 3) education of professionals to disseminate fire safety knowledge. The Fire Safety Institute is incorporated under the laws of the State of Vermont. Fire safety information is scattered among many sources ranging from manufacturer's literature to scientific libraries to foreign periodicals to shelved government documents. The Institute focuses on making this information available to the fire safety specialist in a form that is usable and which will promote the application of available technology to the reduction of fire losses. Information services of the Fire Safety Institute cover: development of guides to fire safety literature; compilation of bibliographies and state-of-the-art papers utilizing on-line computer databases; and editorial and review services for technical fire safety journals. Specific ongoing information activities supported by the Fire Safety Institute include Editor, *Fire Technology*, the quarterly journal of fire safety science and engineering published by the National Fire Protection Association (NFPA); Editorial Board, *Journal of Fire Protection Engineering*, Society of Fire Protection Engineers (SFPE); Editor, *Handbook of Fire Protection Engineering*, Society of Fire Protection Engineers (SFPE) and National Fire Protection Association (NFPA).

Fire safety research takes many forms. The Fire Safety Institute is concerned with applied research to answer the fundamental fire safety questions: What is important? How do we measure it? and How much is enough? In fire safety, these questions are not often amenable to the laboratory experimentation of physical science research, but require more subtle methods of scientific inquiry such as systems analysis, operations research, and decision science. Primary areas of fire safety research pursued by the Fire Safety Institute are design of methods to synthesize the art and science of fire safety decision making; investigation of probabilistic approaches to the evaluation of fire safety; development of concepts of fire risk analysis and fire risk assessment; assessment of alternative approaches to fire safety evaluation.

Fire risk analysis techniques are among the useful information the site has to offer. Risk has always been a part of human endeavor, but we increasingly expect protection against risk, thus governments around the world are mandating risk analysis in areas of health and safety. Computations of the odds of harm are becoming a powerful force in decisions about activities involving risk. These decisions have here-to-fore been largely politically based, but we are learning to debate from a more scientific and quantitative perspective. Risk is generically defined as the uncertainty of loss. Fire loss is usually measured as number of deaths or dollars of property damage, but includes significant intangible losses such as business interruption, mission failure, degradation of the environment, and destruction of irreplaceable cultural artifacts. The concept of safety itself is one of uncertainty. Absolute safety does not exist. Human activity will always and unavoidably involve risks. The concept of fire is also uncertain. Unwanted combustion is perhaps the least predictable common physical phenomenon. Reliability of manufactured or fabricated systems for fire suppression and confinement is another source of uncertainty or risk. Hence, to make meaningful decisions regarding these risks, it is necessary that different scenarios be analyzed. As pointed out by the Institute's director - in fire safety we most often rely on empiricism and intuitive heuristics to make decisions. Increasing computational capabilities and modeling techniques from fields such as decision analysis, management science, operations research, and systems safety now allow us to identify the framework or structure of our decision making process, with varying levels of mathematical sophistication. Fire Risk Analysis is a generic phrase that covers many approaches to decision making about the uncertainties of losses from fire. Within this general structure are techniques for both qualitative and quantitative fire risk analysis. The approach may be as simple as a check list of fire safety features or it may involve mathematically complex probabilistic analysis. Application is variable according to the nature of the risks or hazards involved and according to the experience of the analyst. Each application needs individually to consider the level of mathematical sophistication appropriate to meet objectives. The Fire Safety Institute strives to define and improve the relationships among fire risk analysis, fire modeling, fire risk management, and fire protection engineering. Its purpose here is to enhance the application of fire risk analysis so that it provides an efficient and effective approach to finding solutions to fire safety problems and for selecting among alternative actions or designs.

The following is a list of publications that can be accessed from the site:

Watts, John M., Jr., "Fire Risk Assessment in Cultural Resource Facilities," *Proceedings: Fire Risk & Hazard Assessment Symposium,* National Fire Protection Research Foundation, Quincy MA, 1996, pp. 508-522.

Watts, John M., Jr., "Angle of Exit Remoteness," *Fire Technology,* Vol. 32, No. 1, 1996, pp. 76-82.

Watts, John M., Jr., " Performance- Based Life Safety Code," *Proceedings:*

International Conference on Performance-Based Codes and Fire Safety Design Methods, Society of Fire Protection Engineers, Boston, 1997, pp. 159-169.

Watts, John M., Jr., and Marilyn E. Kaplan, "Performance-Based Approach to Protecting Our Heritage," *Proceedings: International Conference on Performance-Based Codes and Fire Safety Design Methods,* Society of Fire Protection Engineers, Boston, 1997, pp. 339-347.

Watts, John M., Jr., "Systems Concepts for Building Fire Safety," Section 1, Chapter 3 *Fire Protection Handbook, 18th ed.,* NFPA, Quincy MA, 1997, pp. 1-34 to 1-41.

Watts, John M., Jr., "Assessing Life Safety in Buildings," Section 9, Chapter *Protection Handbook, 18th ed.,* NFPA, Quincy MA, 1997, pp. 9-1 to 9-12. , *Fire*

Watts, John M., Jr., "Probabilistic Fire Models," Section 11, Chapter 6, *Fire Protection Handbook, 18th ed.,* NFPA, Quincy MA, 1997, pp. 11 -62 to 11 -69.

Barry, Thomas F. and John M. Watts, Jr., "Simplified Fire Hazard and Risk Calculations," Section 11, Chapter 1, *Fire Protection Handbook, 18th ed.,* NFPA, Quincy MA, 1997, pp. 11 -108 to 11 -118.

Watts, John M., Jr., "Fire Risk Assessment Using Multiattribute Evaluation," *Fire Safety Science, Proceedings of the Fifth International Symposium,* International Association of Fire Safety Science, 1997, pp. 679-690.

Budnick, Edward K., Lawrence A. McKenna, Jr,, and John M. Watts, Jr., "Quantifying Fire Risk for Telecommunications Network Integrity," *Fire Safety Science, Proceedings of the Fifth International Symposium,* International Association of Fire Safety Science, 1997, pp. 691-700.

Watts, John M., Jr., "Communicating Fire Suppression and Detection Research to Users," *Proceedings: Fire Suppression and Detection Symposium,* National Fire Protection Research Foundation, Quincy MA, 1997, pp. 221-233.

Watts, John M., Jr., and Marilyn E. Kaplan, "Development of an Historic Fire Risk Index", *Proceedings: Fire Risk & Hazard Assessment Research Application Symposium,* National Fire Protection Research Foundation, Quincy MA, 1997, pp. 315-327.

Watts, John M., Jr., "Analysis of the NFPA Fire Safety Evaluation System for Business Occupancies," *Fire Technology,* Vol. 33, No., 3,1997, pp. 276-282.

Watts, John M, Jr., "Fire Risk Evaluation in the Codes: A Comparative Analysis," *Proceedings: Second International Conference on Fire Research and Engineering,* Society of Fire Protection Engineers, Bethesda MID, 1998, pp. 226-237.

Beller, D.K., and Watts, J.M., "Human Behavior Approach to Occupancy

Classification," *Human Behavior in Fire, Proceedings of the First International Symposium,* University of Ulster, 1998, pp. 83-103.

Watts, John M., Jr., "Rehabilitating Existing Buildings," *Fire Protection Engineering,* Vol. 1, No. 2, 1999, pp. 6-8, 10, 12, 14-15.

Chapter 7
SAFETY IN THE
LABORATORY

INTRODUCTION

Laboratory settings present special safety problems because of the large number of different chemicals that are often handled. Many chemicals are often incompatible, and fall into such categories as flammable materials, combustible products, water reactive and unstable, compressed gases, pyrophoric, cryogenic, corrosive, explosive, and toxic. Safe management practices require attention be given to special handling and storing of these materials, as well as proper disposal. This chapter focuses on safe handling practices for common laboratory environments. Examples of specific chemicals and proper handling practices are reviewed, and specific recommendations for a number of common chemicals encountered in laboratories are provided. The Web sites noted in the last chapter are a good source for obtaining the MSDSs of many of the chemicals discussed in this chapter.

COMPRESSED AND LIQUEFIED GASES

Compressed and liquefied gases should be used and handled only by properly trained personnel. Some of the most widely used gases in a laboratory are acetylene, hydrogen, methane, propane, carbon monoxide, and natural gas. This section describes the hazardous properties of these gases and provides recommendations for gas storage and handling. Table 1 provides some properties of common flammable gases. Other sources of information on gases include MSDSs, the Matheson Gas Data Book, Compressed Gas Association pamphlets such as, C-6, "Standards for Visual Inspection of Steel Compressed Gas Cylinders" and C-8, "Standard for Re-qualification of DOT-311t Seamless Steel

Cylinders," and NFPA publications. The MSDSs and the Matheson Gas Data Book deal chiefly with properties and use of gases and components of gas systems; pamphlets from the Compressed Gas Association deal with standards and requirements of the Department of Transportation. The NFPA "National Fuel Gas Code" describes appropriate containers and components, and the NFPA Fire Protection Guide on Hazardous Materials is a complete treatise.

Table 1. Properties of Common Flammable Gases

Gas	Exposure Limit TWA (ppm)	Exposure Limit IDLH (ppm)	Flammable Limits in Air	Ignition Temperature °C	Minimum Spark Ignition Energy (mJ)[a]	Vapor Density [0°C) 1 atm.] air = 1
Acetylene	-b	-	2.5-100[c]	260	0.02	0.91
Ammonia	25	500	16-25	850	14	0.60
Arsine	0.05	6	4.5-64	230[d]	-	2.69[e]
Carbon Monoxide	25	1500	12.5-75	593	-	0.97
Diborane (pyrophoric)	0.1	40	0.9-98	N/A[g]	N/A[g]	0.95
Hydrogen	- b	-	4-75	566	0.02	0.07
Hydrogen Sulfide	10	300	4.3-46	260	-	1.19 [f]
MAPP	1000	15000	3.5-11	-	-	-
Methane	- b	-	5-15	538	0.3	0.55
Natural Gas	- b	-	4-17	538	0.5	0.5
Propane	- b	20000	2-9.5	449	0.3	1.52 [e]
Silane (pyrophoric)	5	-	-	N/A[g]	N/A[g]	1.2 [e]

a - Human body static charge can easily exceed 3 mJ; b - Acts as a simple asphyxiant; c - Can violently decompose under pressure, even in the absence of air; d - Decomposes; e - Reference temperature of 70 °F (21.1 °C); f - Reference temperature of 59 °F (15 °C); g - The chemical ignites spontaneously.

Compressed Gases Storage Areas

The following describes the requirements for compressed gas storage areas. Storage of toxic gases should be evaluated to ensure the safety of building occupants and the public. Storage areas should be prominently posted with the hazard class, or the name of the gas, and with "NO SMOKING" signs where appropriate. Designated storage areas in parts of the laboratory containing hazardous/compressed gases should be prominently posted with the names of the gases being stored. Where hazardous/compressed gases of different types are stored at the same location, containers should be grouped by types of gas. Full and empty containers should be stored separately with the storage layout so planned that containers comprising old stock can be removed first with a minimum handling of other containers. Compressed gas cylinders should not be stored near readily ignitable substances (e.g., gasoline or waste), or near combustibles in bulk, (e.g., oil). Compressed gas cylinders should not be exposed to continuous dampness and should not be stored near salt or other corrosive chemicals or fumes. Corrosion may damage the containers and may cause the valve protection caps to stick. Compressed gas cylinders should be protected from any object that will produce a cut or other abrasion in the surface of the metal. Compressed gas cylinders should not be stored near elevators, gangways, or unprotected platform edges, or in locations where heavy moving objects may strike or fall on them. Users should store compressed gas cylinders standing upright where they are not likely to be knocked over. Properly secured gas cylinders less than 305 in^3 water volume may be stored horizontally. Compressed gas cylinders in public areas should be protected against tampering.

Outdoor: Compressed gas cylinders may be stored in the open but should be protected from the ground beneath to prevent bottom corrosion. Containers may be stored in the sun; however, if the supplier recommends storage in the shade for a particular gas, the recommendation should be observed. Outdoor storage areas should have a minimum of 25 percent of the perimeter open to the atmosphere; the open space may be covered with chain link fence, lattice construction, open block, or similar materials. Storage areas should be kept clear of dry vegetation and combustible materials for a minimum distance of 15 feet. Storage areas should be provided with physical protection from vehicle damage. Storage areas can be covered with canopies of noncombustible construction.

Indoor: Gas cylinders should be separated according to their category. Refer to Table 2. Storage rooms for compressed gases should be well ventilated and dry. Where practicable, storage rooms should be of fire-resistive construction. Floors and shelves should be of noncombustible or limited combustible construction.

Storage room temperatures should not exceed 130°F (54°C). Storage in subsurface locations should be avoided in order to avoid creating hazardous conditions such as confined spaces or difficulty in fire fighting. Compressed gas cylinders stored inside should not be located near exits, stairways, or in areas normally used or intended for the safe exit of people. Maximum size and quantity limitations for compressed or liquefied gas cylinders in laboratory work areas are summarized in Table 3.

Table 2. Separation of Gas Cylinders by Hazard

Gas Hazard	Non-flammable	Oxidizing	Flammable	Pyrophoric	Toxic
Toxic	Compatible	20 ft*	20 ft*	20 ft*	-
Pyrophoric	Compatible	20 ft*	20 ft*	-	20 ft*
Flammable	Compatible	20 ft*	-	20 ft*	20 ft*
Oxidizing	Compatible	-	20 ft*	20 ft*	20 ft*
Non-flammable	-	Compatible	Compatible	Compatible	Compatible

Note - Compatible cylinders may be stored adjacent to each other. * = The 20 ft distance may be reduced without limit when the cylinders are separated by a barrier of noncombustible materials at least 5 feet high having a fir resistance rating of at least ½ hour.

Table 3. Size/Quantity Limitations for Compressed or Liquefied Gas Cylinders

Type Material	Maximum Cylinder Size ᵃ	Maximum number of cylinders per 500 ft² or less	
		Non-sprinklered	Sprinklered
Flammable Gases or Oxygen	10×50	3	6
Liquefied Flammable Gases	9×30	2	3
Gases with Health Rating of 3 or 4 (NFPA)	5×15	3	-

a - Approximate dimensions in inches (Diameter × Length)

Compressed and Liquefied Gas Cylinders

The following are recommended requirements for compressed gas cylinders. Cylinders should be:

- Labeled and marked in accordance with DOT and 29 CFR 1910.1200
- Maintained by trained personnel.

Cylinders should also be inspected by users for the following: legible marking and labeling; absence of defects; within the hydrostatic test date. Users should not modify, tamper with, or repair any part of a container or cylinder. The cylinders should have visual and other inspections performed by gas plant personnel. Gas plant personnel should also ensure that safety relief devices for compressed gas cylinders are properly installed and maintained. Cylinders should be used to contain and use the contents. They should not be used for any other purpose. A static pressure test and soapy water can be used to determine the presence and location of leaks from piping systems.

Manifolds should be of proper design for temperatures, pressures, and flows of the materials they contain. Operations involving experimental manifolds must have established SOPs (standard operating procedures).

Cylinders should be secured in place to prevent falling. Cylinders should not be placed where they might become part of an electric circuit. Cylinders should not be exposed to temperature extremes, i.e., 125°F. If ice or snow accumulate on a container, thaw at room temperature or with water not exceeding 125°F. Cylinders that are not necessary for current laboratory operations should be stored in a safe area outside the laboratory work area.

Cylinders should be visually inspected by users for signs of physical damage. Containers that have been damaged, or are leaking, defective, or corroded, should be returned to the gas plant. Valve protection caps should be used at all times except when containers are secured and in use. Valves should be kept closed at all times except when in use. Valve outlets should be pointed away from personnel when being opened. Hand-wheel valves should be opened slowly. Those valves without hand wheels should be opened with a wrench designed or approved for that purpose. Cylinders containing toxic gases having a health rating of 3 or 4, or a health rating of 2 and no warning properties, should be kept in a continuously, mechanically ventilated enclosure. There should be no more than 3 cylinders per enclosure, or fewer depending upon the ventilation capacity. Gas tight valves on poison gas containers should be checked and tightened prior to return to the gas plant. Cylinders in storage containing gases that are corrosive to cylinder valves or that may become unstable while stored in the cylinder should have a maximum retention period of 6 months, unless a shorter period is otherwise specified by the manufacturer. Transfer of compressed gases from one container to another should only be performed by trained, authorized personnel. Empty and full cylinders must be segregated. Empty cylinders must be tagged "empty." Nonliquefied compressed gas containers should not be emptied below the operating pressure of the system, or not less than 20 psig to prevent backflow of atmosphere or other contaminants. All liquefied gas containers, except those designed for horizontal use, should be stored and used valve end up. Nonflammable liquefied gas containers may be inverted for use if secured and dispensed with an apparatus designed for inverted use.

Regulators and Check-Valves: Threads on regulator connections or other equipment should match those on container valve outlets. Connections that do not fit should not be forced. Check valves should be used, inspected, and regularly maintained where a container may be contaminated by backflow of process materials. "Needle" valves are not designed to control the full cylinder pressure. When finished dispensing the gas, it is good practice to leave the needle valve open after closing the cylinder valve. In the event of a leak through the cylinder valve, this arrangement prevents build-up of dangerous pressure in the regulator and needle valve assembly. *Note, however, that the released gas may pose other hazards.* All connections must be gas tight. Tightening, repairs, or removal of the regulator should not be performed while the system is under pressure. Regulators used for toxic gases that are not in service should be stored in plastic bags and labeled with the name of the gas they were used to regulate. Only regulators designed for the particular use and gas should be used. Noncryogenic liquefied gases with relatively low vapor pressures at low ambient temperature may require check valves.

Transportation and Disposal: A suitable hand truck, forklift, or appropriate material handling device should be used to move containers/cylinders. Never lift containers/cylinders by using the container cap or magnets. Ropes, chains, or slings should only be used to suspend containers that have been designed as such. Compressed gas containers/cylinders that are not at or near atmospheric pressure should be handled as hazardous waste and disposed of in accordance with local, state and federal regulations.

Compressed and Liquefied Gas Piping Systems

Piping systems for compressed and liquefied gases should comply with the requirements of applicable NFPA standards: - NFPA 50, "Standard for Bulk

Oxygen Systems at Consumer Sites." - NFPA 50a, "Standard for Gaseous Hydrogen Systems at Consumer Sites." - NFPA 50b, "Standard for Liquefied Hydrogen Systems at Consumer Sites." - NFPA 51, "Standard for the Design and Installation of Oxygen-Fuel Gas Systems for Welding, Cutting, and Allied Processes." - NFPA 54, "National Fuel Gas Code." - NFPA 58, "Standard for the Storage and Handling of Liquefied Petroleum Gases." Systems for other compressed gases and for cryogenic materials should comply with the manufacturer's design and specifications. Manual shut-off valves should be provided at each point of supply and each point of use.

Exception: The following are exceptions to the above recommended guidelines:

• If containers supplying the piping system are equipped with shut-off valves, a separate valve is not required.

• A valve at the point of use is not required if there is a supply shut-off valve within immediate reach of the point of use. All portions of a piping system should have uninterruptible pressure relief.

• Piping designed for a pressure greater than the maximum system pressure that can be developed under normal conditions. Pressure relief systems should be designed to provide a discharge rate sufficient to avoid further pressure increase and should vent to a safe location. Permanent piping should be identified at the supply point and at each discharge point with the name of the compressed or liquefied gas being transported. Piping systems, including regulators, should not be used for gases other than those for which they are designed and identified.

• A piping system can be converted from one gas service to another after a thorough review of design specifications, materials of construction, service compatibility, and other appropriate modifications.

Acetylene [C_2H_2]

Acetylene is highly ignitable and explosive. Explosive limits are 2.5 to 100%. Note that the gas not only has a low value LEL, but has a wide range - making it extremely dangerous. Acetylene is highly unstable at high pressures and may decompose into hydrogen and carbon with explosive violence if subjected to sparks, heat, or friction. Commercial grade acetylene has a characteristic garlic-like odor.

Acetylene cylinders should be stored and used with the valve end up. Storage of acetylene cylinders valve end up will minimize the possibility of solvent (acetone or another suitable acetylene solvent) being discharged. ("Valve end up" includes conditions where the container axis may be inclined as much as 45 degrees from the vertical.) Protect against lightning and static electricity. Isolate from oxidizing gases, especially chlorine. Laboratory handling, storage, and utilization of acetylene in cylinders should be in accordance with Compressed Gas Association (CGA) Pamphlet G-1. Piping systems for transfer and distribution of acetylene should be designed, installed, maintained, and operated in accordance with CGA Pamphlet G-1.3. Generation and charging (filling) of acetylene cylinders should be in accordance with CGA Pamphlet G-1.4.

Hydrogen [H_2]

Hydrogen includes the isotopes deuterium [2 H_2 or D_2] and tritium [3 H_2 or T_2]. Hydrogen has a wide range of explosive limits (4-75%) and readily disperses and stagnates in the vicinity of the ceiling. Hydrogen may explode by sunlight when mixed with halogen gases. Hydrogen burns with an (almost) invisible flame so a fire is difficult to detect. Hydrogen rapidly diffuses through porous materials. It can leak out of systems that are considered gastight for air or other common gases at equivalent pressures.

Hydrogen can cause flaking, hydrogen embrittlement, or delayed brittle fracture when absorbed into steel. Trace amounts of hydrogen sulfide, cyanide, and arsenic can greatly increase the amount of hydrogen that becomes absorbed by steel. Under corrosive conditions, these substances can contribute to severe hydrogen damage, leading to loss of strength in the steel.

Tritium is the radioactive isotope of hydrogen, and is used in research laboratories for specialty applications. Tritium is a low energy emitter (0.018 MeV) with a half-life of 12.26 years (approx. 5% loss each year). It readily exchanges with hydrogen in water, forming tritiated water that disperses uniformly throughout body tissue.

The use and storage of hydrogen are outlined under 29 CFR 1910.103 for gaseous and liquid hydrogen. Hydrogen gas detectors are most effective when placed near the ceiling, since this is where the gas will collect.

FLAMMABLES AND COMBUSTIBLES

The classification of flammable and combustible liquids is based primarily on the fire characteristics of the chemical, particularly the flashpoint. Quantity limitations, classes of flammable and combustible liquids, maximum allowable size of containers and portable tanks, and common flammable and combustible compounds are detailed in Tables 4 through 7. Attention should also be paid to such fire properties as the autoignition temperature, flammability range, upper and lower explosion limits. Also, users should focus on the compatibility of different chemicals, since combinations of certain chemicals can lead to spontaneous combustion situations or other fire hazard risks.

Table 4. Recommended Quantity Limitations

Offices: All storage of flammable and combustible liquids prohibited except for that required for maintenance and operation of the building or operation of equipment. In such cases, liquid should be kept in closed metal containers and stored in a storage cabinet, a safety can, or in an inside storage room that does not have a door opening into the portion of the building used by the public.
Laboratories: 5 gal/100 ft² of floor space for flammable liquids and combustible liquids not in safety cans and flammable liquid storage cabinets; 10 gal/100 ft² (includes quantities in safety cans and flammable liquid storage cabinets)
Other Experimental, Operation, and Warehouse Areas: 25 gal (does not include storage in flammable liquid storage cabinets)
Outdoor Areas: 20 drums or 1100 gal. of flammable and/or combustible liquid

Table 5. Classes of Flammable and Combustible Liquids

Flammable			Combustible		
Flashpoint < 100 °F (37.8 °C) Vapor Pressure < 40 psia			Flashpoint ≥100 °F (37.8 °C)		
Class 1A	Class 1B	Class IC	Class II	Class IIIA	Class IIIB
Flashpoint < 73 °F (22.8 °C) and Boiling point < 100 °F (37.8 °C)	Flashpoint < 73 °F (22.8 °C) and Boiling point ≥100 °F (37.8 °C)	Flashpoint 73 °F (22.8 °C) and < 100 °F (37.8 °C)	Flashpoint 100 °F (37.8 °C) and < 140 °F (60 °C)	Flashpoint 140 °F (60 °C) and < 200 °F (93.3 °C)	Flashpoint ≥200 °F (93.3 °C)

Table 6. Maximum Allowable Size of Containers and Portable Tanks.

Container Type	Flammable Liquids			Combustible Liquids	
	IA	IB	IC	II	III
Glass or approved plastic	1 pt*	1 qt*	1 gal	1 gal	1 gal
Metal (other than DOT drums)	1 gal	5 ga.	5 gal	5 gal	5 gal
Safety cans	2 gal	5 gal	5 gal	5 gal	5 gal
Metal drums (DOT spec.)	60 gal	60 gal	60 gal	60 gal	60 gal
Approved portable tanks	600 gal	600 gal	600 gal	600 gal	600 gal

Glass, 1 gal containers are generally prohibited for class IA or IB flammable liquids, but they may be used if (a) a metal or approved plastic container would contaminate the liquid; (b) the liquid would corrode a metal container; or (c) a single assay lot or more than the listed amount is required.

In general, it should be noted that:

1. Outdoor container and portable tank storage should comply with Tables H-16 and H-17 of 29 CFR 1910.106, respectively.

2. Indoor container and portable tank storage should comply with Tables H-14 and H-15 of 29 CFR 1910.106, respectively.

3. Storage in inside rooms should comply with requirements of Table H-13 of 29 CFR 1910.106.

4. Supply piping should comply with NFPA 30, "Flammable and Combustible Liquids Code."

For specific storage criteria the reader should carefully review 29 CFR 1910.106. Table 7 provides the reader with the classification system used in the United States for flammable and combustible liquids. More extensive data for a large number of different chemical compounds can be found in the *Handbook of Industrial Toxicology and Hazardous Materials*, Marcel Dekker, Inc.(1999).

Table 7. Flammable/Combustible Classification of Common Liquids.

Chemical Name	CASRN*	Flashpoint °F	Flashpoint °C	Class
Acetone	67-64-1	-2	-19	IB
Acetic Acid	64-19-7	109	43	II
Acetonitrile	75-05-8	43	6	IB
Aniline	62-53-3	158	70	IIIA
Benzene	71-43-2	12	-11	IB
Butyl Alcohol,	71-36-3	84	29	IC
sec-	78-92-2	75	24	IC
tert-	75-65-0	52	11	IB
Carbon Disulfide	75-15-0	-22	-30	IB
Carbon Tetrachloride	56-23-5	-	-	Nonflam. Noncomb.
Chloroform	67-66-3	-	-	Nonflam. Noncomb.
Diethyl Ether [Ethyl Ether]	60-29-7	-49	45	IA
Diisopropyl Ether	108-20-3	-18	-28	IB
N,N'-Dimethyl-formamide [DMF]	68-12-2	135	57	II
1,4-Dioxane	123-91-1	54	12	IB
Ethyl Alcohol	64-17-5	54	12	IB
Ethylene Glycol	107-21-1	232	111	IIIB
containing 6% methanol	50-00-0	162	72	IIIA
containing 10% methanol	50-00-0	147	64	IIIA
containing 15% methanol	50-00-0	122	50	II
Isoamyl Alcohol-Primary [Isopentanol]	123-51-3	109	43	II

Table 7. Continued

Chemical Name	CASRN*	Flashpoint		Class
		°F	°C	
Isopropyl Alcohol	67-63-0	54	12	IB
Methyl Alcohol	67-56-1	52	11	IB
Methyl Ethyl Ketone	78-93-3	21	-6	IB
Pentane	109-66-0	-40	40	IA
Petroleum Ether [Petroleum Spirits]	64475-85-0	< 0	< -18	IB
Tetrahydrofuran [THF]	109-99-9	· 7	-14	IB
Toluene	108-88-3	39	4	IB
1,1,1-trichloroethane [Methylchloroform]	71-55-6	30	-1	IB
1, 1,2-trichloroethane	79-00-5	90	32	IC
1, 1, 1 -trichloroethylene	79-01-6	90	32	IC
1,1,1-Trichloromethane [Chloroform]	67-66-3	-		Nonflam.
Xylene, Mixed (m-, o-, p-)	1330-20-7	81-90	27-32	IC
m-	108-38-3	77	25	IC
o-	95-47-6	63	17	IB
p-	106-42-3	77	25	IC

* - *Chemical Abstracts Service Registry Number*

Transfer of Class I Liquids

Class I liquids should not be stored or transferred from one vessel to another in any exit way. The transfer of Class I liquids to smaller containers from bulk stock containers (not exceeding 5 gallons [18.9 liters] in capacity) inside a laboratory building or laboratory work area should be made under the following set of conditions: (a) in a laboratory hood, (b) in an area provided with ventilation adequate to prevent accumulation of flammable vapor/air mixtures from exceeding 25% of the lower flammable (explosive) limit, or (c) in separate inside flammable storage areas. Transfer of Class I liquids from containers of 5 gallons (18.9 liters) or more capacity should be carried out in a separate area outside the building, or in a separate area inside the building that meets the requirements of NFPA 30 for inside storage areas. Class I liquids should not be transferred between metal containers unless the containers are electrically interconnected by direct bonding, or by indirect bonding through a common ground system in the room. The maximum resistance of the bonding should not exceed 6 ohms.

Tank Storage, Design and Use

Tanks used for storage of flammable and combustible liquids should meet the following requirements: Tanks should be constructed of steel or approved noncombustible material. Tanks may be constructed of materials other than steel if appropriate or required by the properties of the liquid stored inside. Good engineering design and practice should be used when designing tanks made of these other materials. Atmospheric tanks should be built and used in accordance with acceptable good standards of design. These standards are provided by Underwriters' Laboratories, Inc.(ULI) and the American Petroleum Institute (API). These types of tanks should not be used for storage of a flammable or combustible liquid at a temperature at or above its boiling point. Low pressure tanks should be built and used in accordance with acceptable good standards of design, set forth by API, the American Society of Mechanical Engineers (ASME) Boiler and Pressure Vessels Code for Unfired Pressure Vessels, and ULI. Pressure vessels should be built in accordance with the ASME Boiler and Pressure Vessels Code for Unfired Pressure Vessels. Pressure vessels are addressed in AR 14-1 and TB 1401. Normal and emergency relief venting and vent piping for atmospheric tanks, low pressure tanks, and pressure vessels should be adequate and in accordance with the design of the vessel. They must, as a minimum, be equipped with a device or other means to prevent overflow into the building. Each connection to a tank inside of buildings through which liquid can normally flow should be provided with an internal or an external valve located as close as practical to the shell of the tank. Flammable or combustible liquid tanks located inside buildings, except for buildings designed and protected for flammable or combustible liquid storage, should be provided with an automatic-closing heat-actuated valve on each withdrawal connection below the liquid level, except for connections used for emergency disposal, to prevent continued flow in the event of fire in the vicinity of the tank. Fill pipes should be designed and installed in accordance with 29 CFR 1910.106, and other applicable requirements. Piping, valves, and fittings should be designed and tested for the expected working pressures and structural stresses, and should conform with the applicable provisions of Pressure Piping, ANSI B31 series.

Metal tanks should be welded, riveted, and caulked, brazed (using appropriate filler metal), or bolted, or constructed by a combination of these methods. Provisions for internal corrosion within tanks, e.g., suitable coatings or linings, should be made as necessary. Supports, foundations, and anchorage for all tanks should be of concrete, masonry, or protected steel, i.e., steel protected with a water spray or fire resistance rating of not less than 2 hours. All tanks should be strength and tightness tested prior to use, in accordance with the applicable paragraphs of the code under which they were built. Outside aboveground tanks used for flammable or combustible liquid storage should be spaced at least 3 feet apart, or not less than one-sixth the sum of their diameters, whichever is greater. If stored liquids are unstable, e.g., operating pressure or emergency venting pressure that exceeds 2.5 psig, the distance between such tanks should not be less than one-half the sum of their diameters. Adequate spacing to ensure accessibility for fire-fighting should also be provided.

Minimum separation between a liquefied petroleum gas (LPG) container of greater than 125 gallon capacity, and a flammable or combustible liquid storage tank of greater than 550 gallon capacity, should be 20 feet. Suitable means should be taken to prevent the accumulation of flammable or combustible liquids under adjacent LPG containers, e.g., diversion curbs or grading. If the tanks are diked, the LPG gas containers should be outside the diked area, and at least 10 feet away from the centerline of the wall of the diked area. Drainage, dikes, and walls for aboveground tanks should be designed and provided to prevent accidental discharge of liquid from endangering adjoining property or reaching waterways.

Underground tanks should be located at appropriate distances from existing structures, at appropriate depth and cover, and be provided with corrosion protection. Location, arrangement, and size of vents and vent piping for Class I. II, and III flammable/combustible liquids should be in accordance with 1910.106 and other requirements, e.g., Environmental Protection Agency (EPA). Tanks should be protected from flood waters, earthquakes, and sources of ignition. Tank openings other than vents should be kept closed with liquid-tight and vapor-tight caps as appropriate, when not in use.

Dip Tanks Containing Flammable or Combustible Liquids: Operations where workpieces; are dipped in, passed through, or coated by flammable or combustible liquids should meet the requirements of 29 CFR 1910.108. The dipping process consists of conveying the workpiece to the tank, immersing it in the fluid, removing it, and passing it over a drainboard where excess liquid will flow back to the tank. Due to the large quantities of flammable liquids used in these processes, the potential for ignition is high. Control of flammable vapors is typically accomplished with properly designed ventilation and exhaust systems. In addition to removing common sources of ignition in the work area, it is also necessary to use equipment designed to be nonsparking, particularly electrical equipment.

The requirements for design, construction, and ventilation of dipping operations address maintaining the concentrations of flammable vapors below the LEL, but do not apply to maintaining operator exposure to within the OELs (Occupational Exposure Limits). Note that an OEL is generally defined by the laboratory or company, and is typically more stringent than an REL or even a PEL. Where a health hazard has been established, controls and modifications may be required and could include increasing the air velocity beyond that specified in the regulation.

Cabinets: Requirements for cabinets used for storage of flammables and combustibles include the following: Not more than 120 gallons of Class I, Class II, and Class III liquids should be stored in a storage cabinet. Of this total, not more than 60 gallons should be Class I and Class II liquids. Not more than three cabinets should be located in a single fire area. If this is necessary, they should be separated from other cabinets or groups of cabinets by at least 100 feet. Storage cabinets should be designed and tested in accordance with NFPA 251 when feasible. Metal and wooden cabinets should be constructed in accordance with Chapter 4 of NFPA 30 when feasible. Although cabinets come with vent openings, cabinets are not required to be vented for fire protection. If the cabinet is not vented, the vent openings should be sealed with the metal bungs provided with the cabinet. If the cabinet is vented for other reasons, it should be vented outdoors to an appropriate location, and preferably in an isolated and well marked area.

CORROSIVES

Corrosive materials cause visible destruction of, or irreversible alterations in living tissue by chemical action at the site of contact. Strong acids and bases, dehydrating agents, halogens, and oxidizing agents are commonly considered to be corrosive materials. Potential accidents with corrosives in which the material may splash onto the skin or eyes are quite common in a laboratory setting. Effects can be immediate or delayed, reversible or irreversible. The eyes are particularly vulnerable to injury, with effects ranging from painful irritation to permanent blindness. Skin injuries, which can range from superficial to deep-seated burns, may be very slow to heal. When inhaled, corrosive mists or gases can cause injury to the respiratory system ranging from moderate irritation to severe injury and death. Ingestion of corrosive materials can cause immediate injury to the mouth, throat, and stomach and, in severe cases, can lead to death. In general, inorganic acids are more dangerous than organic acids. Skin contact with a strong base may be less painful than a comparable exposure to acids. As a result, the damage may extend to greater depths because the injured victim may not be aware of the seriousness of the incident. In addition to the health hazards, some corrosive materials are reactive, water-reactive, or are oxidizers. Inorganic acids, for example, react with metals to release hydrogen gas. Common inorganic acids used in many laboratories include nitric, hydrochloric, and sulfuric. Common bases used in laboratories include calcium and sodium hydroxide.

Container sizes and quantities of corrosive materials should be kept as small as possible, consistent with the rate of use. Since there are incompatibilities among corrosive materials (e.g., strong acids and bases), each class of corrosive material should be stored separately. Corrosives should be stored in a storage cabinet provided with protection against the effects of corrosion. Corrosives should not be stored above the chin level of the using organization's employees, to prevent injuries form handling these containers. MSDSs should be checked for general and specific incompatibilities. For example, hydrochloric acid and formaldehyde should not be stored near each other because the vapors can react in air to form bis(chloromethyl)ether, a known human carcinogen (specifically regulated by OSHA). Specific corrosives used routinely in laboratories are detailed below.

Hydrofluoric Acid [HF]

Hydrofluoric acid can affect the body if it is inhaled, comes in contact with the eyes or skin, or is swallowed. It may enter the body through the skin. Hydrofluoric acid liquid or vapor causes severe irritation and deep-seated burns of the eye and eye lids if it comes in contact with the eyes. If the chemical is not removed immediately, permanent visual damage (e.g., blindness) may result. When lower concentrations (20% of less) come in contact with the skin, the resulting burns do not usually become apparent for several hours. Skin contact with higher concentrations is usually apparent in a much shorter period, if not immediately. The skin burns may be very severe and painful. Fluoride ions readily penetrate skin and tissue and, in extreme cases, may result in necrosis of the subcutaneous tissue, which eventually may become gangrenous. If the penetration is sufficiently deep, decalcification of the bones may result. Hydrofluoric acid is corrosive to the nose, throat, and lungs. Severe exposure to this chemical causes rapid inflammation and congestion of the lungs. Breathing difficulties may not occur until some hours after exposure has ceased. Death may occur from breathing this chemical. If swallowed, hydrofluoric acid will immediately cause severe damage to the throat and stomach. Hydrofluoric acid should be stored in containers that resist the corrosive action of the acid. Lead, platinum, wax, polyethylene, polypropylene, polymethylpentane, and Teflon will resist the corrosive action of hydrofluoric acid. Hydrofluoric acid attacks glass, concrete, and many metals (especially cast iron). It also attacks carbonaceous natural materials (e.g., wood materials), animal products (e.g., leather), and other natural materials used in the laboratory (e.g., rubber).

Uranium Hexafluoride [UF$_6$]

Uranium hexafluoride is a soluble uranium compound which is corrosive to the eyes, skin, and respiratory tract. It also causes kidney damage. The injurious effects observed on the skin, eyes, and respiratory tract are due to the formation of BF; the uranium is responsible for the kidney damage. Repeated or prolonged skin exposure to soluble uranium compounds can cause radiation damage to the skin. Uranium hexaflouride is reported to be capable of penetrating intact skin. Dermatitis has occurred as a result of handing uranium hexafluoride. Uranium hexafluoride reacts vigorously with water to form hydrofluoric acid and uranyl fluoride. Its reaction with water is as follows:

$$UF_6 + 2H_2O \rightarrow 4HF + UF_2O_2$$

Uranium hexafluoride will attack some forms of plastics, rubber, and coatings. Uranium hexafluoride may ignite other combustible materials and may explode when exposed to heat or fire. Therefore, UF$_6$ should be stored away from heat/fire sources and combustible materials.

UF$_6$ also can react vigorously with benzene, toluene, and xylene, and violently with ethanol and water. Proper storage and handling of UF$_6$ is determined by its corrosivity as well as its radioactive properties. As such, extreme caution should be practiced when handling this chemical.

Lithium Hydride [LiH]

Lithium hydride is an odorless, light bluish gray crystalline solid or white powder. It is corrosive and severely irritating to the eyes, mucous membranes, and respiratory tract. Lithium hydride is corrosive due to the formation of lithium hydroxide on contact with moist surfaces. Where eye or skin contact occurs, remove contaminated clothing and rinse with water for at least 15 minutes. Follow up with medical care. Its reaction with water is as follows:

$$LiH + H_2O \rightarrow LiOH + H_2$$

Lithium hydride is a flammable solid and dangerous fire risk. It evolves flammable hydrogen and ignites on contact with water. It may reignite after the fire is initially extinguished. Do not leave extinguished fires unattended. Do not use water, carbon dioxide or halogenated extinguishing agents. Use approved Class D extinguishers, or smother with dry sand, dry clay, or dry ground limestone. One should handle lithium hydride out of contact with air and moisture. Open containers only in inert atmospheres or in a room with very low humidity (less than 50% relative humidity). It is important to store this product in a well-ventilated areas. Protect against dust inhalation. Protect containers against physical damage, and keep water from entering the storage area.

Perchloric Acid [HCℓO₄]

Perchloric acid is a strong acid, and contact with the skin, eyes, or respiratory tract will produce severe bums. Perchloric acid is a colorless, fuming, oily liquid. When cold, its properties are those of a strong acid but when hot, the acid acts as a strong oxidizing agent. Aqueous perchloric acid can cause violent explosions if misused, or when in concentrations greater than the normal commercial strength (72%).

Anhydrous perchloric acid is unstable even at room temperatures and ultimately decomposes spontaneously with a violent explosion. Contact with oxidizable material (e.g., wood, paper, rubber) can cause an immediate explosion. Perchloric acid is known to have caused fires and explosions due to the following:

- The instability of aqueous or of pure anhydrous perchloric acid under various conditions.

- The dehydration of aqueous perchloric acid by contact with dehydrating agents such as concentrated sulfuric acid, phosphorous pentoxide, or acetic anhydride.

- The reaction of perchloric acid with other substances to form unstable material.

- Combustible materials, (e.g., sawdust, excelsior, wood, paper, burlap bags, cotton waste, rags, grease, oil and most organic compounds) contaminated with perchloric acid solution are highly flammable and dangerous. Such materials may explode on heating, in contact with flame, by impact or friction, or may ignite spontaneously.

The acid should be inspected monthly for discoloration; discolored acid should be discarded. Perchloric acid should be used only in specially-designed perchloric acid hoods (these hoods are designed and equipped with water wash down systems). The ventilation ducts should also be designed and equipped for water wash down. There are numerous anecdotes about ducts catching fire or even exploding during maintenance or renovation work because of contamination with accumulated perchloric acid salts resulting from repeated failure to wash down the system. Perchloric acid should be stored on an epoxy-painted shelf, and preferably away from other organic materials and flammable products. The reader should refer to the *Handbook of Industrial Toxicology and Hazardous Materials*, Marcel Dekker Publishers (1999) for more in-depth information on corrosive chemicals.

ETHERS AND OTHER PEROXIDE-FORMING CHEMICALS

Ethers represent a class of materials which can become more dangerous upon prolonged storage because they tend to form explosive peroxides with age. Exposure to light and air enhances the formation of the peroxides. Many ethers tend to absorb and react with oxygen from the air to form unstable peroxides which may detonate with extreme violence when they become concentrated by evaporation or distillation, when combined with other compounds that give a detonable mixture, or when disturbed by unusual heat, shock, or friction.

Peroxides formed in compounds by auto-oxidation have caused many accidents, including the unexpected explosions of the residue of solvents after distillation. Peroxides may form in freshly-distilled, undistilled, and unstabilized ethers within less thaN two weeks; exposure to air, as in opened and partially emptied containers, accelerates formation of peroxides. While the peroxide formation potential in ethers is the primary hazard, it must be noted that inhalation hazards exist. Lower weight ethers are powerful narcotics which in large doses can cause death.

Ethers should be stored in amber bottles or other opaque containers and under a blanket of inert gas, such as nitrogen or argon, or over a reducing agent to inhibit formation of peroxides. It is preferable to use small containers that can be completely emptied, rather than take small amounts from a large container over time. Containers of ether and other peroxide-forming chemicals should be marked with the date they are opened, and marked with the date of required disposal. Table 8 lists some common peroxide-forming chemicals. For additional information about peroxides and peroxide-forming chemicals, refer to NFPA 43B, "Code for the Storage of Organic Peroxide Formulations" and Data Sheet 1-655, Rev. 87, "Recognition and Handling of Peroxidizable Compounds" (National Safety Council). The reader should refer to "Prudent Practices for Disposal of Chemicals from Laboratories, Appendix I," National Academy Press, Washington, DC, 1983, pp. 245-246 for specific recommendations on safe handling of these materials. Also, refer to the *Handbook of Industrial Toxicology and Hazardous Materials*, Marcel Dekker Publishers (1999) for more in-depth information on ethers and peroxides.

Table 8. Common Peroxide Forming Chemicals

Name	CASRN*
The following chemicals are severe peroxide hazards on storage with exposure to air. You should discard within 3 months.	
Diisopropyl ether (isopropyl ether)	108-20-3
Divinylacetylene '	821-08-9
Potassium metal	7440-09-7
Potassium amide	17242-52-3
Sodium amide (sodamide)	7782-92-5
Vinylidene chloride (1,1-dichloroethylene)'	75-35-4

Table 8. Continued

Name	CASRN*
The following chemicals are peroxide hazards on concentration; do not distill or evaporate without first testing for the presence of peroxides. Discard or test for peroxides after 6 months.	
Acetaldehyde diethyl acetal (acetal)	105-57-7
Cumene (isopropylbenzene)	98-82-3
Cyclohexene	110-83-8
Cyclopentene	142-29-0
Decalin (decahydronaphthalene)	91-17-8
Diacetylene (butadiene)	106-99-0
Dicyclopentadiene	77-73-6
Diethyl ether (ether)	60-29-7
Diethylene glycol dimethyl ether (diglyme)	111-96-6
Dioxane	123-91-1
Ethylene glycol dimethyl ether (glyme)	629-14-1
Ethylene glycol ether acetates	-
Ethylene glycol monoethers (cellosolves)	-
Furan	110-00-9
Methylacetylene	74-99-7
Methylcyclopentane	96-37-7
Methyl isobutyl ketone	108-10-1
Tetrahydrofuran	109-99-9
Tetralin (tetrahydronaphthalene)	119-64-2
Vinyl ethers	-
These chemicals pose hazards from rapid polymerization initiated by internally formed peroxides. [a]	
1. Normal liquids; discard or test for peroxides after 6 months. [b]	
Chloroprene (2-chloro-1,3-butadiene) [c]	126-99-8
Styrene	100-42-5
Vinyl acetate	108-05-4

Table 8. Continued

Name	CASRN*
Vinylpyridine	-
2. Normal gases; discard after 12 months.	
Butadiene [c]	106-99-0
Tetrafluoroethylene [c]	116-14-3
Vinylacetylene [c]	689-97-4
Vinyl chloride	75-01-4

* - *Chemical Abstracts Service Registry Number*

a - Polymerizable monomers should be stored with a polymerization inhibitor from which the monomer can be separated by distillation just before use.

b - Although air will not enter a gas cylinder in which gases are stored under pressure, these gases are sometimes transferred from the original cylinder to another in the laboratory, and it is difficult to be sure that there is no residual air in the receiving cylinder. An inhibitor should be put into any such secondary cylinder before one of these gases is transferred into it; the supplier can suggest inhibitors to be used. The hazard posed by these gases is much greater if there is a liquid phase in such a secondary container, and even inhibited gases that have been put into a secondary container under conditions that create a liquid phase should be discarded within 12 months.

c - The hazard from peroxides in these compounds is substantially greater when they are stored in the liquid phase and, if so stored without inhibitors, they should be discarded after 3 months.

OXIDIZERS

Oxidizers represent a significant hazard because of their tendency under appropriate conditions to undergo vigorous reactions when they come into contact with easily oxidized materials. Easily oxidized materials include metal powders and organic materials such as wood, paper, and other organic compounds. Most oxidizing materials increase the rate at which they decompose and release oxygen with temperature. Due to this ability to furnish increased amounts of oxygen with temperature, the reaction rate of most oxidizing agents is significantly enhanced with increasing temperature and concentrations. Containers of oxidizing agents may explode if they are involved in a fire. Table 9 lists common, powerful oxidizing reagents.

Oxidizers in quantities exceeding 10 pounds for solids or liquids or 100 pounds for gases are subject to more restrictive storage (refer to NFPA 43A, "Code for the Storage of Liquid and Solid Oxidizers"). In general, oxidizing agents should never be stored with reducing agents. However, small quantities of oxidizing agents can be safely stored in proximity to reducing agents provided sufficient secondary containment exists. Strong oxidizing agents should be stored and used in glass or other inert containers. Corks and rubber stoppers should not be used. refer to the *Handbook of Industrial Toxicology and Hazardous Materials*, Marcel Dekker Publishers (1999) for more in-depth information on oxidizing agents.

Table 9. Common, Powerful Oxidizing Reagents

Name	CASRN*
Ammonium perchlorate	7790-98-9
Ammonium permanganate	13446-10-1
Barium peroxide	1304-29-6
Bromine	7726-95-6
Calcium chlorate	10137-74-3
Calcium hypochlorite	7778-54-3
Chlorine trifluoride	7790-91-2
Chromium anhydride	1333-82-0
Chromic acid	7738-94-5 13530-68-2
Dibenzoyl peroxide	94-36-0
Fluorine	7782-41-4
Hydrogen peroxide	7722-84-1
Magnesium perchlorate	10034-81-8
Nitric acid	7697-37-2
Nitrogen peroxide	10102-44-0

Table 9. Continued

Name	CASRN*
Nitrogen trioxide	12033-49-7 10544-73-7
Perchloric acid	7601-90-3
Potassium bromate	7758-01-2
Potassium chlorate	3811-04-9
Potassium perchlorate	7778-74-7
Potassium peroxide	17014-71-0
Propyl nitrate	627-13-4
Sodium chlorate	7775-09-9
Sodium chlorite	7758-19-2
Sodium perchlorate	7601-89-0
Sodium peroxide	1313-60-6

* - *Chemical Abstracts Service Registry Number*

CARCINOGENS, HIGHLY TOXIC CHEMICALS, AND CONTROLLED SUBSTANCES

Carcinogens

Carcinogens are materials which have been shown to be capable of causing cancer in laboratory animals or humans. They are generally categorized as either confirmed or suspected human carcinogens. Cancer is an effect of long-term (chronic) overexposure to carcinogens. There is typically a 15 to 40 year period between overexposure to carcinogens and the development of disease (called the latency period). Routes of exposure to carcinogens include inhalation, ingestion, absorption, and injection.

Highly toxic materials are solids, liquids, or gases which can produce injury or death upon contact with the cells of the body. Specifically, a highly toxic material is a chemical which has: 1) an oral median lethal dose (LD_{50}) of 50 milligrams or less per kilogram of body weight, 2) a dermal LD_{50} of 200 milligrams or less per kilogram of body weight, or 3) a median lethal concentration (LC_{50}) of 200 parts per million or less of gas or vapor, or 2 milligrams per cubic meter or less of dust, mist, or fume. Health effects can be immediate or delayed, reversible or irreversible, local or systemic. Toxic effects are dependent on the magnitude and duration of exposure. Routes of exposure to carcinogens include:

- inhalation,
- ingestion,
- absorption, and
- injection.

Controlled substances are narcotic, hallucinogenic, stimulant, or depressive drugs, some of which have a high potential for abuse.

Carcinogens should be kept in sealed, unbreakable containers inside a secure cabinet. Highly toxic materials should be stored in secure cabinets. Both carcinogens and highly toxic materials should be used only in a fume hood appropriately designed for the intended operation. Open containers should be closed after use and unneeded reagents should be returned to storage. Methods to prevent overexposure to carcinogens and highly toxic materials include:

- engineering controls (e.g., ventilation system design),
- personal protective equipment (e.g., gloves, respirators), and
- training.

The major concern regarding handling and storage of controlled substances is security. Controlled substances should be kept in a storage unit which is sufficiently strong to prevent forced entry. The storage unit should either be sufficiently heavy (750 pounds or more) or rigidly bolted to a floor or wall to prevent the entire storage unit from being carried away. A careful inventory of all controlled substances must be maintained.

Shock-Sensitive Chemicals

Shock-sensitive compounds include materials classified as explosives and chemicals that can act as explosives. If the material is maintained in inventory for its explosive properties, it must be treated as an explosive, and the U.S. DOE

Explosive Safety Manual applies. Further details may also be found in the CRC "Handbook of Chemical Safety."

Tables 10 and 11 provide a list of potentially shock-sensitive compounds and highly reactive and/or heat sensitive materials, respectively. Further information can be obtained from *Prudent Practices for Disposal of Chemicals from Laboratories*, Appendix F, National Academy Press, Washington, D.C. 1983, and the *Handbook of Industrial Toxicology and Hazardous Materials*, N. P. Cheremisinoff, Marcel Dekker Publishers (1999) and to *Safety Management Practices for Hazardous Materials*, Marcel Dekker Publishers (1996).

Table 10. Shock-Sensitive Compounds

Acetylenic compounds	Fulminates
Acyl nitrates	Polynitroaromatic compounds
Alkyl nitrates	N-Halogen compounds
Alkyl and acyl nitrites	N-nitro compounds
Alkyl perchlorates	Oxo salts of nitrogenous bases
Amine metal oxosalts	Perchlorate salts
Azides	Peroxides and hydroperoxides
Chlorite salts of metals	Diazonium salts, when dry
Diazo compounds	Polynitroalkyl compounds
Picrates, especially picric acid when dry [creanine picric reagent or trinitrile phenol]	Hydrogen peroxide in concentrations above 30%

Quantities of shock-sensitive materials should be kept to a minimum by maintaining proper inventory consistent with the rate of use. Inventory control is also important in order to dispose of chemicals which tend to form unstable materials with age, such as ethers, or materials which become dangerous when they become dehydrated, such as perchloric and picric acids.

Shock-sensitive materials should be stored in a cool, dry area, protected from heat and shock. During storage, the materials should be segregated from incompatible materials including flammables and corrosives. Materials which are used specifically because of their explosive properties should be treated as an explosive of the appropriate class and kept in a magazine or the equivalent. Further information can be obtained from the *Handbook of Industrial Toxicology and Hazardous Materials*, Marcel Dekker Publishers (1999).

Table 11. Highly Reactive or Heat-Sensitive Materials

Name	CASRN	Name	CASRN
Ammonium perchlorate	7790-98-9	Dibenzoyl peroxide	94-36-0
Ammonium permanganate	13446-10-1	Diisopropyl peroxydicarbonate	105-64-6
Anhydrous perchloric acid	7601-90-3	ortho-Dinitrobenzene	25154-54-5
Butyl hydroperoxide	75-91-2	Ethyl methyl ketone peroxide	1338-23-4
Butyl perbenzoate	614-46-9	Ethyl nitrate	625-58-1
tert-Butyl peroxyacetate	107-71-1	Hydroxylamine	7803-49-8
tert-Butyl peroxypivalate	927-07-1	Peroxyacetic acid	79-21-0
1-Chloro-2,4-dinitrobenzene	97-00-7	Picric acid	88-89-1
Cumeme hydroperoxide	80-15-9	Trinitrobenzene	99-35-4 25377-32-6
Diacetyl peroxide	110-22-5	Trinitrotoluene [TNT]	118-96-7

Pyrophoric Materials

Pyrophoric materials are those that are capable of spontaneous combustion in the presence of air. Spontaneous ignition or combustion takes place when these substances reach ignition temperature without application of external heat. Ignition may be delayed or only occur if the material is finely divided or spread as a diffuse layer (titanium powder is an example in the first case, and mixed tributyl phosphine isomers is an example in the second). Or, on the other hand, ignition could be essentially instantaneous, the time delay being measured in milliseconds as with trimethylaluminum. The following classes of compounds are prone to pyrophoricity: finely divided metals (calcium, zirconium); alkali metals (sodium, potassium); metal hydrides or nonmetal hydrides (germane, diborane, sodium hydride, lithium aluminum hydride); Grignard reagents (compounds of the form RmgX); partially or fully alkylated derivatives of metal and nonmetal hydrides (diethylaluminum hydride, trimethylphosphine);alkylated metal alkoxides or nonmetal halides (diethylethoxyaluminum, dichloro(methyl)silane); Metal carbonyls (pentacarbonyliron, octacarbonyldicobalt, nickel carbonyl); used hydrogenation catalysts (especially hazardous because of the adsorbed hydrogen); phosphorus (white). A more extensive list of pyrophoric compounds can be found in Bretherick's *Handbook of Reactive Chemical Hazards.* Pyrophoric materials should be stored in tightly closed containers under an inert atmosphere or liquid. All transfers and manipulations of them must also be carried out under an inert atmosphere or liquid. These materials may not be disposed of in a landfill because of their characteristic reactivity.

Reactive Metals

Lithium, sodium, and potassium react vigorously with moisture as well as many other substances. In the reaction with water, the corresponding hydroxide is formed along with hydrogen gas, which will ignite. The generalized reaction is as follows:

$$2M + 2H_2O \rightarrow 2MOH + H_2$$

where M = Li, Na, or K.

Lithium and sodium should be stored under mineral oil or other hydrocarbon liquids that are free of oxygen and moisture. Potassium should be stored under dry xylene. Lithium, sodium, or potassium may also be stored in an inert-atmosphere glovebox. For more information, refer to the DOE Handbook (HDBK1081-94), "Primer on Spontaneous Heating and Pyrophoricity." Sprinkler protection is undesirable and combustible materials should not be stored in the same area with these metals.

Cryogens

Cryogens in common use include the liquid forms of nitrogen, helium, argon, carbon dioxide, oxygen, hydrogen, and ammonia. These cryogenic fluids exist as liquids only at temperatures considerably below ambient conditions and at volumes typically 700-800 times less than their gaseous volumes. The chemical properties and reaction rates of substances are changed under cryogenic conditions. Liquid oxygen, for example, will react explosively with materials usually considered to be noncombustible. Cryogens have a significant potential for creating an oxygen deficiency. When expelled to the atmosphere at room temperatures, they evaporate and expand on the order of 700-800 times their liquid volume. Consequently, leaks of even small quantities of cryogenic fluids can expand to displace large amounts of oxygen. Some cryogens (carbon dioxide, ammonia, fluorine) can be toxic at high concentrations. Cryogens can also create oxygen-enriched conditions. Cryogenic fluids can condense oxygen out of the air if exposed to the atmosphere, thereby accumulating oxygen as an unwanted contaminant. If the cryogenic fluid is permitted to evaporate, oxygen concentrations of 50% can be reached. Under conditions of oxygen enrichment, there is an increased potential for violent reactions (rapid combustion or explosions). The extreme cold of cryogens presents the potential of frostbite and hypothermia. When cryogens are spilled, a thin gaseous layer forms next to the skin. This layer protects the tissue from freezing if the quantity of liquid is small and the exposure is brief. However, having wet skin or exposures to larger quantities of cryogens for extended periods of time can produce freezing of the tissue or cooling of the internal organs of the body. Since cryogenic storage vessels are pressurized containers, there exists the potential for leaks, cracks, or ruptures of the vessels.

Design of systems and apparatus used for cryogenic fluids should be in accordance with NFPA 45. Pressure relief must be provided to cryogenic storage vessels to permit routine off-gassing of the vapors generated by external heat flux into the cryogenic system. The use of pressure relief devices and storage vessel materials which are compatible with cryogenic temperatures are required to prevent leaks, cracks, or rupture of the vessels. The space in which cryogenic systems are located should be ventilated commensurate with the specific cryogenic fluid in use.

Oxygen sensors should be installed where there is sufficient danger of the development of an oxygen-deficient atmosphere. The oxygen sensors should be placed at the lowest point in a given area because the cold, dense, escaping gases will be heavier than the warmer ambient air, at least initially. A simple emergency plan should be developed to guide personnel during malfunctions or accidents. The plan should cover shut-down, alarms, and evacuation procedures for foreseeable emergencies.

Epoxies

Epoxies are multi-component resin systems that begin with a plastic material that is cured or hardened after application. Epoxies are made by mixing two or more reactive components, a base of one or more resins, and one or more curing agents. In general, epoxy components tend to be toxic; they are also sensitizers and irritants. The curing agents and, to a lesser extent, the uncured resins are often sensitizers. Individuals can unknowingly become sensitized. Subsequent exposures to small amounts may produce an adverse health effect. Improperly cured epoxies contain unreacted components and therefore have toxicity similar to the components themselves. For this reason, careful adherence to manufacturer's instructions on the use of epoxies is important.

The less time a sensitizer is in contact with the skin, the less likely it is to cause sensitization. For this reason, personal hygiene and protective clothing, such as gloves, are a first line of defense against the hazards presented by epoxies. Any skin contact with uncured epoxy components should be carefully avoided. When mixing resins and hardeners, workers should use a ventilated hood.

Incompatible Chemicals

A wide variety of chemicals react dangerously when mixed with certain other materials. Some of the more widely-used incompatible chemicals are listed in Table 12, but the absence of a chemical from this list should not be taken to indicate that it is safe to mix it with any other chemical.

Table 12. List of Incompatible Chemicals

Chemical	Incompatible Chemicals
Acetic acid	Chromic acid, ethylene glycol, nitric acid, hydroxyl compounds, perchloric acid, peroxides, permanganates
Acetone	Concentrated sulphuric and nitric acid mixtures
Acetylene	Chlorine, bromine, copper, fluorine, silver, mercury
Alkali and alkaline earth metals	Water, chlorinated hydrocarbons, carbon dioxide, halogens, alcohols, aldehydes, ketones, acids

Table 12. Continued

Chemical	Incompatible Chemicals
Aluminium (powdered)	Chlorinated hydrocarbons, halogens, carbon dioxide, organic acids
Anhydrous ammonia	Mercury, chlorine, calcium hypochlorite, iodine, bromine, hydrofluoric acid
Ammonium nitrate	Acids, metal powders, flammable liquids, chlorates, nitrites, sulphur, finely divided organic combustible materials
Aniline	Nitric acid, hydrogen peroxide
Arsenic compounds	Reducing agents
Azides	Acids
Bromine	Ammonia, acetylene, butadiene, hydrocarbons, hydrogen, sodium, finely-divided metals, turpentine, other hydrocarbons
Calcium carbide	Water, alcohol
Calcium oxide	Water
Carbon, activated	Calcium hypochlorite, oxidizing agents
Chlorates	Ammonium salts, acids, metal powders, sulphur, finely divided organic or combustible materials
Chromic acid	Acetic acid, naphthalene, camphor, glycerin, turpentine, alcohols, flammable liquids in general
Chlorine	See bromine
Chlorine dioxide	Ammonia, methane, phosphine, hydrogen sulphide
Copper	Acetylene, hydrogen peroxide
Cumene hydroperoxide	Acids, organic or inorganic

Table 12. Continued

Chemical	Incompatible Chemicals
Cyanides	Acids
Flammable liquids	Ammonium nitrate, chromic acid, hydrogen peroxide, nitric acid, sodium peroxide, halogens
Hydrocarbons	Fluorine, chlorine, bromine, chromic acid, sodium peroxide
Hydrocyanic acid	Nitric acid, alkali
Hydrofluoric acid	Aqueous or anhydrous ammonia
Hydrogen peroxide	Copper, chromium, iron, most metals or their salts, alcohols, acetone, organic materials, aniline, nitromethane, flammable liquids, oxidizing gases
Hydrogen sulphide	Fuming nitric acid, oxidizing gases
Hypochlorites	Acids, activated carbon
Iodine	Acetylene, ammonia (aqueous or anhydrous), hydrogen
Mercury	Acetylene, fulminic acid, ammonia
Mercuric oxide	Sulphur
Nitrates	Sulphuric acid
Nitric acid (concentrated)	Acetic acid, aniline, chromic acid, hydrocyanic acid, hydrogen sulphide, flammable liquids, flammable gases
Oxalic acid	Silver, mercury
Perchloric acid	Acetic anhydride, bismuth and its alloys, ethanol, paper, wood
Peroxides (organic)	Acids, avoid friction or shock
Phosphorus (white)	Air, alkalies, reducing agents, oxygen

Table 12. Continued

Chemical	Incompatible Chemicals
Potassium	Carbon tetrachloride, carbon dioxide, water
Potassium chlorate	Acids
Potassium perchlorate	Acids
Potassium permanganate	Glycerin, ethylene glycol, benzaldehyde, sulphuric acid
Selenides	Reducing agents
Silver	Acetylene, oxalic acid, tartaric acid, ammonium compounds, fulminic acid
Sodium	Carbon tetrachloride, carbon dioxide, water
Sodium nitrate	Ammonium salts
Sodium nitrite	Ammonium salts
Sodium peroxide	Ethanol, methanol, glacial acetic acid, acetic anhydride, benzaldehyde, carbon disulphide, glycerin, ethylene glycol, ethyl acetate, methyl acetate, furfural
Sulphides	Acids
Sulphuric acid	Potassium chlorate, potassium perchlorate, potassium permanganate (or compounds with similar light metals, such as sodium, lithium, etc.)
Tellurides	Reducing agents
Zinc powder	Sulphur

NFPA HAZARD RATINGS

The National Fire Protection Association, NFPA, a private nonprofit organization, is the leading authoritative source of technical background, data, and consumer advice on fire protection, problems and prevention. Their web site is http://www.nfpa.gov/. The primary goal of NFPA is to reduce the worldwide burden of fire and other hazards on the quality of life by providing and advocating scientifically-based consensus codes and standards, research, training, and education.

The NFPA has over 300 codes worldwide which are for sale through their web site. These codes cover every conceivable topic including basic fire safety, the **National Electrical Code,** and life safety. These codes are developed and updated through an open process, ensuring their broad acceptance.

While NFPA codes cover several aspects of flammable materials pertinent to MSDS's, perhaps the most significant is the NFPA 704 Hazard Identification ratings system (the familiar NFPA "hazard diamond" shown on the right) for health, flammability, and instability. "NFPA 325: *Guide to Fire Hazard Properties of Flammable Liquids, Gases, and Volatile Solids*, 1994 Edition" can be purchased on-line for $27.00 and includes basic fire hazard properties of approximately 1,300 chemicals.

The diamond is broken into four sections. Numbers in the three colored sections range from 0 (least severe hazard) to 4 (most severe hazard). The fourth (white) section is left blank and is used to denote special fire fighting procedures. Table 13 provides the NFPA hazard definitions corresponding to the number designations in Figure 1.

Figure 1. *NFPA hazard warning label.*

Table 13. NFPA Hazard Ratings (Refer to Figure 1).

HEALTH HAZARD	
4	Very short exposure could cause death or serious residual injury even though prompt medical attention was given.
3	Short exposure could cause serious temporary or residual injury even though prompt medical attention was given.
2	Intense or continued exposure could cause temporary incapacitation or possible residual injury unless prompt medical attention is given.
1	Exposure could cause irritation buy only minor residual injury even if no treatment is given.
0	Exposure under fire conditions would offer no hazard beyond that of ordinary combustible materials.
FLAMMABILITY	
4	Will rapidly or completely vaporize at normal pressure and temperature, or is readily dispersed in air and will burn readily.
3	Liquids and solids that can be ignited under almost all ambient conditions.
2	Must be moderately heated or exposed to relatively high temperature before ignition can occur.
1	Must be preheated before ignition can occur.
0	Materials that will not burn.
REACTIVITY	
4	Readily capable of detonation or of explosive decomposition or reaction at normal temperatures and pressures.
3	Capable of detonation or explosive reaction, but requires a strong initiating source or must be heated under confinement before initiation, or reacts explosively with water.
2	Normally unstable and readily undergo violent decomposition but do not detonate. Also: may react violently with water or may form potentially explosive mixtures with water.

Table 13. Continued

1	Normally stable, but can become unstable at elevated temperatures and pressures for may react with water with some release of energy, but not violently.
0	Normally stable, even under fire exposure conditions, and are not reactive with water.

SPECIAL HAZARDS
This section is used to denote unusual reactivity with water. The letter W with a horizontal line through it indicates a potential hazard using water to fight a fire involving this material. Other symbols may also appear here to indicate, for example, radioactivity, proper extinguishing agent, protective equipment etc.

Target Organ Effects: Target organ effects indicate which bodily organs are most likely to be affected by exposure to a substance. When working with chemicals that have target organ effects it is critical to prevent exposure. This is especially true if you have a pre-existing condition, disease or injury to that particular organ.

Read the MSDS to find out what the most effective personal protection equipment (PPE) for dealing with the chemical and be certain to minimize release of the chemical in the first place. Some terms used when describing target organ effects are listed in Table 14.

It is an employer's responsibility to make sure that employees understand and know how to use a material safety data sheet. Most companies maintain these in a central file where workers can gain ready access to them. Today, there are a multitude of software products available that can maintain and manage MSDSs, and the ready access to the Internet sites where literally thousands of product specific MSDSs can be obtained makes the management of this information a little easier. The MSDS should always be explained to workers through a Right-to-Know training program. Nearly all workplaces are required under OSHA regulations to provide basic training on the hazards associated with handling chemicals and the workplace. Laboratory environments in particular should be particularly sensitive to providing adequate training for technicians. This training should be specific to the laboratory conditions and the chemicals that are handles and stored on site. Adequate training records should be meticulously maintained for employees and the training should be frequent.

Table 14. Target Organs Affected by Chemicals

Class and Definition	Signs and Symptoms	Examples
Hepatotoxins - produce hepatic (liver) damage	Jaundice liver enlargement	Carbon tetrachloride, nitrosamines
Nephrotoxins - produce kidney damage	Edema, proteinuria	Halogenated hydrocarbons, uranium
Neurotoxins - produce their primary toxic effects on the nervous system	Narcosis, behavioral changes, decrease in motor functions	Mercury, carbon disulfide
Reproductive Toxins - affect the reproductive capabilities	Birth defects, sterility, chromosomal damage (mutations and effects on fetuses (teratogenesis)	Lead, DBCP
Cutaneous Hazards - affect the dermal	Defatting of the skin, layer (skin) of the body rashes, irritation	Ketones, chlorinated compounds
Pulmonary Agents - damage the lung	Cough, tightness in, irritate or damage pulmonary tissue (lung) chest, shortness of breath	Silica, asbestos
Eye Hazard - affect the eye or visual	Conjunctivitis, corneal capacity damage	Organic solvents, acids
Hematopoietic agents - act on the blood or hematopoietic system, decrease hemoglobin function, deprive the body ,tissues of oxygen	Cyanosis loss of consciousness	Carbon monoxide, cyanides

Chapter 8
SELECT OSH FACTS

INTRODUCTION

This final chapter provides highlights and general notes on several OSH (occupational safety and health) subjects addressed by OSHA. Specific standards are referenced so that the reader can refer to more detailed information and data. The safety officer needs to assess whether the specific functions and workplace environment are subject to certain OSHA standards and regulations. Proper assessments can only be made by performing a detailed safety and risk assessment audit of the operations, worker practices, company policies, and the workplace environment.

BLOODBORNE PATHOGENS STANDARD

In 1991 OSHA issued its final rule on Occupational Exposure to Bloodborne Pathogens (29 CFR Part 1910.1030). This regulation marks the first time OSHA has taken regulatory action in the area of infectious diseases. The standard is aimed at protecting more than 5.6 million workers and prevent more than 9,200 bloodborne infections and 200 deaths each year.

Bloodborne pathogens are microorganisms that exist in human blood and other body fluids. They include the hepatitis B virus (HBV) and the human immunodeficiency virus (HIV), the virus that causes AIDS. When body fluids that are infectious enter the blood stream of another person, they can cause disease. Full implementation of the Bloodborne Pathogens Standard will protect workers from exposure to these diseases.

The standard covers workers who can "reasonably anticipate skin, eye, mucous membrane or parenteral contact (e.g., needlesticks, sharps injuries, human bites, etc.) with blood or other potentially infectious materials including semen, vaginal secretions, cerebrospinal fluid, synovial fluid, pleural fluid, pericardial fluid, peritoneal fluid, amniotic fluid, saliva in dental procedures, any fluid that is visibly contaminated with blood, and all body fluids in situations where it is difficult to differentiate between body fluids.

That means that the standard must be implemented in many places, such as schools and correctional institutions, and not just health care institutions. As with other OSHA regulations, however, the Bloodborne Pathogens Standard does not cover all workers. Public employees in states that do not have federally approved state OSHA plans are not covered.

Employers and employees should become familiar with the Bloodborne Pathogens Standard. Although reading it may appear intimidating at first, it is important for employees who are exposed to blood to become familiar with their rights and the employer's obligations. A summary of the main points of the standard follows, but you should also read the standard itself. For those who want more detail on how the standard is enforced, OSHA has prepared a publication CPL 2-2.44C, *Enforcement Procedures for the Occupational Exposure to Bloodborne Pathogens Standard* which provides detailed instructions to OSHA inspectors on how to enforce the standard. This document may be requested from the AFSCME Research Department. The key components of the standard are outlined below:

Exposure Control Plan 1910.1030(c)

The standard requires employers to have a *written* exposure control plan that identifies workers who might be at risk. As part of the plan, employers must identify, in writing, tasks and procedures as well as job classifications where occupational exposure to blood occurs. The plan must establish a schedule for implementing other provisions of the standard and specify the procedure for evaluating the causes of exposure incidents. The written Exposure Control Plan must be accessible to employees and updated at least annually or when new or revised procedures are implemented.

Methods Of Compliance 1910.1030(d)

Universal Precautions: The standard mandates "universal precautions." This means that *all* blood and body fluids must be treated as it they are infectious, whether or not you know if the patient is infected with HIV or hepatitis B.

Engineering and Work Practice Controls: The standard emphasizes engineering and work practice controls to prevent exposure. Engineering controls include such items as self-sheathing syringes. These would reduce needlestick injuries and thereby control potential exposures to blood.

Work practice controls include:

- handwashing: handwashing facilities must be readily accessible to employees, or where this is not feasible, the employer must provide an antiseptic hand cleaner.
- procedures to minimize needlesticks: bending, recapping or removing needles or sharps is forbidden (except where this is not possible). Sharps must be placed in a puncture resistant, leakproof container that is labeled or color-coded and located in the area where needles and sharps can be expected to be found. This includes hospital laundries, as well as patient rooms.

Personal Protective Equipment: The standard requires that when engineering or work practice controls cannot eliminate exposures the employer provide personal protective equipment to the employee. Personal protective equipment must be available in appropriate sizes and must be readily accessible. This equipment includes:

- *gloves* must be worn by workers who can reasonably anticipate that their hands will be exposed to blood. They must replaced when contaminated or torn or punctured, must be appropriate for its intended use and be available in a variety of sizes. Gloves must be changed after each patient contact.
- *masks, eye protection and face shields* must be used whenever splashes, spray spatter or droplets of blood may contaminate the eye, nose or mouth.
- *gowns, aprons, lab coats* and other protective body clothing must be worn in occupational exposure situations. Such clothing that is used as personal protective equipment must be laundered by the employer and not sent home with the employee for cleaning.
- *surgical caps or hoods and/or shoe covers or boots* must be worn where gross contamination can be expected (such as during autopsies or orthopaedic surgery).

Housekeeping: Employers must ensure that the worksite is maintained in a clean and sanitary condition. Sharps containers must be kept clean, easily accessible and not allowed to overfill. All equipment and working surfaces must be clean and decontaminated after exposure to blood.

Laundry: Contaminated laundry should be handled as little as possible with a minimum of agitation. Contaminated laundry must be bagged or containerized at the location where it was used and shall not be sorted or rinsed in the location of use. Contaminated laundry must be bagged in labeled or color-coded bags. Employees who have contact with contaminated laundry must wear gloves.

Post-exposure Evaluation and Follow-up 1910.1030(f)

The employer is required to develop, and have available to employees, the procedures to be followed in the event of an occupational exposure. Follow-up procedures must include a confidential medical evaluation documenting the circumstances of exposure and identifying and testing the source individual if feasible. (Information on the source individual must be given to the employee, but not the employer.) Confidential testing of the exposed employee's blood if he/she consents, post-exposure treatment, and counseling are also required. The only information that is required to be given to the employer is the health care professional's written opinion for whether or not a hepatitis B vaccination should be given (if the employee has not received a vaccination), the healthcare professional's assurance that the employee has been informed of the results of the evaluation and told about any medical conditions resulting from the exposure that may require future follow-up. No test results, diagnoses, or other findings may be given to the employer.

Hepatitis B Vaccination 1910.1030(f)(2)

The standard requires that employers make the hepatitis B vaccination available, at no cost to the employee, to all workers who can reasonably anticipate occupational exposure to blood. Vaccinations must be made available during regular work hours. If the vaccination requires travel, the employer must bear the cost. Any employee who chooses not to be vaccinated must sign a declination form, but may later choose to receive the vaccine at no cost.

Hazard Communication 1910.1030(g)(1)

Warning labels, including the orange or orange-red biohazard symbol, must be affixed to containers of regulated waste, refrigerators, freezers and other containers that are used to store or transport blood or other potentially infectious materials. Red bags may be used instead of labeling.

Employee Training 1910.1030(g)(2)

The standard requires that training be provided to all personnel who may potentially be exposed to blood or body fluids. The training must be specific to the tasks the workers perform and be in the literacy and language level of the employees. If the employee only speaks a foreign language, the training must be done in that language. The training must include information about the risks posed by bloodborne pathogens and other potentially infectious materials. The training must include information about modes of transmission and effective controls. Training must also cover the employer's responsibility to offer free hepatitis B vaccination. Training sessions must be done for all affected employees and must be repeated annually. For new employees, training must be done at the time of hire. There must be an opportunity for questions and answers and the trainer must be knowledgeable in the subject matter. A video tape alone is not adequate.

Recordkeeping 1910.1030(h)

Medical records must be kept for each employee with occupational exposure for the duration of employment plus 30 years. These records must be kept confidential. Training records must be maintained for three years and must include dates, contents of training program or summary, trainer's name and qualifications and names and job titles of all persons attending the sessions.

Reviewing the Exposure Control Plan

Employees should make sure that the Exposure Control Plan includes all affected employees and methods of compliance. Inspect the workplace to check for the location of sharps disposal containers, personal protective equipment, biohazards warning labels, and the use of safer medical devices. Educate your co-workers about the bloodborne pathogens standard. Also review the employer's written protocol for handling needlestick injuries and other exposure incidents. If gloves are not available or the right size gloves are not supplied, if potentially exposed workers have not been offered the hepatitis B vaccine, if sharps containers are not accessible in all patient care areas, or if other violations of the standard exist, your employer should be asked to correct the situation and comply with the law. All requests should be filed in writing and the local health and safety committee should keep copies of the complaint and records of management's response. If your employer does not comply, use the health and safety language in your contract to file a grievance. Strong health and safety language is especially important for public employees in states with no state OSHA plan.

An OSHA inspection should be called for only after management has been notified of the problem and refused to take appropriate action. The easiest way to file a formal complaint is to use a complaint form requesting an inspection. These forms are available from the state OSHA office nearest you. Or you can write a letter requesting an inspection. Telephone complaints will not be accepted unless the situation poses an imminent danger. To avoid retaliation against individual members, it may be better for a health and safety committee member or a steward to file the OSHA complaint, rather than individual members. When the inspector comes out, a member designated by the union has a right to meet alone with the inspector and accompany him or her on the inspection. It is very important for the union representative to be present in order to point out to the inspector where the violations occur. The OSHA inspector should also interview employees from all appropriate areas of the facility to verify the accuracy of the facility's injury and illness records and the effectiveness of the infection control program. If possible, photographs of unsafe conditions should be taken to give to the inspector. Written copies of union requests to management to clean up the workplace should also be supplied to the inspector. The inspector should also look at records in employee health to see which employees have been vaccinated against hepatitis B, the number of injuries from needlesticks and cuts, as well as records of employees who have contracted hepatitis B or other infectious diseases on the job. The inspector has the option to decide that no inspection is necessary or that no violation has occurred.

MERCURY FACT SHEET

Mercury is probably best know as the silver liquid in thermometers. However, it has over 3000 industrial uses. Desirable properties such as the ability to alloy with most metals, liquidity at room temperature, ease of vaporizing and freezing, and electrical conductivity make mercury an important industrial metal. In 1973, U.S. consumption of mercury was 1900 metric tons. Primary among its over 3000 industrial uses are battery manufacturing and chlorine-alkali production. Paints and industrial instruments have also been among the major uses. Until paint manufacturers agreed to eliminate the use of mercury in interior paints, 480,000 pounds of mercury in paints and coatings were produced each year. Table 1 provides a list of mercury uses.

Mercury and its compounds are widely distributed in the environment as a result of both natural and man-made activities. The utility, and the toxicity, of mercury have been known for centuries. New studies demonstrate that even low levels of mercury exposure may be hazardous. Mercury occurs naturally in the environment as mercuric sulfide, also known as cinnabar.

It is also present in some fossil fuels. Cinnabar has been refined for its mercury content since the 15th or 16th century B.C. Its health hazards have been known at least since the Roman conquest of Spain. Due to the toxicity of mercury in cinnabar, criminals sentenced to work in quicksilver mines by the Romans had a life expectancy of only 3 years.

Table 1. Common Uses of Mercury

Barometers	Cells for Caustic Soda and Chlorine Production
Use In Boilers/Turbines for Electricity Generation	Synthetic Silk
Thermometers and Manometers	Photography
Textile Production	Metal Plating
Tanning and Dyeing	Mercury Vapor Lamps
Solder	Laboratory Reagent
Medicines	Investment Casting
Electrical Instruments	Fungicides/Preservatives (most uses now banned)
Dental Amalgams	Catalysts And Pigments

Mercury is present in numerous chemical forms. Elemental mercury itself is toxic and cannot be broken down into less hazardous compounds. Elemental or inorganic forms can be transformed into organic (especially methylated) forms by biological systems. Not only are these methylated mercury compounds toxic, but highly bioaccumulative as well. The increase in mercury as it rises in the aquatic food chain results in relatively high levels of mercury in fish consumed by humans.

Widespread poisoning of Japanese fisherman and their families occurred in Minamata, Japan in the 1950's as a result of consumption of methyl mercury contaminated fish. Today, we continue to be exposed to mercury in our diets, primarily from fish and shellfish. As a result, the U.S. Food and Drug Administration (FDA) has an action level for mercury of 1 part per million (ppm) in fish and the Michigan Department of Public Health issues fish consumption advisories to anglers when mercury levels exceed 0.5 ppm in fish tissue. Widespread industrial production of mercury, along with lack of careful handling and disposal practices, has contributed to environmental contamination. The U.S. Environmental Protection Agency (EPA) has made efforts to regulate the continued release of mercury into the environment. EPA regulates industrial discharges to air and water, as well as regulating some aspects of mercury waste disposal. In 1976, EPA banned most pesticide uses of mercury - with the exceptions of fungicidal uses in paints and outdoor fabrics, and for control of Dutch Elm disease. In 1990, mercury use as a fungicide in interior latex paint was halted by the EPA. This action stemmed from requests by Michigan officials after a child was poisoned from over formulated mercury-containing paint used in his home. More recently, the use of mercury compounds in exterior latex paint has also been halted. In addition to the early workers in the cinnabar mines, modern workers in industries using mercury are at risk from overexposure. The Occupational Safety and Health Administration (OSHA) has been reviewing the current occupational exposure standard of 0.1 mg/m^3 (milligrams per cubic meter of air) to determine if they should reduce the 8 hour acceptable exposure limit to 0.05 mg/m^3. Although no regulatory limit exists for airborne exposure to mercury outside of an occupational setting, the EPA suggests that 0.3 μg/m^3 (micro-grams per cubic meter of air) of mercury is a no-effect level (or reference dose = Rfd) for chronic inhalation exposure.

Humans come in contact with mercury through environmental, occupational or accidental exposure scenarios. An estimated 80% of utilized mercury is eventually released back into the environment. Because it is easily vaporized, air around chlorine-alkali plants, smelters, municipal incinerators, sewage treatment plants and even contaminated soils may contain increased levels of mercury. A primary route of exposure is through transport into surface waters, where mercury becomes biomagnified in fish tissues.

Workplace exposure to mercury occurs through inhalation of contaminated air, direct skin contact with liquid mercury, or oral exposure through contaminated hands, food, etc. A recent edition of the television show 60 Minutes highlighted concerns about mercury exposure in patients receiving silver dental fillings with mercury-containing amalgam. Insufficient scientific evidence exists at this time to either support or refute the claims that dental fillings may result in harmful exposure to mercury. Accidents have resulted in several cases of mercury poisoning in Michigan in the past two years. Four members of a Lincoln Park family were killed after one member attempted to refine dental amalgam in his home while attempting to recover silver. High levels of mercury were found throughout the house, including wrapped food inside the freezer. The entire house had to be demolished and disposed of in a hazardous waste landfill. A number of children have developed mercury poisoning after playing with small vials of mercury which they found at home or school. These children were hospitalized when symptoms became so severe that they could not longer walk. One contamination incident involved closing a school for weeks and entailed environmental investigation of residences, cars, school buses and day care centers.

Exposure to mercury can occur through inhalation, ingestion or dermal absorption. the amount of mercury absorbed by the body, and thus the degree of toxicity is dependent upon the chemical form of mercury. For instance, ingested elemental mercury is only 0.01% absorbed, but methyl mercury is nearly 100% absorbed from the gastrointestinal tract. The biological half-life of mercury is 60 days. Thus, even though exposure is reduced, the body burden will remain for at least a few months. Elemental mercury is most hazardous when inhaled. Only about 25% of an inhaled dose is exhaled. Skin absorption of mercury vapor occurs, but at low levels (e.g., 2.2% of the total dose). Dermal contact with liquid mercury can significantly increase biological levels. The primary focus of this article is elemental mercury, since that is the form of exposure to health care workers involved with mercury-containing instrument accidents. In the human body, mercury accumulates in the liver, kidney, brain, and blood. Mercury may cause acute or chronic health effects. Acute exposure (i.e., short term, high dose) is not as common today due to greater precautions and decreased handling. However, severe acute effects may include severe gastrointestinal damage, cardiovascular collapse, or kidney failure, all of which could be fatal. Inhalation of 1-3 mg/m^3 for 2-5 hours may cause headaches, salivation, metallic taste in the mouth, chills, cough, fever, tremors, abdominal cramps, diarrhea, nausea, vomiting, tightness in the chest, difficulty breathing, fatigue, or lung irritation. Symptoms may be delayed in onset for a number of hours.

Chronic effects include central nervous system effects, kidney damage and birth defects. Genetic damage is also suspected. Nervous system effects are the most critical effects of chronic mercury exposure from adult exposure as they are consistent and pronounced. Some elemental mercury is dissolved in the blood and may be transported across the blood/brain barrier, oxidized and retained in brain tissue. Elimination from the brain is slow, resulting in nerve tissue accumulation. Symptoms of chronic mercury exposure on the nervous system include: Increased excitability, mental instability, tendency to weep, fine tremors of the hands and feet, and personality changes. The term "Mad as a Hatter" came from these symptoms which were a result of mercury exposure in workers manufacturing felt hats using a mercury-containing process.

Kidney damage includes increased protein in the urine and may result in kidney failure at high dose exposure. Neurologic damage from methyl mercury leads to birth defects. The manifestations of mild exposure include delayed developmental milestones, altered muscle tone and tendon reflexes, and depressed intelligence. Mercury exposure in children can cause a severe form of poisoning termed acrodynia. Acrodynia is evidenced by pain in the extremities, pinkness and peeling of the hands, feet and nose, irritability, sweating, rapid heartbeat and loss of mobility.

Substitutes for mercury-containing medical devices should be used whenever possible, e.g., thermometers and sphygmomanometers. When mercury devices must be used, special precautions should be taken. These devices should never be used on a cloth surface, such as upholstered chair or in a room with a carpeted floor. If a spill occurred in such an area, the upholstery or carpeting would need to be discarded as it could not be effectively decontaminated. Children should never be left unattended near these devices. If mercury thermometers are used, a mercury spill kit should be kept readily accessible. The kit should contain a sulfur powder to suppress volatilization and a collection device.

If a spill occurs, evacuate the immediate area and ventilate as well as possible. An environmental consultant will need to be contacted for clean-up and disposal. Do not attempt to clean-up a mercury spill using rags or an ordinary vacuum. This will only serve to disperse the mercury and encourage volatilization. For further assistance, contact your local health department and/or the Michigan Department of Public Health, Division of Health Risk Assessment. For assistance with a large spill, call the Fire Department for assistance.

BERYLLIUM

Pure beryllium is a hard, grayish metal. In nature, beryllium can be found in compounds in mineral rocks, coal, soil, and volcanic dust. Beryllium compounds are commercially mined, and the beryllium purified for use in electrical parts, machine parts, ceramics, aircraft parts, nuclear weapons, and mirrors. Beryllium compounds have no particular smell. Beryllium can be harmful if you breathe it. The effects depend on how much you are exposed to and for how long. High levels of beryllium in air cause lung damage and a disease that resembles pneumonia. If you stop breathing beryllium dust, the lung damage may heal. Some people become sensitive to beryllium. This is called a hypersensitivity or allergy. These individuals develop an inflammatory reaction to low levels of beryllium. This condition is called chronic beryllium disease, and can occur long after exposure to small amounts of beryllium. This disease can make you feel weak and tired, and can cause difficulty in breathing. Both the short-term, pneumonia-like disease and the chronic beryllium disease can cause death. Swallowing beryllium has not been reported to cause effects in humans because very little beryllium can move from the stomach and intestines into the bloodstream. Beryllium contact with scraped or cut skin can cause rashes or ulcers. The Department of Health and Human Services (DHHS) has determined that beryllium and certain beryllium compounds may reasonably be anticipated to be carcinogens. This determination is based on animal studies and studies in workers. None of the studies provide conclusive evidence, but when taken as a whole, they indicate that long-term exposure to beryllium in the air results in an increase in lung cancer. Tests can measure beryllium in the urine and blood. The amount of beryllium in blood or urine may not indicate how much or how recently you were exposed. Small amounts of human lung and skin can also be removed from the body and examined for beryllium. These tests can be done in a doctor's office or in a hospital. One test uses blood cells washed out of the lung. If these cells start growing in the presence of beryllium, you are probably sensitive to beryllium and may have chronic beryllium disease. The Environmental Protection Agency (EPA) restricts the amount of beryllium that industries may emit into the environment to 10 grams (g) in a 24-hour period, or to an amount that would result in atmospheric levels of 0.01 micrograms of beryllium per cubic meter of air (0.01 $\mu g/m^3$), averaged over a 30-day period. The National Institute for Occupational Safety and Health (NIOSH) recommends a standard for occupational exposure of 0.5 $\mu g/m^3$ of beryllium in workroom air during an 8-hour shift to protect workers from potential cancer. The Occupational Safety and Health Administration (OSHA) sets a limit of 2 $\mu g/m^3$ of beryllium in workroom air for an 8-hour work shift. Important references that the reader can refer to for detailed information on health risks are:

1.   Agency for Toxic Substances and Disease Registry (ATSDR). 1993. Toxicological profile for beryllium; Atlanta: U.S. Department of Health and Human Services, Public Health Service.

2.   Agency for Toxic Substances and Disease Registry (ATSDR). 1993. Case studies in environmental medicine: Beryllium toxicity. Atlanta: U.S. Department of Health and Human Services, Public Health Service.

Table 2 provides a summary of health risks associated with exposure to beryllium. Table 3 provides a description of safe handling and labeling procedures. Table 4 provides an abridged MSDS on beryllium.

Table 2. Major Health Risks Associated with Beryllium

TYPES OF HAZARD/ EXPOSURE	ACUTE HAZARDS/ SYMPTOMS	PREVENTION	FIRST AID/ FIRE FIGHTING
FIRE	Combustible	No open flames.	Special powder, dry sand - No other agents.
EXPLOSION	Finely dispersed particles form explosive mixtures in air	Prevent deposition of dust; closed system, dust explosion-proof electrical equipment and lighting	
EXPOSURE		Prevent dispersion of dust! Avoid all contact!	In all cases consult a doctor
INHALATION	Cough. Shortness of breath. Sore throat. Weakness. Symptoms may be delayed	Local exhaust. Breathing protection	Fresh air, rest. Refer for medical attention
SKIN	Redness	Protective gloves. Protective clothing	Remove contaminated clothes. Rinse skin with plenty of water or shower
EYES	Redness and pain	Face shield or eye protection in combination with breathing protection if powder	First rinse with plenty of water for several minutes (remove contact lenses if easily possible), then take to a doctor
INGESTION		Do not eat, drink, or smoke during work. Wash hands before eating	Rinse mouth. DO NOT induce vomiting. Seek medical attention

Table 3. Handling and Labeling Procedures

SPILLAGE DISPOSAL: Evacuate danger area! Consult an expert! Carefully collect the spilled substance into containers; if appropriate moisten first, then remove to safe place. Do NOT let this chemical enter the environment (extra personal protection: complete protective clothing including self-contained breathing apparatus).
STORAGE: Separated from strong acids, bases, chlorinated solvents, food and feedstuffs.
PACKAGING & LABELING: Unbreakable packaging; put breakable packaging into closed unbreakable container. Do not transport with food and feedstuffs. UN Hazard Class: 6.1; UN Subsidiary Risks: 4.1; UN Packing Group: II

Table 4. Abridged MSDS on Beryllium

PHYSICAL STATE; APPEARANCE: Grey to white metal or powder. Boiling point: above 2500°C; Melting point: 1287°C; Relative density (water = 1): 1.9 Solubility in water: none	**ROUTES OF EXPOSURE:** The substance can be absorbed into the body by inhalation of its aerosol and by ingestion.
PHYSICAL DANGERS: Dust explosion possible if in powder or granular form, mixed with air.	**INHALATION RISK:** Evaporation at 20°C is negligible; a harmful concentration of airborne particles can, however, be reached quickly when dispersed.
CHEMICAL DANGERS: Reacts with strong acids and strong bases forming combustible gas. Forms shock sensitive mixtures with some chlorinated solvents, such as carbon tetrachloride and trichloroethylene.	**EFFECTS OF SHORT-TERM EXPOSURE:** The aerosol of this substance irritates the respiratory tract. Inhalation of dust or fumes may cause chemical pneumonitis. Exposure may result in death. The effects may be delayed. Medical observation is indicated.
OCCUPATIONAL EXPOSURE LIMITS (OELs): TLV (as TWA): ppm; 0.002 mg/m^3 A2 (Suspected Human Carcinogen) (ACGIH 1994-1995).	**EFFECTS OF LONG-TERM OR REPEATED EXPOSURE:** Repeated or prolonged contact may cause skin sensitization. Lungs may be affected by repeated or prolonged exposure to dust particles , resulting in chronic beryllium disease (cough, weight loss, weakness). This substance is carcinogenic to humans.
ENVIRONMENTAL DATA: The substance is very toxic to aquatic organisms.	

LEAD

OSHA's standard on lead is standard number 1926.62 (Subpart Number D). The standard applies to all construction work where an employee may be occupationally exposed to lead. All construction work excluded from coverage in the general industry standard for lead by 29 CFR 1910.1025(a)(2) is covered by this standard. Construction work is defined as work for construction, alteration and/or repair, including painting and decorating. It includes but is not limited to the following:

- Demolition or salvage of structures where lead or materials containing lead are present;
- Removal or encapsulation of materials containing lead;
- New construction, alteration, repair, or renovation of structures, substrates, or portions thereof, that contain lead, or materials containing lead;
- Installation of products containing lead;
- Lead contamination/emergency cleanup;
- Under ..1926.62(a)(6), Transportation, disposal, storage, or containment of lead or materials containing lead on the site or location at which construction activities are performed, and
- Under (a)(7), Maintenance operations associated with the construction activities.

Some important definitions in the standard are as follows:

- "Action level" means employee exposure, without regard to the use of respirators, to an airborne concentration of lead of 30 micrograms per cubic meter of air (30 $\mu g/m^3$) calculated as an 8-hour time-weighted average (TWA).
- "Competent person" means one who is capable of identifying existing and predictable lead hazards in the surroundings or working conditions and who has authorization to take prompt corrective measures to eliminate them.
- "Permissible exposure limit" (PEL) - The employer shall assure that no employee is exposed to lead at concentrations greater than fifty micrograms per cubic meter of air (50 $\mu g/m^3$) averaged over an 8-hour period.

The following highlights some, but not all of the important sections of the standard. Employers engaged in lead handling operations or in the use of contractors dealing with lead remediation activities should review the standard closely. There are very specific requirements for PPE and respirators, employee training, medical surveillance, recordkeeping, in addition to the highlights offered below.

If an employee is exposed to lead for more than 8 hours in any work day the employees' allowable exposure, as a time weighted average (TWA) for that day, shall be reduced according to the following relationship:

Allowable employee exposure (in units of $\mu g/m^3$)
= 400 ÷ (hours worked in the day)

When respirators are used to limit employee exposure, employee exposure may be considered to be at the level provided by the protection factor of the respirator for those periods the respirator is worn. Those periods may be averaged with exposure levels during periods when respirators are not worn to determine the employee's daily TWA exposure. Each employer who has a workplace or operation covered by this standard shall initially determine if any employee may be exposed to lead at or above the action level.

The employer shall collect personal samples representative of a full shift including at least one sample for each job classification in each work area either for each shift or for the shift with the highest exposure level. Full shift personal samples shall be representative of the monitored employee's regular, daily exposure to lead. With respect to the lead related tasks, until the employer performs an employee exposure assessment, the employer shall treat the employee as if the employee were exposed above the PEL, and not in excess of ten (10) times the PEL, and shall implement employee protective measures.

Where the employer has previously monitored for lead exposures, and the data were obtained within the past 12 months during work operations conducted under workplace conditions closely resembling the processes, type of material, control methods, work practices, and environmental conditions used and prevailing in the employer's current operations, the employer may rely on such earlier monitoring results to satisfy the requirements of the standards.

Where the employer has objective data, demonstrating that a particular product or material containing lead or a specific process, operation or activity involving lead cannot result in employee exposure to lead at or above the action level during processing, use, or handling, the employer may rely upon such data instead of implementing initial monitoring. It is also the employer's responsibility to establish and maintain an accurate record documenting the nature and relevancy of objective data, where used in assessing employee exposure in lieu of exposure monitoring.

If there is a possibility of any employee exposure at or above the action level, the employer shall conduct monitoring which is representative of the exposure for each employee in the workplace who is exposed to lead. Where the employer has previously monitored for lead exposure, and the data were obtained within the past 12 months during work operations conducted under workplace conditions closely resembling the processes, type of material, control methods, work practices, and environmental conditions used and prevailing in the employer's current operations, the employer may rely on such earlier monitoring results to satisfy the requirements of the standard. If the initial determination reveals employee exposure to be below the action level, further exposure determination need not be repeated. If the initial determination or subsequent determination reveals employee exposure to be at or above the action level but at or below the PEL the employer shall perform monitoring at least every 6 months. The employer shall continue monitoring at the required frequency until at least two consecutive measurements, taken at least 7 days apart, are below the action level at which time the employer may discontinue monitoring for that employee.

If the initial determination reveals that employee exposure is above the PEL the employer shall perform monitoring quarterly. The employer shall continue monitoring at the required frequency until at least two consecutive measurements, taken at least 7 days apart, are at or below the PEL but at or above the action level at which time the employer shall repeat monitoring for that employee at the frequency specified in the standard. The employer shall continue monitoring at the required frequency until at least two consecutive measurements, taken at least 7 days apart, are below the action level at which time the employer may discontinue monitoring for that employee.

Whenever there has been a change of equipment, process, control, personnel or a new task has been initiated that may result in additional employees being exposed to lead at or above the action level or may result in employees already exposed at or above the action level being exposed above the PEL, the employer shall conduct additional monitoring.

Within 5 working days after completion of the exposure assessment the employer shall notify each employee in writing of the results which represent that employee's exposure. Whenever the results indicate that the representative employee exposure, without regard to respirators, is at or above the PEL the employer shall include in the written notice a statement that the employees exposure was at or above that level and a description of the corrective action taken or to be taken to reduce exposure to below that level.

The employer shall use a method of monitoring and analysis which has an accuracy (to a confidence level of 95 percent) of not less than plus or minus 25 percent for airborne concentrations of lead equal to or greater than 30 $\mu g/m^3$.

Employers are required to implement engineering and work practice controls, including administrative controls, to reduce and maintain employee exposure to lead to or below the permissible exposure limit to the extent that such controls are feasible. Wherever all feasible engineering and work practices controls that can be instituted are not sufficient to reduce employee exposure to or below the permissible exposure limit, the employer shall nonetheless use them to reduce employee exposure to the lowest feasible level and shall supplement them by the use of respiratory protection. Prior to commencement of the job each employer shall establish and implement a written compliance program to achieve compliance. Written plans for these compliance programs shall include at least the following:

- A description of each activity in which lead is emitted; e.g., equipment used, material involved, controls in place, crew size, employee job responsibilities, operating procedures and maintenance practices;
- A description of the specific means that will be employed to achieve compliance and, where engineering controls are required engineering plans and studies used to determine methods selected for controlling exposure to lead;
- A report of the technology considered in meeting the PEL;
- Air monitoring data which documents the source of lead emissions;
- A detailed schedule for implementation of the program, including documentation such as copies of purchase orders for equipment, construction contracts, etc.;
- A work practice program;
- An administrative control schedule, if applicable;
- A description of arrangements made among contractors on multi-contractor sites with respect to informing affected employees of potential exposure to lead and with respect to responsibility for compliance with this section as set-forth in 1926.16.

The compliance program shall provide for frequent and regular inspections of job sites, materials, and equipment to be made by a competent person. Written programs shall be submitted upon request to any affected employee or authorized employee representatives, to OSHA, and shall be available at the worksite for examination and copying by the Assistant Secretary and the Director. Written programs shall be revised and updated at least every 6 months to reflect the current status of the program.

When ventilation is used to control lead exposure, the employer shall evaluate the mechanical performance of the system in controlling exposure as necessary to maintain its effectiveness.

If administrative controls are used as a means of reducing employees TWA exposure to lead, the employer shall establish and implement a job rotation schedule which includes:

- Name or identification number of each affected employee;
- Duration and exposure levels at each job or work station where each affected employee is located; and
- Any other information which may be useful in assessing the reliability of administrative controls to reduce exposure to lead.

For employees who use respirators required by the standard, the employer must provide respirators that comply with the requirements. Respirators must be used during:

- Periods when an employee's exposure to lead exceeds the PEL.
- Work operations for which engineering and work-practice controls are not sufficient to reduce employee exposures to or below the PEL.
- Periods when an employee requests a respirator.

If an employee has breathing difficulty during fit testing or respirator use, the employer must provide the employee with a medical examination to determine whether or not the employee can use a respirator while performing the required duty. The employer must select the appropriate respirator or combination of respirators from Table 5. The employer must provide a powered air-purifying respirator when an employee chooses to use such a respirator and it will provide adequate protection to the employee. Where an employee is exposed to lead above the PEL without regard to the use of respirators, where employees are exposed to lead compounds which may cause skin or eye irritation (e.g., lead arsenate, lead azide), and as interim protection for employees performing tasks, the employer shall provide at no cost to the employee and assure that the employee uses appropriate protective work clothing and equipment that prevents contamination of the employee and the employee's garments such as, but not limited to:

- Coveralls or similar full-body work clothing;
- Gloves, hats, and shoes or disposable shoe coverlets; and
- Face shields, vented goggles, or other appropriate protective equipment which complies with 1910.133

It is the employer's responsibility to assure that contaminated protective clothing which is to be cleaned, laundered, or disposed of, is placed in a closed container in the change area which prevents dispersion of lead outside the container. The employer shall inform in writing any person who cleans or launders protective clothing or equipment of the potentially harmful effects of exposure to lead. The employer shall assure that the containers of contaminated protective clothing and equipment are labeled as follows:

CAUTION:
CLOTHING CONTAMINATED WITH LEAD. DO NOT REMOVE DUST BY BLOWING OR SHAKING. DISPOSE OF LEAD CONTAMINATED WASH WATER IN ACCORDANCE WITH APPLICABLE LOCAL, STATE, OR FEDERAL REGULATIONS.

Table 5. OSHA Recommended Respiratory Protection for Lead Aerosols.

Lead Airborne Concentration	Required Type of Respirator
Not in excess of 500 $\mu m/m^3$	½ mask air purifying respirator with high efficiency filters. ½ mask supplied air respirator operated in demand (negative pressure) mode.
Not in excess of 1,250 $\mu m/m^3$	Loose fitting hood or helmet powered air purifying respirator with high efficiency filters. Hood or helmet supplied respirator operated in a continuous flow mode, e.g., type CE abrasive blasting respirators operated in a continuous flow mode.
Not in excess of 2,500 $\mu m/m^3$	Full facepiece air purifying respirator with high efficiency filters. Tight fitting powered air purifying respirator with high efficiency filters. Full facepiece supplied air respirator operated in demand mode. ½ mask or full facepiece air respirator operated in continuous flow mode. Full facepiece self-contained breathing apparatus (SCBA) operated in pressure demand mode.
Not in excess of 50,000 $\mu m/m^3$	½ mask supplied air respirator operated in pressure demand or other positive-pressure mode.
Not in excess of 100,000 $\mu m/m^3$	Full facepiece supplied air respirator operated in pressure demand or other positive-pressure mode, e.g., type CE abrasive blasting respirator operated in pressure demand or in a positive-pressure mode.
Greater than 100,000 $\mu m/m^3$. Unknown concentration or fire fighting situation.	Full facepiece SCBA operated in a positive pressure demand or other positive-pressure mode.

Medical Surveillance: The employer shall make available initial medical surveillance to employees occupationally exposed on any day to lead at or above the action level. Initial medical surveillance consists of biological monitoring in the form of blood sampling and analysis for lead and zinc protoporphyrin levels. The employer shall institute a medical surveillance program for all employees who are or may be exposed by the employer at or above the action level for more than 30 days in any consecutive 12 months. All medical examinations and procedures are to be performed by or under the supervision of a licensed physician. The employer shall make available biological monitoring in the form of blood sampling and analysis for lead and zinc protoporphyrin levels to each employee covered under the standard. This must be done at least every 2 months for the first 6 months and every 6 months thereafter. For each employee whose last blood sampling and analysis indicated a blood lead level at or above 40 μg/dl, at least every two months. This frequency shall continue until two consecutive blood samples and analyses indicate a blood lead level below 40 μg/dl; and for each employee who is removed from exposure to lead due to an elevated blood lead level at least monthly during the removal period. Whenever the results of a blood lead level test indicate that an employee's blood lead level exceeds the numerical criterion for medical removal, the employer shall provide a second (follow-up) blood sampling test within two weeks after the employer receives the results of the first blood sampling test.

FACTS ABOUT HEAT STRESS

Operations involving high air temperatures, radiant heat sources, high humidity, direct physical contact with hot objects, or strenuous physical activities have a high potential for inducing heat stress in employees engaged in such operations. Such places include: iron and steel foundries, nonferrous foundries, brick-firing and ceramic plants, glass products facilities, rubber products factories, electrical utilities (particularly boiler rooms), bakeries, confectioneries, commercial kitchens, laundries, food canneries, chemical plants, mining sites, smelters, and steam tunnels. Outdoor operations conducted in hot weather, such as construction, refining, asbestos removal, and hazardous waste site activities, especially those that require workers to wear semipermeable or impermeable protective clothing, are also likely to cause heat stress among exposed workers. The principle causal factors are: age, weight, degree of physical fitness, degree of acclimatization, metabolism, use of alcohol or drugs, and a variety of medical conditions such as hypertension all affect a person's sensitivity to heat. However, even the type of clothing worn must be considered. Prior heat injury predisposes an individual to additional injury.

It is difficult to predict just who will be affected and when, because individual susceptibility varies. In addition, environmental factors include more than the ambient air temperature. Radiant heat, air movement, conduction, and relative humidity all affect an individual's response to heat.

The American Conference of Governmental Industrial Hygienists (1992) states that workers should not be permitted to work when their deep body temperature exceeds 38°C (100.4°F). Heat is a measure of energy in terms of quantity. A calorie is the amount of heat required to raise 1 gram of water 1°C (based on a standard temperature of 16.5 to 17.5°C). Conduction is the transfer of heat between materials that contact each other. Heat passes from the warmer material to the cooler material. For example, a worker's skin can transfer heat to a contacting surface if that surface is cooler, and vice versa. Convection is the transfer of heat in a moving fluid. Air flowing past the body can cool the body if the air temperature is cool. On the other hand, air that exceeds 35°C (95°F) can increase the heat load on the body. Evaporative cooling takes place when sweat evaporates from the skin. High humidity reduces the rate of evaporation and thus reduces the effectiveness of the body's primary cooling mechanism. Radiation is the transfer of heat energy through space. A worker whose body temperature is greater than the temperature of the surrounding surfaces radiates heat to these surfaces. Hot surfaces and infrared light sources radiate heat that can increase the body's heat load. Globe temperature is the temperature inside a blackened, hollow, thin copper globe. Metabolic heat is a byproduct of the body's activity. Natural wet bulb (NWB) temperature is measured by exposing a wet sensor, such as a wet cotton wick fitted over the bulb of a thermometer, to the effects of evaporation and convection. The term natural refers to the movement of air around the sensor. Dry bulb (DB) temperature is measured by a thermal sensor, such as an ordinary mercury-in-glass thermometer, that is shielded from direct radiant energy sources.

Heat Disorders and Health Effects: Heat stroke occurs when the body's system of temperature regulation fails and body temperature rises to critical levels. This condition is caused by a combination of highly variable factors, and its occurrence is difficult to predict. Heat stroke is a medical emergency. The primary signs and symptoms of heat stroke are confusion; irrational behavior; loss of consciousness; convulsions; a lack of sweating (usually); hot, dry skin; and an abnormally high body temperature, e.g., a rectal temperature of 41°C (105.8°F). If body temperature is too high, it causes death. The elevated metabolic temperatures caused by a combination of work load and environmental heat load, both of which contribute to heat stroke, are also highly variable and difficult to predict.

If a worker shows signs of possible heat stroke, professional medical treatment should be obtained immediately. The worker should be placed in a shady area and the outer clothing should be removed. The worker's skin should be wetted and air movement around the worker should be increased to improve evaporative cooling until professional methods of cooling are initiated and the seriousness of the condition can be assessed. Fluids should be replaced as soon as possible. The medical outcome of an episode of heat stroke depends on the victim's physical fitness and the timing and effectiveness of first aid treatment. Regardless of the worker's protests, no employee suspected of being ill from heat stroke should be sent home or left unattended unless a physician has specifically approved such an order.

The signs and symptoms of **heat exhaustion** are headache, nausea, vertigo, weakness, thirst, and giddiness. Fortunately, this condition responds readily to prompt treatment. Heat exhaustion should not be dismissed lightly, however, for several reasons. One is that the fainting associated with heat exhaustion can be dangerous because the victim may be operating machinery or controlling an operation that should not be left unattended; moreover, the victim may be injured when he or she faints. Also, the signs and symptoms seen in heat exhaustion are similar to those of heat stroke, a medical emergency. Workers suffering from heat exhaustion should be removed from the hot environment and given fluid replacement. They should also be encouraged to get adequate rest.

Heat cramps are usually caused by performing hard physical labor in a hot environment. These cramps have been attributed to an electrolyte imbalance caused by sweating. It is important to understand that cramps can be caused by both too much and too little salt. Cramps appear to be caused by the lack of water replenishment. Because sweat is a hypotonic solution ($\pm 0.3\%$ NaCl), excess salt can build up in the body if the water lost through sweating is not replaced. Thirst cannot be relied on as a guide to the need for water; instead, water must be taken every 15 to 20 minutes in hot environments. Under extreme conditions, such as working for 6 to 8 hours in heavy protective gear, a loss of sodium may occur. Recent studies have shown that drinking commercially available carbohydrate-electrolyte replacement liquids is effective in minimizing physiological disturbances during recovery.

Heat collapse ("fainting") is a condition where the brain does not receive enough oxygen because blood pools in the extremities. As a result, the exposed individual may lose consciousness. This reaction is similar to that of heat exhaustion and does not affect the body's heat balance. However, the onset of heat collapse is rapid and unpredictable. To prevent heat collapse, the worker should gradually become acclimatized to the hot environment.

Heat rashes are the most common problem in hot work environments. Prickly heat is manifested as red papules and usually appears in areas where the clothing is restrictive. As sweating increases, these papules give rise to a prickling sensation. Prickly heat occurs in skin that is persistently wetted by unevaporated sweat, and heat rash papules may become infected if they are not treated. In most cases, heat rashes will disappear when the affected individual returns to a cool environment. A factor that predisposes an individual to **heat fatigue** is lack of acclimatization. The use of a program of acclimatization and training for work in hot environments is advisable. The signs and symptoms of heat fatigue include impaired performance of skilled sensorimotor, mental, or vigilance jobs. There is no treatment for heat fatigue except to remove the heat stress before a more serious heat-related condition develops. The following are guidelines for evaluating employee heat stress, and closely follow those found in the 1992-1993 ACGIH publication, *Threshold Limit Values for Chemical Substances and Physical Agents and Biological Exposure Indices*.

- The inspector will review the OSHA 200 Log and, if possible, the OSHA 101 forms for indications of prior heat stress problems. Following are some questions for employer interviews: What type of action, if any, has the employer taken to prevent heat stress problems? What are the potential sources of heat? What employee complaints have been made? Following are some questions for employee interviews: What heat stress problems have been experienced? What type of action has the employee taken to minimize heat stress? What is the employer's involvement, i.e., does employee training include information on heat stress?

- During a walk-around inspection, the investigator will: determine building and operation characteristics; determine whether engineering controls are functioning properly; verify information obtained from the employer and employee interviews; and perform temperature measurements and make other determinations to identify potential sources of heat stress. Investigators may wish to discuss any operations that have the potential to cause heat stress with engineers and other knowledgeable personnel. The walk-around inspection should cover all affected areas. Heat sources, such as furnaces, ovens, and boilers, and relative heat load per employee should be noted.

- Under conditions of high temperature and heavy workload, the CSHO (Certified Safety and Health Officer) should determine the work-load category of each job. Work-load category is determined by averaging metabolic rates for the tasks and then ranking them as follows: Light work: up to 200 kcal/hour; Medium work: 200-350 kcal/hour; Heavy work: 350-500 kcal/hour.

Where heat conditions in the rest area are different from those in the work area, the metabolic rate (M) should be calculated using a time-weighted average, as follows:

$$M_{average} = \{M_1t_1 + M_2t_2 + ... + M_nt_n\}/\{t_1 + t_2 + ... + t_n\}$$

where M is the metabolic rate and t is time in minutes.

Although instruments are available to estimate deep body temperature by measuring the temperature in the ear canal or on the skin, these instruments are not sufficiently reliable to use in compliance evaluations. Environmental heat measurements should be made at, or as close as possible to, the specific work area where the worker is exposed.

When a worker is not continuously exposed in a single hot area but moves between two or more areas having different levels of environmental heat, or when the environmental heat varies substantially at a single hot area, environmental heat exposures should be measured for each area and for each level of environmental heat to which employees are exposed. Wet Bulb Globe Temperature (WBGT) should be calculated using the appropriate formula. The WBGT for continuous all-day or several hour exposures should be averaged over a 60-minute period. Intermittent exposures should be averaged over a 120-minute period. These averages should be calculated using the following formulae:

For indoor and outdoor conditions with no solar load, WBGT is calculated as:

$$WBGT = 0.7NWB + 0.3GT$$

For outdoors with a solar load, WBGT is calculated as:

$$WBGT = 0.7NWB + 0.2GT + 0.1DB$$

where: WBGT = Wet Bulb Globe Temperature Index; NWB = Nature Wet-Bulb Temperature; DB = Dry-Bulb Temperature; and GT = Globe temperature.

The exposure limits in Table 6 are valid for employees wearing light clothing. They must be adjusted for the insulation from clothing that impedes sweat evaporation and other body cooling mechanisms. Use Table 7 to correct Table 6 for various kinds of clothing. Use of Table 6 requires knowledge of the WBGT and approximate workload.

Portable heat stress meters or monitors are used to measure heat conditions. These instruments can calculate both the indoor and outdoor WBGT index according to established ACGIH Threshold Limit Value equations. With this information and information on the type of work being performed, heat stress meters can determine how long a person can safely work or remain in a particular hot environment.

Table 6. Permissible Heat Exposure Threshold Limit Value

Work Load*			
Work/rest regime	Light	Moderate	Heavy
Continuous work	30.0°C (86°F)	26.7°C (80°F)	25.0°C (77°F)
75% Work, 25% rest, each hour	30.6°C (87°F)	28.0°C (82°F)	25.9°C (78°F)
50% Work, 50% rest, each hour	31.4°C (89°F)	29.4°C (85°F)	27.9°C (82°F)
25% Work, 75% rest, each hour	32.2°C (90°F)	31.1°C (88°F)	30.0°C (86°F)

Values are in °C and °F, WBGT.

Table 7. WGBT Correction Factors in °C (Source: ACGIH, 1992)

Clothing Type	Clo-Value	WBGT Correction
Summer. lightweight working clothing	0.6	0
Cotton coveralls	1.0	-2
Winter work clothing	1.4	-4
Water barrier, permeable	1.2	-6

TLVs are based on the assumption that nearly all acclimatized, fully clothed workers with adequate water and salt intake should be able to function effectively under the given working conditions without exceeding a deep body temperature of 38°C (100.4° F). They are also based on the assumption that the WBGT of the resting place is the same or very close to that of the workplace. Where the WBGT of the work area is different from that of the rest area, a time-weighted average should be used [consult the ACGIH 1992-1993 Threshold Limit Values for Chemical Substances and Physical Agents and Biological Exposure Indices (1992)]. These TLV's apply to physically fit and acclimatized individuals wearing light summer clothing. If heavier clothing that impedes sweat or has a higher insulation value is required, the permissible heat exposure TLVs in Table 6 must be reduced by the corrections shown in Table 7.

The Effective Temperature index (ET) combines the temperature, the humidity of the air, and air velocity. This index has been used extensively in the field of comfort ventilation and air-conditioning. ET remains a useful measurement technique in mines and other places where humidity is high and radiant heat is low.

The Heat-Stress Index (HSI) was developed by Belding and Hatch in 1965. Although the HSI considers all environmental factors and work rate, it is not completely satisfactory for determining an individual worker's heat stress and is also difficult to use.

Ventilation, air cooling, fans, shielding, and insulation are the five major types of engineering controls used to reduce heat stress in hot work environments. Heat reduction can also be achieved by using power assists and tools that reduce the physical demands placed on a worker. However, for this approach to be successful, the metabolic effort required for the worker to use or operate these devices must be less than the effort required without them. Another method is to reduce the effort necessary to operate power assists. The worker should be allowed to take frequent rest breaks in a cooler environment.

Acclimation: The human body can adapt to heat exposure to some extent. This physiological adaptation is called acclimatization. After a period of acclimatization, the same activity will produce fewer cardiovascular demands. The worker will sweat more efficiently (causing better evaporative cooling), and thus will more easily be able to maintain normal body temperatures. A properly designed and applied acclimatization program decreases the risk of heat-related illnesses. Such a program basically involves exposing employees to work in a hot environment for progressively longer periods. NIOSH (1986) says that, for workers who have had previous experience in jobs where heat levels are high enough to produce heat stress, the regimen should be 50% exposure on day one, 60% on day two, 80% on day three, and 100% on day four. For new workers who will be similarly exposed, the regimen should be 20% on day one, with a 20% increase in exposure each additional day.

Fluid Replacement: Cool (50°-60°F) water or any cool liquid (except alcoholic beverages) should be made available to workers to encourage them to drink small amounts frequently, e.g., one cup every 20 minutes. Ample supplies of liquids should be placed close to the work area. Although some commercial replacement drinks contain salt, this is not necessary for acclimatized individuals because most people add enough salt to their summer diets.

Engineering Controls: General ventilation is used to dilute hot air with cooler air (generally cooler air that is brought in from the outside). This technique clearly works better in cooler climates than in hot ones. A permanently installed ventilation system usually handles large areas or entire buildings. Portable or local exhaust systems may be more effective or practical in smaller areas. **Air treatment/air cooling** differs from ventilation because it reduces the temperature of the air by removing heat (and sometimes humidity) from the air. **Air conditioning** is a method of air cooling, but it is expensive to install and operate. An alternative to air conditioning is the use of chillers to circulate cool water through heat exchangers over which air from the ventilation system is then passed; chillers are more efficient in cooler climates or in dry climates where evaporative cooling can be used. **Local air cooling** can be effective in reducing air temperature in specific areas. Two methods have been used successfully in industrial settings. One type, cool rooms, can be used to enclose a specific workplace or to offer a recovery area near hot jobs. The second type is a portable blower with built-in air chiller. The main advantage of a blower, aside from portability, is minimal set-up time.

Another way to reduce heat stress is to increase the air flow or **convection** using fans, etc. in the work area (as long as the air temperature is less than the worker's skin temperature). Changes in air speed can help workers stay cooler by increasing both the convective heat exchange (the exchange between the skin surface and the surrounding air) and the rate of evaporation. Because this method does not actually cool the air, any increases in air speed must impact the worker directly to be effective. If the dry bulb temperature is higher than 35°C (95°F), the hot air passing over the skin can actually make the worker hotter. When the temperature is more than 35°C and the air is dry, evaporative cooling may be improved by air movement, although this improvement will be offset by the convective heat. When the temperature exceeds 35°C and the relative humidity is 100%, air movement will make the worker hotter. Increases in air speed have no effect on the body temperature of workers wearing vapor-barrier clothing.

Heat conduction methods include insulating the hot surface that generates the heat and changing the surface itself. Simple engineering controls, such as shields, can be used to reduce radiant heat, i.e., heat coming from hot surfaces within the worker's line of sight. Surfaces that exceed 35°C (95°F) are sources of infrared radiation that can add to the worker's heat load. Flat black surfaces absorb heat more than smooth, polished ones. Having cooler surfaces surrounding the worker assists in cooling because the worker's body radiates heat toward them. With some sources of radiation, such as heating pipes, it is possible to use both insulation and surface modifications to achieve a substantial reduction in radiant heat. Instead of reducing radiation from the source, shielding can be used to interrupt the path between the source and the worker. Polished surfaces make the best barriers, although special glass or metal mesh surfaces can be used if visibility is a problem. Shields should be located so that they do not interfere with air flow, unless they are also being used to reduce convective heating. The reflective surface of the shield should be kept clean to maintain its effectiveness.

Administrative Controls And Work Practices: Training is the key to good work practices. Unless all employees understand the reasons for using new, or changing old, work practices, the chances of such a program succeeding are greatly reduced. NIOSH (1986) states that a good heat stress training program should include at least the following components:

- Knowledge of the hazards of heat stress;
- Recognition of predisposing factors, danger signs, and symptoms;
- Awareness of first-aid procedures for, and the potential health effects of, heat stroke;
- Employee responsibilities in avoiding heat stress;
- Dangers of using drugs, including therapeutic ones, and alcohol in hot work environments;
- Use of protective clothing and equipment; and
- Purpose and coverage of environmental and medical surveillance programs and the advantages of worker participation in such programs.

Hot jobs should be scheduled for the cooler part of the day, and routine maintenance and repair work in hot areas should be scheduled for the cooler seasons of the year.

Worker Monitoring Programs: Every worker who works in extraordinary conditions that increase the risk of heat stress should be personally monitored. These conditions include wearing semipermeable or impermeable clothing when the temperature exceeds 21°C (69.8°F), working at extreme metabolic loads (greater than 500 kcal/hour), etc. Personal monitoring can be done by checking the heart rate, recovery heart rate, oral temperature, or extent of body water loss. To check the heart rate, count the radial pulse for 30 seconds at the beginning of the rest period. If the heart rate exceeds 110 beats per minute, shorten the next work period by one third and maintain the same rest period. The recovery heart rate can be checked by comparing the pulse rate taken at 30 seconds (P_1) with the pulse rate taken at 2.5 minutes (P_3) after the rest break starts. Oral temperature can be checked with a clinical thermometer after work but before the employee drinks water. If the oral temperature taken under the tongue exceeds 37.6°C, shorten the next work cycle by one third. Body water loss can be measured by weighing the worker on a scale at the beginning and end of each work day. The worker's weight loss should not exceed 1.5% of total body weight in a work day. If a weight loss exceeding this amount is observed, fluid intake should increase. The following administrative controls can be used to reduce heat stress:

- Reduce the physical demands of work, e.g., excessive lifting or digging with heavy objects;
- Provide recovery areas, e.g., air-conditioned enclosures and rooms;

- Use shifts, e.g., early morning, cool part of the day, or night work;
- Use intermittent rest periods with water breaks;
- Use relief workers;
- Use worker pacing; and
- Assign extra workers and limit worker occupancy, or the number of workers present, especially in confined or enclosed spaces.

Personal Protective Equipment

Reflective clothing, which can vary from aprons and jackets to suits that completely enclose the worker from neck to feet, can stop the skin from absorbing radiant heat. However, since most reflective clothing does not allow air exchange through the garment, the reduction of radiant heat must more than offset the corresponding loss in evaporative cooling. For this reason, reflective clothing should be worn as loosely as possible. In situations where radiant heat is high, auxiliary cooling systems can be used under the reflective clothing.

Commercially available ice vests, though heavy, may accommodate as many as 72 ice packets, which are usually filled with water. Carbon dioxide (dry ice) can also be used as a coolant. The cooling offered by ice packets lasts only 2 to 4 hours at moderate to heavy heat loads, and frequent replacement is necessary. However, ice vests do not encumber the worker and thus permit maximum mobility. Cooling with ice is also relatively inexpensive.

Wetted clothing is another simple and inexpensive personal cooling technique. It is effective when reflective or other impermeable protective clothing is worn. The clothing may be wetted terry cloth coveralls or wetted two-piece, whole-body cotton suits. This approach to auxiliary cooling can be quite effective under conditions of high temperature and low humidity, where evaporation from the wetted garment is not restricted.

Water-cooled garments range from a hood, which cools only the head, to vests and "long johns," which offer partial or complete body cooling. Use of this equipment requires a battery-driven circulating pump, liquid-ice coolant, and a container. Although this system has the advantage of allowing wearer mobility, the weight of the components limits the amount of ice that can be carried and thus reduces the effective use time. The heat transfer rate in liquid cooling systems may limit their use to low-activity jobs; even in such jobs, their service time is only about 20 minutes per pound of cooling ice. To keep outside heat from melting the ice, an outer insulating jacket should be an integral part of these systems.

Circulating air is the most highly effective, as well as the most complicated, personal cooling system. By directing compressed air around the body from a supplied air system, both evaporative and convective cooling are improved. The greatest advantage occurs when circulating air is used with impermeable garments or double cotton overalls.

One type, used when respiratory protection is also necessary, forces exhaust air from a supplied-air hood ("bubble hood") around the neck and down inside an impermeable suit. The air then escapes through openings in the suit. Air can also be supplied directly to the suit without using a hood in three ways: by a single inlet; by a distribution tree; or by a perforated vest. In addition, a vortex tube can be used to reduce the temperature of circulating air. The cooled air from this tube can be introduced either under the clothing or into a bubble hood.

The use of a vortex tube separates the air stream into a hot and cold stream; these tubes also can be used to supply heat in cold climates. Circulating air, however, is noisy and requires a constant source of compressed air supplied through an attached air hose. One problem with this system is the limited mobility of workers whose suits are attached to an air hose. Another is that of getting air to the work area itself. These systems should therefore be used in work areas where workers are not required to move around much or to climb. Another concern with these systems is that they can lead to dehydration. The cool, dry air feels comfortable and the worker may not realize that it is important to drink liquids frequently.

It should be noted that the weight of a self-contained breathing apparatus (SCBA) increases stress on a worker, and this stress contributes to overall heat stress. Chemical protective clothing such as totally encapsulating chemical protection suits will also add to the heat stress problem.

Investigating Heat Stress in the Workplace

A compliance officer or a Health and Safety Officer (HSO) has two approaches to minimizing the risks of heat stress. Both approaches should be applied. First, workers should be trained in the symptoms and causes in order to recognize early warnings. The second approach is to conduct a site investigation or an audit, which may identify worksite conditions and operations that can lead to the risk of heat stress. Table 8 provides sample questions that the HSO may wish to consider when investigating heat stress in the workplace. By conducting a thorough analysis, corrective actions can be taken to minimize the health risks of employees and also reduce the liabilities of employers.

Table 8. Sample Questions for Investigating Heat Stress Problems at a Workplace

Workplace description		
1	Type of business	
2	Heat-producing equipment or processes used	
3	Previous history (if any) of heat-related problems	
4	At "hot" spots:	Is the heat steady or intermittent?
		Number of employees exposed?
		For how many hours per day?
		Is potable water available?
		Are supervisors trained to detect/evaluate heat stress symptoms?
Are exposures typical for a workplace in this industry?		
5	Weather at Time of Review	Temperature
		Humidity
		Air velocity
6	Is Day Typical of Recent Weather Conditions? (Get information from the Weather Bureau)	
7	Heat-Reducing Engineering Controls	Ventilation in place?
		Ventilation operating?
		Air conditioning in place?
		Air conditioning operating?
		Fans in place?
		Fans operating?
		Shields or insulation between sources and employees?
		Are reflective faces of shields clean?

Work practices to detect, evaluate, and prevent or reduce heat stress		
8	Training program?	Content?
		Where given?
		For whom?
9	Liquid replacement program?	
10	Acclimatization program?	
11	Work/rest schedule?	
12	Scheduling of work (during cooler parts of shift, cleaning and maintenance during shut-downs, etc.)	
13	Cool rest areas (including shelter at outdoor worksites)?	
14	Heat monitoring program?	
15	Personal Protective Equipment	Reflective clothing in use?
		Ice and/or water-cooled garments in use?
		Wetted undergarments (used with reflective or impermeable clothing) in use?
		Circulating air systems in use?
16	First Aid Program	Trained personnel?
		Provision for rapid cool-down?
		Procedures for getting medical attention?
		Transportation to medical facilities readily available for heat stroke victims?
17	Medical Screening and Surveillance Program	Content?
		Who manages program?

INDEX

A

abrasive blasting, 87
absorber bottoms, 133
absorber tail gas, 73
accident investigations, 319-324
acclimation, 408
accumulation, 120
acetylene, 356
ACGIH, 33
ACGIH skin designations, 88
ACGIH TLVs, 177
acid precipitation, 17
acid rain, 16
acid sludge, 73, 137
acids, definition, 32
acute hazards events, 300
acute toxicity, 33
administrative controls, 409
afterburners, 62
air cleaning strategies, 182
air contaminants, in petroleum
    refining, 98
air emissions, 141
air emissions, from valves, 137
air monitoring practices, 212
air pollutants, 31
air pollution, 15-20
air quality measurement
    techniques, 187
air quality profiles, 156
air sampling methods, 206
air sampling statistics, 215
airborne dusts, 144
airflow measurements, 185
air-purifying respirator, 269, 275
aldehydes, 102

aldrin, 83
aliphatics, 100
alkanes, 101
alkenes, 102
alkyd resins, 55
allergies, 5
Ames Test, 33
amino resins, 55
anesthesia, 34
anhydride, 34
anorexia, 34
anosmia, 34
anoxia, 35
ANSI, 35
ANSI standards, 311
aqueous, 35
Arrhenius, 32
asbestos, 105, 279-288
asbestos sampling, 195, 280
ASHRAE, 145, 161
ASHRAE standards, 151
asphysia, 36
asphyxiant, 35
asphyxiation, 37
asthma, 37
ASTM standards, 311
atmosphere-supplying respirator, 269
atrophy, 37
auditing practices, indoor air
    quality, 153
autoignition, 37
automatic flow control, 137
automobile exhaust, 217

B

back and neck injuries, 11
backpressure, 119, 121
back-up power systems, 315